Lecture Notes in Computer Science 3001

Commenced Publication in 1973
Founding and Former Series Editors:
Gerhard Goos, Juris Hartmanis, and Jan van Leeuwen

Springer

Berlin
Heidelberg
New York
Hong Kong
London
Milan
Paris
Tokyo

Alois Ferscha   Friedemann Mattern (Eds.)

Pervasive Computing

Second International Conference, PERVASIVE 2004
Linz/Vienna, Austria, April 18-23, 2004
Proceedings

 Springer

Volume Editors

Alois Ferscha
Johannes Kepler Universität Linz
Institut für Pervasive Computing
Altenberger Str. 69, 4040 Linz, Austria
E-mail: ferscha@soft.uni-linz.ac.at

Friedemann Mattern
ETH Zürich (Swiss Federal Institute of Technology)
Department of Computer Science, Institute for Pervasive Computing
Haldeneggsteig 4, 8092 Zürich, Switzerland
E-mail: mattern@inf.ethz.ch

Library of Congress Control Number: 2004103931

CR Subject Classification (1998): C.2.4, C.3, C.5.3, D.4, H.3-5, K.4, K.6.5, J.7

ISSN 0302-9743
ISBN 3-540-21835-1 Springer-Verlag Berlin Heidelberg New York

Springer-Verlag is a part of Springer Science+Business Media

springeronline.com

© Springer-Verlag Berlin Heidelberg 2004
Printed in Germany

Typesetting: Camera-ready by author, data conversion by PTP-Berlin, Protago-TeX-Production GmbH
Printed on acid-free paper   SPIN: 10996911   06/3142   5 4 3 2 1 0

Preface

Welcome to the proceedings of PERVASIVE 2004, the 2nd International Conference on Pervasive Computing and the premier forum for the presentation and appraisal of the most recent and most advanced research results in all foundational and applied areas of pervasive and ubiquitous computing. Considering the half-life period of technologies and knowledge this community is facing, PERVASIVE is one of the most vibrant, dynamic, and evolutionary among the computer-science-related symposia and conferences.

The research challenges, efforts, and contributions in pervasive computing have experienced a breathtaking acceleration over the past couple of years, mostly due to technological progress, growth, and a shift of paradigms in computer science in general. As for technological advances, a vast manifold of tiny, embedded, and autonomous computing and communication systems have started to create and populate a pervasive and ubiquitous computing landscape, characterized by paradigms like autonomy, context-awareness, spontaneous interaction, seamless integration, self-organization, ad hoc networking, invisible services, smart artifacts, and everywhere interfaces. The maturing of wireless networking, miniaturized information-processing possibilities induced by novel microprocessor technologies, low-power storage systems, smart materials, and technologies for motors, controllers, sensors, and actuators envision a future computing scenario in which almost every object in our everyday environment will be equipped with embedded processors, wireless communication facilities, and embedded software to perceive, perform, and control a multitude of tasks and functions. Since many of these objects are already able to communicate and interact with global networks and with each other, the vision of context-aware "smart appliances" and "smart spaces" has already become a reality. Service provision is based on the ability of being aware of the presence of other objects or users, and systems can be designed in order to be sensitive, adaptive, and responsive to their needs, habits, and even emotions. With pervasive computing technology embodied into real-world objects like furniture, clothing, crafts, rooms, etc., those artifacts also become the interface to "invisible" services and allow them to mediate between the physical and digital (or virtual) world via natural interaction – away from desktop displays and keyboards. All these observations pose serious challenges to the conceptual architectures of computing, and the related engineering disciplines in computer science. PERVASIVE rises to those challenges.

A program committee of 30 leading scientists, together with the help of external expert reviewers, shaped the PERVASIVE 2004 scientific program, the incarnation of which you now hold in your hands. Upon the call for papers, 278 submissions were received for consideration in the conference program – 212 for the *paper* track (including 8 tech-notes), 49 for the *hot spot paper* track, and 17 for the *video paper* track. In the *paper* track, each submission was assigned for review to at least three program committee members, who in turn often involved further experts in the review process, so that each paper received at least three (on average 3.27, at most 8) independent reviews. After a lively discussion in the program committee meeting on December 13, 2003, assessing the scientific quality and merits of each individual submission on top of the scoring it received from reviewers, 27 papers were accepted for presentation at PERVASIVE 2004 (12.7% acceptance). One accepted paper had to be withdrawn by the authors for restricted corporate reasons. Out of the 27 papers 19 were accepted in the category *regular papers* and 8 in the category *tech-notes*. Tech-notes are not to be understood as short papers condensed into fewer pages, but are intended to present pointed results at a high level of technicality in a very focused and compact format.

The PERVASIVE 2004 venue and presentation schedule was to some extent experimental, but appealing and promising. While an international doctoral colloquium preceded the main conference on April 18–19 at the University of Linz, tutorials and workshops opened the PERVASIVE 2004 activities in Vienna on April 20. The workshop topics expressed a good blend of topical research issues emerging under the pervasive computing umbrella: Gaming Applications in Pervasive Computing Environments (W1), Toolkit Support for Interaction in the Physical World (W2), Memory and Sharing of Experiences (W3), Computer Support for Human Tasks and Activities (W4), Benchmarks and a Database for Context Recognition (W5), SPPC: Security and Privacy in Pervasive Computing (W6), and Sustainable Pervasive Computing (W7). Technical paper sessions were scheduled from April 21 through April 23, highlighted by two very distinguished keynote speeches, and an inspiring banquet speech. A special PERVASIVE 2004 Video Night event presented video contributions in a lively format in a marvelous, historic place: the festival hall of the University of Vienna. All video clips are included in the PERVASIVE 2004 Video DVD. All doctoral colloquium papers, hot spot papers, and video papers are published in the "Advances in Pervasive Computing" book of the OCG (Vol. 176, ISBN 3-85403-176-9).

We want to thank all the people on the program committee and the volunteer reviewers (listed on the following pages) with sincere gratitude for their valuable assistance in this very difficult task of reviewing, judging, and scoring the technical paper submissions, as well as for their upright and factual contributions to the final decision process. We particularly wish to thank Albrecht Schmidt (Ludwig-Maximilians-Universität München) for being a very pragmatic workshop chair; Gabriele Kotsis (Johannes Kepler University Linz) for chairing the doctoral colloquium and for her pioneering work in making the colloquium ECTS credible; Horst Hörtner from the AEC (Ars Electronica Center) Future Lab for chairing the video track, as well as his team for the support in getting the PERVASIVE 2004 Video DVD produced; Rene Mayrhofer and Simon Vogl (both Johannes Kepler University Linz) for chairing the tutorials track; and Karin Anna Hummel (University of Vienna) and Rene Mayrhofer for their excellent work as publicity co-chairs.

From the many people who contributed to make PERVASIVE 2004 happen, our special thanks go to Gabriele Kotsis, president of the OCG (Oesterreichische Computergesellschaft), and her team headed by Eugen Mühlvenzl for co-organizing this event. As in many previous events of this nature, she was the real "organizational memory" behind everything – PERVASIVE 2004 would not have come to happen without her help. Warmest thanks go to both the Rektor of the University of Linz, Rudolf Ardelt, and the Rektor of the University of Vienna, Georg Winckler, for hosting PERVASIVE 2004. For their invaluable support making PERVASIVE 2004 a first-rank international event we thank Reinhard Göbl (Austrian Ministry of Transport, Innovation and Technology), Erich Prem (Austria's FIT-IT Embedded Systems Program), and Günter Haring (University of Vienna). Jörgen Bang Jensen, CEO of Austria's mobile communications provider ONE, Florian Pollack (head of ONE Mobile Living), and Florian Stieger (head of ONE Smart Space) generously helped in facilitating PERVASIVE 2004 and hosted the program committee meeting in ONE's smart space. Finally, we are grateful for the cooperative interaction with the organizers of the UbiComp conference series and their helpful support in finding the right time slot for this and future PERVASIVE conferences – PERVASIVE is planned to happen annually in spring, UbiComp in fall. Particular thanks go to Gregory Abowd (Georgia Institute of Technology), Hans-Werner Gellersen (Lancaster University), Albrecht Schmidt (Ludwig-Maximilians-Universität München), Lars Erik Holmquist (Viktoria Institute), Tom Rodden (Nottingham University), Anind Dey (Intel Research Berkeley), and Joe McCarthy (Intel Research Seattle) for their mentoring efforts – we look forward to a lively and sisterly interaction with UbiComp.

Finally, this booklet would not be in your hands without the hard work and selfless contributions of Rene Mayrhofer, our technical editor, and the patience and professional support of Alfred Hofmann and his team at Springer-Verlag. Last but not least we would like to express our sincere appreciation to the organizing committee at the Institute for Pervasive Computing at the University of Linz, in particular Monika Scholl, Sandra Derntl, and Karin Haudum, as well as Rene Mayrhofer, Simon Vogl, Dominik Hochreiter, Volker Christian, Wolfgang Narzt, Hans-Peter Baumgartner, Clemens Holzmann, Stefan Oppl, Manfred Hechinger, Günter Blaschek, and Thomas Scheidl.

The numerous authors who submitted papers, expressing their interest in PERVASIVE as the outlet for their research work, deserve our deepest thanks. It is their work – very often conducted in selfless and expendable efforts – that gives PERVASIVE its special vitality. We wish to strongly encourage the authors not presenting this year to continue their endeavors, and the participants new to PERVASIVE to remain part of it by submitting next year. We all hope that this year's program met with your approval, and we encourage you to actively contribute to (and thus steer) future PERVASIVE events.

February 2004   Alois Ferscha
                                                        Friedemann Mattern

Organization

PERVASIVE 2004, the second in a series of international conferences on Pervasive Computing, took place in Linz and Vienna, Austria from April 18 to 23, 2004. It was organized by the Department of Pervasive Computing, Johannes Kepler University of Linz, in cooperation with the Oesterreichische Computergesellschaft.

Executive Committee

General Chair	Friedemann Mattern
	(ETH Zurich, Switzerland)
Program Chair	Alois Ferscha
	(Johannes Kepler University Linz, Austria)
Doctoral Colloquium Chair	Gabriele Kotsis
	(Johannes Kepler University Linz, Austria)
Workshops Chair	Albrecht Schmidt
	(Ludwig-Maximilians-Universität, Germany)
Video Chair	Horst Hörtner
	(Ars Electronica Center, Austria)
Tutorial Co-chairs	Rene Mayrhofer
	(Johannes Kepler University Linz, Austria) and
	Simon Vogl
	(Johannes Kepler University Linz, Austria)
Organization Chair	Eugen Mühlvenzl
	(Oesterreichische Computergesellschaft, Austria)
Finance Co-chairs	Alois Ferscha
	(Johannes Kepler University Linz, Austria) and
	Gabriele Kotsis
	(Oesterreichische Computergesellschaft, Austria)
Technical Editor	Rene Mayrhofer
	(Johannes Kepler University Linz, Austria)
Publicity Co-chairs	Karin Anna Hummel
	(University of Vienna, Austria) and
	Rene Mayrhofer
	(Johannes Kepler University Linz, Austria)

Program Committee

Gregory Abowd, Georgia Institute of Technology, USA
Michael Beigl, TecO, University of Karlsruhe, Germany
Mark Billinghurst, University of Washington, USA
David De Roure, University of Southampton, UK

Anind Dey, Intel Research, USA
Elgar Fleisch, Universität St. Gallen, Switzerland
Hans Werner Gellersen, Lancaster University, UK
Lars Erik Holmquist, Viktoria Institute of Technology, Sweden
Horst Hörtner, Ars Electronica Center, Austria
Tim Kindberg, Hewlett-Packard Laboratories Bristol, UK
Gerd Kortuem, University of Lancaster, UK
Gabriele Kotsis, Johannes Kepler University Linz, Austria
Antonio Krüger, Saarland University, Germany
Marc Langheinrich, ETH Zurich, Switzerland
Max Mühlhäuser, Technische Universität Darmstadt, Germany
Joe Paradiso, MIT Media Laboratory, USA
Tom Pfeifer, TSSG, Waterford Institute of Technology, Ireland
Jun Rekimoto, Sony Computer Science Laboratories, Japan
Thomas Rist, DFKI Standort Saarbrücken, Germany
Tom Rodden, Nottingham University, UK
Anthony Savidis, ICS Forth, Greece
Bernt Schiele, ETH Zurich, Switzerland
Dieter Schmalstieg, TU Vienna, Austria
Albrecht Schmidt, Ludwig-Maximilians-Universität Munich, Germany
Vincent Stanford, NIST, USA
Thad Starner, Georgia Institute of Technology, USA
Franco Zambonelli, University of Modena and Reggio Emilia, Italy
Albert Zomaya, University of Sydney, Australia

Reviewers

Erwin Aitenbichler, TU Darmstadt, Germany
Stavros Antifakos, ETH Zurich, Switzerland
Gerhard Austaller, TU Darmstadt, Germany
Yuji Ayatsuka, Sony Computer Science Laboratories, Japan
Jonathan Bachrach, MIT Computer Science & Artificial Intelligence Lab, USA
Stephan Baldes, DFKI Saarbrücken, Germany
Istvan Barakonyi, TU Vienna, Austria
Steven Bathiche, Microsoft, USA
Martin Bauer, University of Stuttgart, Germany
Hans-Peter Baumgartner, Johannes Kepler University Linz, Austria
Jörg Baus, Saarland University, Germany
Ari Benbasat, MIT Media Laboratory, USA
Alastair Beresford, University of Cambridge, UK
Ansgar Bernardi, DFKI Kaiserslautern, Germany
Aggelos Bletsas, MIT Media Laboratory, USA
Juergen Bohn, ETH Zurich, Switzerland
Gaetano Borriello, University of Washington, USA

Boris Brandherm, Saarland University, Germany
Elmar Braun, TU Darmstadt, Germany
Sonja Buchegger, EPFL, Switzerland
Andreas Butz, Saarland University, Germany
Giacomo Cabri, University of Modena and Reggio Emilia, Italy
Eduardo Carrillo, Universidad de Valencia, Spain
Brian Clarkson, Sony Computer Science Laboratories, Japan
Christian Decker, TecO, University of Karlsruhe, Germany
Klaus Dorfmueller-Ulhaas, Universität Augsburg, Germany
Christoph Endres, Saarland University, Germany
Mattias Esbjörnsson, Interactive Institute, Sweden
Petra Fagerberg, Swedish Institute of Computer Science, Sweden
Mikael Fernström, University of Limerick, Ireland
Luca Ferrari, University of Modena and Reggio Emilia, Italy
Alois Ferscha, Johannes Kepler University Linz, Austria
Joe Finney, Lancaster University, UK
Christian Floerkemeier, ETH Zurich, Switzerland
Lalya Gaye, Viktoria Institute, Sweden
Matthias Gerlach, Fraunhofer FOKUS Berlin, Germany
Peter Gober, Fraunhofer FOKUS Berlin, Germany
Jacques Govignon, Draper Laboratory, USA
Raphael Grasset, INRIA, France
Thomas Grill, Johannes Kepler University Linz, Austria
Tom Gross, Bauhaus University Weimar, Germany
Susanne Guth, Wirtschafts-Universität Vienna, Austria
Michael Halle, Harvard Medical School, USA
Michael Haller, Fachhochschule Hagenberg, Austria
Andreas Hartl, TU Darmstadt, Germany
Ralf Hauber, Johannes Kepler University Linz, Austria
Mike Hazas, Lancaster University, UK
Manfred Hechinger, Johannes Kepler University Linz, Austria
Andreas Heinemann, TU Darmstadt, Germany
Ken Hinckley, Microsoft Research, USA
Helmut Hlavacs, University of Vienna, Austria
Jörg Hähner, University of Stuttgart, Germany
Maria Håkansson, Viktoria Institute, Sweden
Tobias Höllerer, University of California, USA
Jens Hünerberg, Fraunhofer FOKUS Berlin, Germany
Philipp Hünerberg, Fraunhofer FOKUS Berlin, Germany
Ismail Khalil Ibrahim, Johannes Kepler University Linz, Austria
Sozo Inoue, Kyushu University, Japan
Stephen Intille, MIT Department of Architecture and Planning, USA
Robert Jacob, Tufts University, USA
Anthony Jameson, DFKI Saarbrücken, Germany
Jussi Kangasharju, TU Darmstadt, Germany

Guenter Karjoth, IBM Research, Switzerland
Oliver Kasten, ETH Zurich, Switzerland
Hannes Kaufmann, TU Vienna, Austria
Nicky Kern, ETH Zurich, Switzerland
Engin Kirda, Technical University of Vienna, Austria
Georg Klein, Cambridge University, UK
Gudrun Klinker, Technical University of Munich, Germany
Michimune Kohno, Sony Computer Science Laboratories, Japan
Christian Kray, Lancaster University, UK
Michael Kreutzer, University of Freiburg, Germany
Albert Krohn, TecO, University of Karlsruhe, Germany
Reinhard Kronsteiner, Johannes Kepler University Linz, Austria
John Krumm, Microsoft Research, USA
Alexander Kröner, DFKI Saarbrücken, Germany
Uwe Kubach, SAP AG, Germany
James L. Crowley, INRIA, France
Anton L. Fuhrmann, VRVis Research Center, Austria
Mathew Laibowitz, MIT Media Laboratory, USA
Markus Lauff, SAP AG, Germany
Florian Ledermann, TU Vienna, Austria
Joshua Lifton, MIT Media Laboratory, USA
Tobias Limberger, TU Darmstadt, Germany
Sara Ljungblad, Viktoria Institute, Sweden
Asa MacWilliams, Technische Universität München, Germany
Carl Magnus Olsson, Viktoria Institute, Sweden
Rainer Malaka, European Media Lab, Germany
Marco Mamei, University of Modena and Reggio Emilia, Italy
Heiko Maus, DFKI Kaiserslautern, Germany
Rene Mayrhofer, Johannes Kepler University Linz, Austria
Joe McCarthy, Intel Research, USA
Florian Michahelles, ETH Zurich, Switzerland
Martin Mueller, Universität Augsburg, Germany
Joe Newman, TU Vienna, Austria
Daniela Nicklas, University of Stuttgart, Germany
Ian Oakley, Media Lab Europe, Ireland
Roy Oberhauser, Siemens AG, Germany
Mattias Östergren, Interactive Institute, Sweden
Stefan Oppl, Johannes Kepler University Linz, Austria
Steven Peters, MIT Computer Science & Artificial Intelligence Lab, USA
Mario Pichler, SCCH, Austria
Wayne Piekarski, University of Southern Australia, Australia
Claudio Pinhanez, IBM Research, USA
Ivan Poupyrev, Sony Computer Science Laboratories, Japan
Daniel Prince, Lancaster University, UK
Thomas Psik, TU Vienna, Austria

Sponsoring Institutions

Austrian Ministry of Transport, Innovation and Technology
FIT-IT Embedded Systems
Forschungsförderungsfonds für die Gewerbliche Wirtschaft
Oesterreichische Computergesellschaft
Land Oberösterreich
Stadt Linz
Stadt Wien

Table of Contents

Sensors

Security

Architectures and Systems

Algorithms

New Interfaces

Activity Recognition from User-Annotated Acceleration Data

Ling Bao and Stephen S. Intille

Massachusetts Institute of Technology
1 Cambridge Center, 4FL
Cambridge, MA 02142 USA
intille@mit.edu

Abstract. In this work, algorithms are developed and evaluated to detect physical activities from data acquired using five small biaxial accelerometers worn simultaneously on different parts of the body. Acceleration data was collected from 20 subjects without researcher supervision or observation. Subjects were asked to perform a sequence of everyday tasks but not told specifically where or how to do them. Mean, energy, frequency-domain entropy, and correlation of acceleration data was calculated and several classifiers using these features were tested. Decision tree classifiers showed the best performance recognizing everyday activities with an overall accuracy rate of 84%. The results show that although some activities are recognized well with subject-independent training data, others appear to require subject-specific training data. The results suggest that multiple accelerometers aid in recognition because conjunctions in acceleration feature values can effectively discriminate many activities. With just two biaxial accelerometers – thigh and wrist – the recognition performance dropped only slightly. This is the first work to investigate performance of recognition algorithms with multiple, wire-free accelerometers on 20 activities using datasets annotated by the subjects themselves.

1 Introduction

One of the key difficulties in creating useful and robust ubiquitous, context-aware computer applications is developing the algorithms that can detect context from noisy and often ambiguous sensor data. One facet of the user's context is his physical activity. Although prior work discusses physical activity recognition using acceleration (e.g. [17,5,23]) or a fusion of acceleration and other data modalities (e.g. [18]), it is unclear how most prior systems will perform under real-world conditions. Most of these works compute recognition results with data collected from subjects under artificially constrained laboratory settings. Some also evaluate recognition performance on data collected in natural, out-of-lab settings but only use limited data sets collected from one individual (e.g. [22]). A number of works use naturalistic data but do not quantify recognition accuracy. Lastly, research using naturalistic data collected from multiple subjects has focused on

A. Ferscha and F. Mattern (Eds.): PERVASIVE 2004, LNCS 3001, pp. 1–17, 2004.

recognition of a limited subset of nine or fewer everyday activities consisting largely of ambulatory motions and basic postures such as sitting and standing (e.g. [10,5]). It is uncertain how prior systems will perform in recognizing a variety of everyday activities for a diverse sample population under real-world conditions.

In this work, the performance of activity recognition algorithms under conditions akin to those found in real-world settings is assessed. Activity recognition results are based on acceleration data collected from five biaxial accelerometers placed on 20 subjects under laboratory and semi-naturalistic conditions. Supervised learning classifiers are trained on labeled data that is acquired without researcher supervision from subjects themselves. Algorithms trained using only user-labeled data might dramatically increase the amount of training data that can be collected and permit users to train algorithms to recognize their own individual behaviors.

2   Background

Researchers have already prototyped wearable computer systems that use acceleration, audio, video, and other sensors to recognize user activity (e.g. [7]). Advances in miniaturization will permit accelerometers to be embedded within wrist bands, bracelets, adhesive patches, and belts and to wirelessly send data to a mobile computing device that can use the signals to recognize user activities.

For these applications, it is important to train and test activity recognition systems on data collected under naturalistic circumstances, because laboratory environments may artificially constrict, simplify, or influence subject activity patterns. For instance, laboratory acceleration data of walking displays distinct phases of a consistent gait cycle which can aide recognition of pace and incline [2]. However, acceleration data from the same subject outside of the laboratory may display marked fluctuation in the relation of gait phases and total gait length due to decreased self-awareness and fluctuations in traffic. Consequently, a highly accurate activity recognition algorithm trained on data where subjects are told exactly where or how to walk (or where the subjects are the researchers themselves) may rely too heavily on distinct phases and periodicity of accelerometer signals found only in the lab. The accuracy of such a system may suffer when tested on naturalistic data, where there is greater variation in gait pattern.

Many past works have demonstrated 85% to 95% recognition rates for ambulation, posture, and other activities using acceleration data. Some are summarized in Figure 1 (see [3] for a summary of other work). Activity recognition has been performed on acceleration data collected from the hip (e.g. [17,19]) and from multiple locations on the body (e.g. [5,14]). Related work using activity counts and computer vision also supports the potential for activity recognition using acceleration. The energy of a subject's acceleration can discriminate sedentary activities such as sitting or sleeping from moderate intensity activities such as walking or typing and vigorous activities such as running [25]. Recent work

Ref.	Recognition Accuracy	Activities Recognized	No. Subj.	Data Type	No. Sensors	Sensor Placement
[17]	92.85% to 95.91%	ambulation	8	L	2	2 thigh
[19]	83% to 90%	ambulation, posture	6	L	6	3 left hip, 3 right hip
[10]	95.8%	ambulation, posture, typing, talking, bicycling	24	L	4	chest, thigh, wrist, forearm
[10]	66.7%	ambulation, posture, typing, talking, bicycling	24	N	4	chest, thigh, wrist, forearm
[1]	89.30%	ambulation, posture	5	L	2	chest, thigh
[12]	N/A	walking speed, incline	20	L	4	3 lower back 1 ankle
[22]	86% to 93%	ambulation, posture, play	1	N	3	2 waist, 1 thigh
[14]	≈65% to ≈95%	ambulation, typing, stairs shake hands, write on board	1	L	up to 36	all major joints
[6]	96.67%	3 Kung Fu arm movements	1	L	2	2 wrist
[23]	42% to 96%	ambulation, posture, bicycling	1	L	2	2 lower back
[20]	85% to 90%	ambulation, posture	10	L	2	2 knee

Fig. 1. Summary of a representative sample of past work on activity recognition using acceleration. The "No. Subj." column specifies the number of subjects who participated in each study, and the "Data Type" column specifies whether data was collected under laboratory (L) or naturalistic (N) settings. The "No. Sensors" column specifies the number of uniaxial accelerometers used per subject.

with 30 wired accelerometers spread across the body suggests that the addition of sensors will generally improve recognition performance [24].

Although the literature supports the use of acceleration for physical activity recognition, little work has been done to validate the idea under real-world circumstances. Most prior work on activity recognition using acceleration relies on data collected in controlled laboratory settings. Typically, the researcher collected data from a very small number of subjects, and often the subjects have included the researchers themselves. The researchers then hand-annotated the collected data. Ideally, data would be collected in less controlled settings without researcher supervision. Further, to increase the volume of data collected, subjects would be capable of annotating their own data sets. Algorithms that could be trained using only user-labeled data might dramatically increase the amount of training data that can be collected and permit users to train algorithms to recognize their own individual behaviors. In this work we assume that labeled training data is required for many automatic activity recognition tasks. We note, however, that one recent study has shown that unsupervised learning can be used to cluster accelerometer data into categories that, in some instances, map onto meaningful labels [15].

The vast majority of prior work focuses on recognizing a special subset of physical activities such as ambulation, with the exception of [10] which examines nine everyday activities. Interestingly, [10] demonstrated 95.8% recognition rates for data collected in the laboratory but recognition rates dropped to 66.7% for data collected outside the laboratory in naturalistic settings. These results demonstrate that the performance of algorithms tested only on laboratory data or data acquired from the experimenters themselves may suffer when tested on data collected under less-controlled (i.e. naturalistic) circumstances.

Prior literature demonstrates that forms of locomotion such as walking, running, and climbing stairs and postures such as sitting, standing, and lying down can be recognized at 83% to 95% accuracy rates using hip, thigh, and ankle acceleration (see Figure 1). Acceleration data of the wrist and arm are known to improve recognition rates of upper body activities [6,10] such as typing and martial arts movements. All past works with multiple accelerometers have used accelerometers connected with wires, which may restrict subject movement. Based on these results, this work uses data collected from five wire-free biaxial accelerometers placed on each subject's right hip, dominant wrist, non-dominant upper arm, dominant ankle, and non-dominant thigh to recognize ambulation, posture, and other everyday activities. Although each of the above five locations have been used for sensor placement in past work, no work addresses which of the accelerometer locations provide the best data for recognizing activities even though it has been suggested that for some activities that more sensors improve recognition [24]. Prior work has typically been conducted with only 1-2 accelerometers worn at different locations on the body, with only a few using more than 5 (e.g. [19,14,24]).

3   Design

Subjects wore 5 biaxial accelerometers as they performed a variety of activities under two different data collection protocols.

3.1   Accelerometers

Subject acceleration was collected using ADXL210E accelerometers from Analog Devices. These two-axis accelerometers are accurate to ±10 G with tolerances within 2%. Accelerometers were mounted to *hoarder boards* [11], which sampled at 76.25 Hz (with minor variations based on onboard clock accuracy) and stored acceleration data on compact flash memory. This sampling frequency is more than sufficient compared to the 20 Hz frequency required to assess daily physical activity [4]. The hoarder board time stamped one out of every 100 acceleration samples, or one every 1.31 seconds. Four AAA batteries can power the hoarder board for roughly 24 hours. This is more than sufficient for the 90 minute data collection sessions used in this study. A hoarder board is shown in Figure 2a.

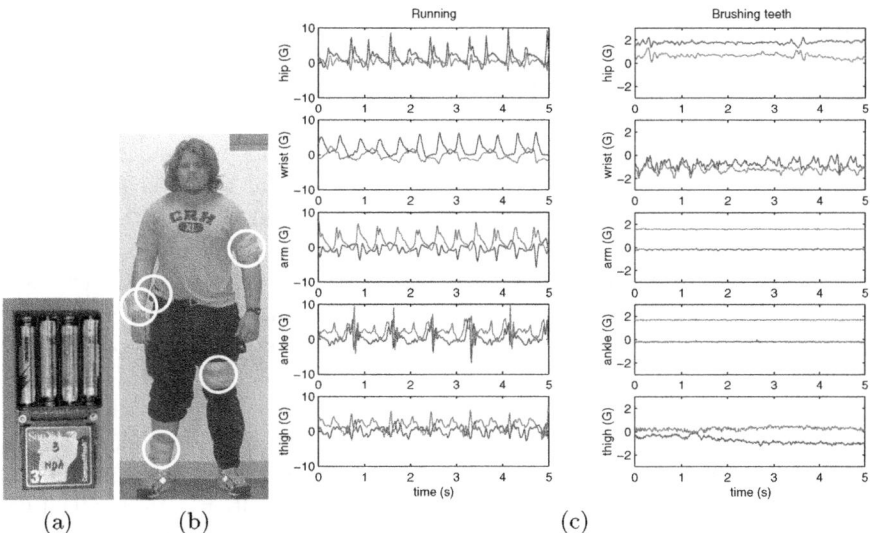

Fig. 2. (a) Hoarder data collection board, which stored data from a biaxial accelerometer. The biaxial accelerometers are attached to the opposite side of the board. (b) Hoarder boards were attached to 20 subjects on the 4 limb positions shown here (held on with medical gauze), plus the right hip. (c) Acceleration signals from five biaxial accelerometers for walking, running, and tooth brushing.

Previous work shows promising activity recognition results from ±2 G acceleration data (e.g. [9,14]) even though typical body acceleration amplitude can range up to 12 G [4]. However, due to limitations in availability of ±12 G accelerometers, ±10 G acceleration data was used. Moreover, although body limbs and extremities can exhibit a 12 G range in acceleration, points near the torso and hip experience a 6 G range in acceleration [4].

The hoarder boards were not electronically synchronized to each other and relied on independent quartz clocks to time stamp data. Electronic synchronization would have required wiring between the boards which, even when the wiring is carefully designed as in [14], would restrict subject movements, especially during whole body activities such as bicycling or running. Further, we have found subjects wearing wiring feel self-conscious when outside of the laboratory and therefore restrict their behavior.

To achieve synchronization without wires, hoarder board clocks were synchronized with subjects' watch times at the beginning of each data collection session. Due to clock skew, hoarder clocks and the watch clock drifted between 1 and 3 seconds every 24 hours. To minimize the effects of clock skew, hoarder boards were shaken together in a fixed sinusoidal pattern in two axes of acceleration at the beginning and end of each data collection session. Watch times were manually recorded for the periods of shaking. The peaks of the distinct sinusoidal patterns at the beginning and end of each acceleration signal were visually aligned between the hoarder boards. Time stamps during the shaking period were also shifted to be consistent with the recorded watch times for shaking. Acceleration time stamps were linearly scaled between these manually aligned start and end points.

To characterize the accuracy of the synchronization process, three hoarder boards were synchronized with each other and a digital watch using the above protocol. The boards were then shaken together several times during a full day to produce matching sinusoidal patterns on all boards. Visually comparing the peaks of these matching sinusoids across the three boards showed mean skew of 4.3 samples with a standard deviation of 1.8 samples between the boards. At a sampling frequency of 76.25 Hz, the skew between boards is equivalent to $.0564 \pm .0236$ s.

A T-Mobile Sidekick phone pouch was used as a carrying case for each hoarder board. The carrying case was light, durable, and provided protection for the electronics. A carrying case was secured to the subject's belt on the right hip. All subjects were asked to wear clothing with a belt. Elastic medical bandages were used to wrap and secure carrying cases at sites other than the hip. Typical placement of hoarder boards is shown in Figure 2b. Figure 2c shows acceleration data collected for walking, running, and tooth brushing from the five accelerometers.

No wires were used to connect the hoarder boards to each other or any other devices. Each hoarder in its carrying case weighed less than 120 g. Subjects could engage in vigorous, complex activity without any restriction on movement or fear of damaging the electronics. The sensors were still visually noticeable. Subjects who could not wear the devices under bulky clothing did report feeling self conscious in public spaces.

3.2   Activity Labels

Twenty activities were studied. These activities are listed in Figure 5. The 20 activities were selected to include a range of common everyday household activities that involve different parts of the body and range in level of intensity. Whole body activities such as walking, predominantly arm-based activities such as brushing of teeth, and predominantly leg-based activities such as bicycling were included as were sedentary activities such as sitting, light intensity activities such as eating, moderate intensity activities such as window scrubbing, and vigorous activities such as running. Activity labels were chosen to reflect the content of the actions but do not specify the style. For instance, "walking" could be parameterized by walking speed and quantized into slow and brisk or other categories.

3.3   Semi-naturalistic, User-Driven Data Collection

The most realistic training and test data would be naturalistic data acquired from subjects as they go about their normal, everyday activities. Unfortunately, obtaining such data requires direct observation of subjects by researchers, subject self-report of activities, or use of the experience sampling method [8] to label subject activities for algorithm training and testing. Direct observation can be costly and scales poorly for the study of large subject populations. Subject self-report recall surveys are prone to recall errors [8] and lack the temporal precision required for training activity recognition algorithms. Finally, the experience sampling method requires frequent interruption of subject activity, which agitates subjects over an extended period of time. Some activities we would like to develop recognition algorithms for, such as folding laundry, riding escalators, and scrubbing windows, may not occur on a daily basis. A purely naturalistic protocol would not capture sufficient samples of these activities for thorough testing of recognition systems without prohibitively long data collection periods.

In this work we compromise and use a semi-naturalistic collection protocol that should permit greater subject variability in behavior than laboratory data. Further, we show how training sets can be acquired from subjects themselves without the direct supervision of a researcher, which may prove important if training data must be collected by end users to improve recognition performance.

For semi-naturalistic data collection, subjects ran an obstacle course consisting of a series of activities listed on a worksheet. These activities were disguised as goals in an obstacle course to minimize subject awareness of data collection. For instance, subjects were asked to "use the web to find out what the world's largest city in terms of population is" instead of being asked to "work on a computer." Subjects recorded the time they began each obstacle and the time they completed each obstacle. Subjects completed each obstacle on the course ensuring capture of all 20 activities being studied. There was no researcher supervision of subjects while they collected data under the semi-naturalistic collection protocol. As subjects performed each of these obstacles in the order given on their worksheet, they labeled the start and stop times for that activity and made any relevant notes about that activity. Acceleration data collected between the start and stop times were labeled with the name of that activity. Subjects were free to rest between obstacles and proceed through the worksheet at their own pace as long as they performed obstacles in the order given. Furthermore, subjects had freedom in how they performed each obstacle. For example, one obstacle was to "read the newspaper in the common room. Read the entirety of at least one non-frontpage article." The subject could choose which and exactly how many articles to read. Many activities were performed outside of the lab. Subjects were not told where or how to perform activities and could do so in a common room within the lab equipped with a television, vacuum, sofa, and reading materials or anywhere they preferred. No researchers or cameras monitored the subjects.

3.4 Specific Activity Data Collection

After completing the semi-naturalistic obstacle course, subjects underwent another data collection session to collect data under somewhat more controlled conditions. Linguistic definitions of activity are often ambiguous. The activity

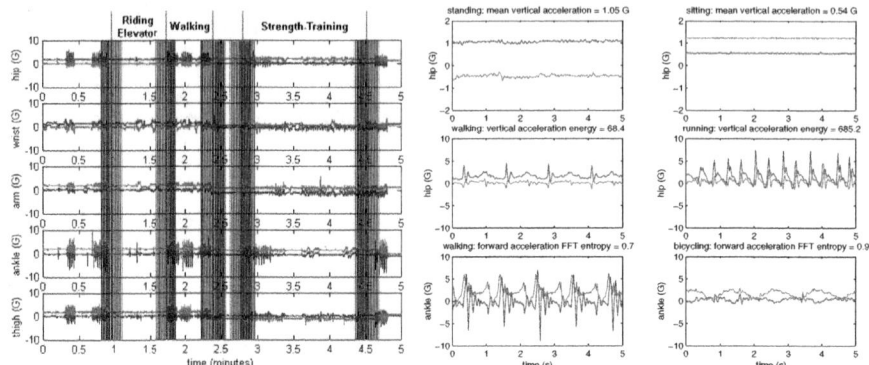

Fig. 3. (a) Five minutes of 2-axis acceleration data annotated with subject self-report activity labels. Data within 10s of self-report labels is discarded as indicated by masking. (b) Differences in feature values computed from FFTs are used to discriminate between different activities.

"scrubbing," for example, can be interpreted as window scrubbing, dish scrubbing, or car scrubbing. For this data collection session, subjects were therefore given short definitions of the 20 activity labels that resolved major ambiguities in the activity labels while leaving room for interpretation so that subjects could show natural, individual variations in how they performed activities. For example, walking was described as "walking without carrying any items in you hand or on your back heavier than a pound" and scrubbing was described as "using a sponge, towel, or paper towel to wipe a window." See [3] for descriptions for all 20 activities.

Subjects were requested to perform random sequences of the 20 activities defined on a worksheet during laboratory data collection. Subjects performed the sequence of activities given at their own pace and labeled the start and end times of each activity. For example, the first 3 activities listed on the worksheet might be "bicycling," "riding elevator," and "standing still." The researcher's definition of each of these activities was provided. As subjects performed each of these activities in the order given on their worksheet, they labeled the start and stop times for that activity and made any relevant notes about that activity such as "I climbed the stairs instead of using the elevator since the elevator was out of service." Acceleration data collected between the start and stop times were labeled with the name of that activity. To minimize mislabeling, data within 10 s of the start and stop times was discarded. Since the subject is probably standing still or sitting while he records the start and stop times, the data immediately around these times may not correspond to the activity label. Figure 3a shows acceleration data annotated with subject self-report labels.

Although data collected under this second protocol is more structured than the first, it was still acquired under less controlled conditions than in most prior work. Subjects, who were not the researchers, could perform their activities anywhere including outside of the laboratory. Also, there was no researcher supervision during the data collection session.

3.5 Feature Computation

Features were computed on 512 sample windows of acceleration data with 256 samples overlapping between consecutive windows. At a sampling frequency of 76.25 Hz, each window represents 6.7 seconds. Mean, energy, frequency-domain entropy, and correlation features were extracted from the sliding windows signals for activity recognition. Feature extraction on sliding windows with 50% overlap has demonstrated success in past works [9,23]. A window of several seconds was used to sufficiently capture cycles in activities such as walking, window scrubbing, or vacuuming. The 512 sample window size enabled fast computation of FFTs used for some of the features.

The DC feature is the mean acceleration value of the signal over the window. The energy feature was calculated as the sum of the squared discrete FFT component magnitudes of the signal. The sum was divided by the window length for normalization. Additionally, the DC component of the FFT was excluded in this sum since the DC characteristic of the signal is already measured by another feature. Note that the FFT algorithm used produced 512 components for each 512 sample window. Use of mean [10,1] and energy [21] of acceleration features has been shown to result in accurate recognition of certain postures and activities (see Figure 1).

Frequency-domain entropy is calculated as the normalized information entropy of the discrete FFT component magnitudes of the signal. Again, the DC component of the FFT was excluded in this calculation. This feature may support discrimination of activities with similar energy values. For instance, biking and running may result in roughly the same amounts of energy in the hip acceleration data. However, because biking involves a nearly uniform circular movement of the legs, a discrete FFT of hip acceleration in the vertical direction may show a single dominant frequency component at 1 Hz and very low magnitude for all other frequencies. This would result in a low frequency-domain entropy. Running on the other hand may result in complex hip acceleration and many major FFT frequency components between 0.5 Hz and 2 Hz. This would result in a higher frequency-domain entropy.

Features that measure correlation or acceleration between axes can improve recognition of activities involving movements of multiple body parts [12,2]. Correlation is calculated between the two axes of each accelerometer hoarder board and between all pairwise combinations of axes on different hoarder boards.

Figure 3b shows some of these features for two activities. It was anticipated that certain activities would be difficult to discriminate using these features. For example, "watching TV" and "sitting" should exhibit very similar if not identical body acceleration. Additionally, activities such as "stretching" could show marked variation from person to person and for the same person at different times. Stretching could involve light or moderate energy acceleration in the upper body, torso, or lower body.

As discussed in the the next section, several classifiers were tested for activity recognition using the feature vector.

4   Evaluation

Subjects were recruited using posters seeking research study participants for compensation. Posters were distributed around an academic campus and were also emailed to the student population. Twenty subjects from the academic community volunteered. Data was collected from 13 males and 7 females. Subjects ranged in age from 17 to 48 (mean 21.8, sd 6.59).

Each subject participated in two sessions of study. In the first session, subjects wore five accelerometers and a digital watch. Subjects collected the semi-naturalistic data by completing an obstacle course worksheet, noting the start and end times of each obstacle on the worksheet. Each subject collected between 82 and 160 minutes of data (mean 104, sd 13.4). Six subjects skipped between one to two obstacles due to factors such as inclement weather, time constraints, or problems with equipment in the common room (e.g. the television, vacuum, computer, and bicycle). Subjects performed each activity on their obstacle course for an average of 156 seconds (sd 50).

In the second session, often performed on a different day, the same subjects wore the same set of sensors. Subjects performed the sequence of activities listed on an activity worksheet, noting the start and end times of these activities. Each subject collected between 54 and 131 minutes of data (mean 96, sd 16.7). Eight subjects skipped between one to four activities due to factors listed earlier.

4.1   Results

Mean, energy, entropy, and correlation features were extracted from acceleration data. Activity recognition on these features was performed using decision table, instance-based learning (IBL or nearest neighbor), C4.5 decision tree, and naive Bayes classifiers found in the Weka Machine Learning Algorithms Toolkit [26].

Classifiers were trained and tested using two protocols. Under the first protocol, classifiers were trained on each subject's activity sequence data and tested on that subject's obstacle course data. This user-specific training protocol was repeated for all twenty subjects. Under the second protocol, classifiers were trained on activity sequence and obstacle course data for all subjects except one. The classifiers were then tested on obstacle course data for the only subject left out of the training data set. This leave-one-subject-out validation process was repeated for all twenty subjects. Mean and standard deviation for classification accuracy under both protocols is summarized in Figure 4.

Overall, recognition accuracy is highest for decision tree classifiers, which is consistent with past work where decision based algorithms recognized lying, sitting, standing and locomotion with 89.30% accuracy [1]. Nearest neighbor is the second most accurate algorithm and its strong relative performance is

Classifier	User-specific Training	Leave-one-subject-out Training
Decision Table	36.32 ± 14.501	46.75 ± 9.296
IBL	69.21 ± 6.822	82.70 ± 6.416
C4.5	71.58 ± 7.438	84.26 ± 5.178
Naive Bayes	34.94 ± 5.818	52.35 ± 1.690

Fig. 4. Summary of classifier results (mean ± standard deviation) using user-specific training and leave-one-subject-out training. Classifiers were trained on laboratory data and tested on obstacle course data.

Activity	Accuracy	Activity	Accuracy
Walking	89.71	Walking carrying items	82.10
Sitting & relaxing	94.78	Working on computer	97.49
Standing still	95.67	Eating or drinking	88.67
Watching TV	77.29	Reading	91.79
Running	87.68	Bicycling	96.29
Stretching	41.42	Strength-training	82.51
Scrubbing	81.09	Vacuuming	96.41
Folding laundry	95.14	Lying down & relaxing	94.96
Brushing teeth	85.27	Climbing stairs	85.61
Riding elevator	43.58	Riding escalator	70.56

Fig. 5. Aggregate recognition rates (%) for activities studied using leave-one-subject-out validation over 20 subjects.

also supported by past prior work where nearest neighbor algorithms recognized ambulation and postures with over 90% accuracy [16,10].

Figure 5 shows the recognition results for the C4.5 classifier. Rule-based activity recognition appears to capture conjunctions in feature values that may lead to good recognition accuracy. For instance, the C4.5 decision tree classified sitting as an activity having nearly 1 G downward acceleration and low energy at both hip and arm. The tree classified bicycling as an activity involving moderate energy levels and low frequency-domain entropy at the hip and low energy levels at the arm. The tree distinguishes "window scrubbing" from "brushing teeth" because the first activity involves more energy in hip acceleration even though both activities show high energy in arm acceleration. The fitting of probability distributions to acceleration features under a Naive Bayesian approach may be unable to adequately model such rules due to the assumptions of conditional independence between features and normal distribution of feature values, which may account for the weaker performance. Furthermore, Bayesian algorithms may require more data to accurately model feature value distributions.

Figure 6 shows an aggregate confusion matrix for the C4.5 classifier based on all 20 trials of leave-one-subject-out validation. Recognition accuracies for stretching and riding an elevator were below 50%. Recognition accuracies for

a	b	c	d	e	f	g	h	i	j	k	l	m	n	o	p	q	r	s	t	< classified as
942	46	0	0	2	0	0	0	8	3	8	1	4	2	7	0	3	8	8	8	a = walking
83	**1183**	9	0	3	2	0	0	8	1	3	8	14	1	16	0	8	53	38	11	b = walking/carry
0	9	**762**	11	0	1	17	3	0	0	0	0	0	0	0	1	0	0	0	0	c = sitting relaxed
0	0	10	**893**	9	1	0	1	0	1	0	0	0	0	1	0	0	0	0	0	d = computer work
0	0	0	7	**774**	11	0	0	0	6	1	2	2	0	4	0	2	0	0	0	e = standing still
0	2	1	0	12	**712**	9	1	0	0	2	1	10	1	18	0	26	1	4	3	f = eating/drinking
0	0	42	21	0	1	**320**	28	0	0	0	0	0	0	0	0	0	0	0	1	g = watching TV
0	0	23	1	1	6	16	**961**	9	0	2	0	0	1	0	1	2	0	2	22	h = reading
14	12	0	0	1	1	0	17	**491**	10	1	1	1	1	1	0	1	3	4	1	i = running
0	1	0	0	5	0	0	0	8	**830**	10	0	1	0	3	0	2	1	0	1	j = bicycling
9	3	2	16	30	22	45	9	3	35	**309**	37	26	21	99	1	38	12	3	26	k = stretching
4	10	0	0	6	5	2	7	0	6	23	**500**	13	2	9	3	6	5	3	2	l = strength train
1	7	0	0	5	10	0	0	0	3	9	9	**403**	11	10	1	26	1	6	4	m = scrubbing
1	0	0	0	3	1	0	0	2	0	1	9	10	**885**	11	0	1	0	2	2	n = vacuuming
1	1	0	0	1	6	0	0	1	4	1	4	7	10	**822**	8	4	0	1	3	o = folding laundry
0	0	4	9	0	2	1	7	0	0	0	1	0	10	0	**791**	8	0	0	0	p = lying down
1	2	0	0	3	32	0	0	1	5	0	18	7	10	9	0	**637**	10	2	10	q = brushing teeth
7	14	0	0	1	1	0	0	3	2	1	1	0	2	0	12	0	**351**	10	5	r = climbing stairs
84	70	0	7	20	60	0	0	8	40	33	11	24	34	40	0	0	59	**502**	160	s = riding elevator
5	2	0	0	5	6	0	1	0	1	0	3	3	1	0	0	3	7	16	**127**	t = riding escalator

Fig. 6. Aggregate confusion matrix for C4.5 classifier based on leave-one-subject-out validation for 20 subjects, tested on semi-naturalistic data.

"watching TV" and "riding escalator" were 77.29% and 70.56%, respectively. These activities do not have simple characteristics and are easily confused with other activities. For instance, "stretching" is often misclassified as "folding laundry" because both may involve the subject moving the arms at a moderate rate. Similarly, "riding elevator" is misclassified as "riding escalator" since both involve the subject standing still. "Watching TV" is confused with "sitting and relaxing" and "reading" because all the activities involve sitting. "Riding escalator" is confused with "riding elevator" since the subject may experience similar vertical acceleration in both cases. "Riding escalator" is also confused with "climbing stairs" since the subject sometimes climbs the escalator stairs.

Recognition accuracy was significantly higher for all algorithms under the leave-one-subject-out validation process. This indicates that the effects of individual variation in body acceleration may be dominated by strong commonalities between people in activity pattern. Additionally, because leave-one-subject-out validation resulted in larger training sets consisting of data from 19 subjects, this protocol may have resulted in more generalized and robust activity classifiers. The markedly smaller training sets used for the user-specific training protocol may have limited the accuracy of classifiers.

To control for the effects of sample size in comparing leave-one-subject-out and user-specific training, preliminary results were gathered using a larger training data set collected for three subjects. These subjects were affiliates of the researchers (unlike the 20 primary subjects). Each of these subjects participated in one semi-naturalistic and five laboratory data collection sessions. The C4.5 decision tree algorithm was trained for each individual using data collected from all five of his laboratory sessions and tested on the semi-naturalistic data. The algorithm was also trained on five laboratory data sets from five random subjects other than the individual and tested on the individual's semi-naturalistic data. The results are compared in Figure 7. In this case, user-specific training resulted in an increase in recognition accuracy of 4.32% over recognition rates for leave-one-subject-out-training. This difference shows that given equal amounts of training data, training on user-specific training data can result in classifiers

Classifier	User-specific Training	Leave-one-subject-out Training
C4.5	77.31 ± 4.328	72.99 ± 8.482

Fig. 7. Summary of classifier results (mean ± standard deviation) using user-specific training and leave-one-subject-out training where both training data sets are equivalent to five laboratory data sessions.

that recognize activities more accurately than classifiers trained on example data from many people. However, the certainty of these conclusions is limited by the low number of subjects used for this comparison and the fact that the three individuals studied were affiliates of the researchers. Nonetheless, these initial results support the need for further study of the power of user-specific versus generalized training sets.

The above results suggest that real-world activity recognition systems can rely on classifiers that are pre-trained on large activity data sets to recognize some activities. Although preliminary results show that user-specific training can lead to more accurate activity recognition given large training sets, pre-trained systems offer greater convenience. Pre-trained systems could recognize many activities accurately without requiring training on data from their user, simplifying the deployment of these systems. Furthermore, since the activity recognition system needs to be trained only once before deployment, the slow running time for decision tree training is not an obstacle. Nonetheless, there may be limitations to a pre-trained algorithm. Although activities such as "running" or "walking" may be accurately recognized, activities that are more dependent upon individual variation and the environment (e.g. "stretching") may require person-specific training [13]).

To evaluate the discriminatory power of each accelerometer location, recognition accuracy using the decision tree classifier (the best performing algorithm) was also computed using a leave-one-accelerometer-in protocol. Specifically, recognition results were computed five times, each time using data from only one of the five accelerometers for the training and testing of the algorithm. The differences in recognition accuracy rates using this protocol from accuracy rates obtained from all five accelerometers are summarized in Figure 8. These results show that the accelerometer placed on the subject's thigh is the most powerful for recognizing this set of 20 activities. Acceleration of the dominant wrist is more useful in discriminating these activities than acceleration of the non-dominant arm. Acceleration of the hip is the second best location for activity discrimination. This suggests that an accelerometer attached to a subject's cell phone, which is often placed at a fixed location such as on a belt clip, may enable recognition of certain activities.

Confusion matrices resulting from leave-one-accelerometer-in testing [3] show that data collected from lower body accelerometers placed on the thigh, hip, and ankle is generally best at recognizing forms of ambulation and posture. Ac-

Accelerometer(s) Left In	Difference in Recognition Accuracy
Hip	-34.12 ± 7.115
Wrist	-51.99 ± 12.194
Arm	-63.65 ± 13.143
Ankle	-37.08 ± 7.601
Thigh	-29.47 ± 4.855
Thigh and Wrist	-3.27 ± 1.062
Hip and Wrist	-4.78 ± 1.331

Fig. 8. Difference in overall recognition accuracy (mean $\pm$ standard deviation) due to leaving only one or two accelerometers in. Accuracy rates are aggregated for 20 subjects using leave-one-subject-out validation.

celerometer data collected from the wrist and arm is better at discriminating activities involving characteristic upper body movements such as reading from watching TV or sitting and strength-training (push ups) from stretching. To explore the power of combining upper and lower body accelerometer data, data from thigh and wrist accelerometers and hip and wrist accelerometers were also used and results are shown in Figure 8. Note that recognition rates improved over 25% for the leave-two-accelerometers-in results as compared to the best leave-one-accelerometer-in results. Of the two pairs tested, thigh and wrist acceleration data resulted in the highest recognition accuracy. However, both thigh and wrist and hip and wrist pairs showed less than a 5% decrease in recognition rate from results using all five accelerometer signals. This suggests that effective recognition of certain everyday activities can be achieved using two accelerometers placed on the wrist and thigh or wrist and hip. Others have also found that for complex activities at least one sensor on the lower and upper body is desirable [14].[1]

4.2   Analysis

This work shows that user-specific training is not necessary to achieve recognition rates for some activities of over 80% for 20 everyday activities. Classification accuracy rates of between 80% to 95% for walking, running, climbing stairs, standing still, sitting, lying down, working on a computer, bicycling, and vacuuming are comparable with recognition results using laboratory data from previous works. However, most prior has used data collected under controlled laboratory conditions to achieve their recognition accuracy rates, typically where data is hand annotated by a researcher. The 84.26% overall recognition rate achieved in this work is significant because study subjects could move about freely outside the lab without researcher supervision while collecting and annotating their own

---

[1] Only the decision tree algorithm was used to evaluate the information content of specific sensors, leaving open the possibility that other algorithms may perform better with different sensor placements.

semi-naturalistic data. This is a step towards creating mobile computing systems that work outside of the laboratory setting.

The C4.5 classifier used mean acceleration to recognize postures such as sitting, standing still, and lying down. Ambulatory activities and bicycling were recognized by the level of hip acceleration energy. Frequency-domain entropy and correlation between arm and hip acceleration strongly distinguished bicycling, which showed low entropy hip acceleration and low arm-hip correlation, from running, which displayed higher entropy in hip acceleration and higher arm-hip movement correlation. Both activities showed similar levels of hip acceleration mean and energy. Working on a computer, eating or drinking, reading, strength-training as defined by a combination of sit ups and push-ups, window scrubbing, vacuuming, and brushing teeth were recognized by arm posture and movement as measured by mean acceleration and energy.

Lower recognition accuracies for activities such as stretching, scrubbing, riding an elevator, and riding an escalator suggest that higher level analysis is required to improve classification of these activities. Temporal information in the form of duration and time and day of activities could be used to detect activities. For instance, standing still and riding an elevator are similar in terms of body posture. However, riding an elevator usually lasts for a minute or less whereas standing still can last for a much longer duration. By considering the duration of a particular posture or type of body acceleration, these activities could be distinguished from each other with greater accuracy. Similarly, adults may be more likely to watch TV at night than at other times on a weekday. Thus, date and time or other multi-modal sensing could be used to improve discrimination of watching TV from simply sitting and relaxing. However, because daily activity patterns may vary dramatically across individuals, user-specific training may be required to effectively use date and time information for activity recognition.

The decision tree algorithm used in this work can recognize the content of activities, but may not readily recognize activity style. Although a decision tree algorithm could potentially recognize activity style using a greater number of labels such as "walking slowly," "walking briskly," "scrubbing softly," or "scrubbing vigorously," the extensibility of this technique is limited. For example, the exact pace of walking cannot be recognized using any number of labels. Other techniques may be required to recognize parameterized activity style.

Use of other sensor data modalities may further improve activity recognition. Heart rate data could be used to augment acceleration data to detect intensity of physical activities. GPS location data could be used to infer whether an individual is at home or at work and affect the probability of activities such as working on the computer or lying down and relaxing. Use of such person-specific sensors such as GPS, however, is more likely to require that training data be acquired directly from the individual rather than from a laboratory setting because individuals can work, reside, and shop in totally different locations.

5   Conclusion

Using decision tree classifiers, recognition accuracy of over 80% on a variety of 20 everyday activities was achieved using leave-one-subject-out-validation on data acquired without researcher supervision from 20 subjects. These results are competitive with prior activity recognition results that only used laboratory data. Furthermore, this work shows acceleration can be used to recognize a variety of household activities for context-aware computing. This extends previous work on recognizing ambulation and posture using acceleration (see Figure 1).

This work further suggests that a mobile computer and small wireless accelerometers placed on an individual's thigh and dominant wrist may be able to detect some common everyday activities in naturalistic settings using fast FFT-based feature computation and a decision tree classifier algorithm. Decision trees are slow to train but quick to run. Therefore, a pre-trained decision tree should be able to classify user activities in real-time on emerging mobile computing devices with fast processors and wireless accelerometers.

Acknowledgements. This work was supported, in part, by National Science Foundation ITR grant #0112900 and the Changing Places/House_n Consortium.

References

1. K. Aminian, P. Robert, E.E. Buchser, B. Rutschmann, D. Hayoz, and M. Depairon. Physical activity monitoring based on accelerometry: validation and comparison with video observation. *Medical & Biological Engineering & Computing*, 37(3):304–8, 1999.
2. K. Aminian, P. Robert, E. Jequier, and Y. Schutz. Estimation of speed and incline of walking using neural network. *IEEE Transactions on Instrumentation and Measurement*, 44(3):743–746, 1995.
3. L. Bao. *Physical Activity Recognition from Acceleration Data under Semi-Naturalistic Conditions*. M.Eng. Thesis, Massachusetts Institute of Technology, 2003.
4. C.V. Bouten, K.T. Koekkoek, M. Verduin, R. Kodde, and J.D. Janssen. A triaxial accelerometer and portable data processing unit for the assessment of daily physical activity. *IEEE Transactions on Bio-Medical Engineering*, 44(3):136–47, 1997.
5. J.B. Bussmann, W.L. Martens, J.H. Tulen, F.C. Schasfoort, H.J. van den Berg-Emons, and H.J. Stam. Measuring daily behavior using ambulatory accelerometry: the Activity Monitor. *Behavior Research Methods, Instruments, & Computers*, 33(3):349–56, 2001.
6. G.S. Chambers, S. Venkatesh, G.A.W. West, and H.H. Bui. Hierarchical recognition of intentional human gestures for sports video annotation. In *Proceedings of the 16th International Conference on Pattern Recognition*, volume 2, pages 1082–1085. IEEE Press, 2002.
7. B.P. Clarkson. *Life Patterns: Structure from Wearable Sensors*. Ph.D. Thesis, Massachusetts Institute of Technology, 2002.
8. M. Csikszentmihalyi and R. Larson. Validity and reliability of the Experience-Sampling Method. *The Journal of Nervous and Mental Disease*, 175(9):526–36, 1987.

9. R.W. DeVaul and S. Dunn. Real-Time Motion Classification for Wearable Computing Applications. Technical report, MIT Media Laboratory, 2001.
10. F. Foerster, M. Smeja, and J. Fahrenberg. Detection of posture and motion by accelerometry: a validation in ambulatory monitoring. *Computers in Human Behavior*, 15:571–583, 1999.
11. V. Gerasimov. Hoarder Board Specifications, Access date: January 15 2002. http://vadim.www.media.mit.edu/Hoarder/Hoarder.htm.
12. R. Herren, A. Sparti, K. Aminian, and Y. Schutz. The prediction of speed and incline in outdoor running in humans using accelerometry. *Medicine & Science in Sports & Exercise*, 31(7):1053–9, 1999.
13. S.S. Intille, L. Bao, E. Munguia Tapia, and J. Rondoni. Acquiring in situ training data for context-aware ubiquitous computing applications. In *Proceedings of CHI 2004 Connect: Conference on Human Factors in Computing Systems*. ACM Press, 2004.
14. N. Kern, B. Schiele, and A. Schmidt. Multi-sensor activity context detection for wearable computing. In *European Symposium on Ambient Intelligence (EUSAI)*. 2003.
15. A. Krause, D.P. Siewiorek, A. Smailagic, and J. Farringdon. Unsupervised, dynamic identification of physiological and activity context in wearable computing. In *Proceedings of the 7th International Symposium on Wearable Computers*, pages 88–97. IEEE Press, 2003.
16. S.-W. Lee and K. Mase. Recognition of walking behaviors for pedestrian navigation. In *Proceedings of 2001 IEEE Conference on Control Applications (CCA01)*, pages 1152–5. IEEE Press, 2001.
17. S.-W. Lee and K. Mase. Activity and location recognition using wearable sensors. *IEEE Pervasive Computing*, 1(3):24–32, 2002.
18. P. Lukowicz, H. Junker, M. Stager, T.V. Buren, and G. Troster. WearNET: a distributed multi-sensor system for context aware wearables. In G. Borriello and L.E. Holmquist, editors, *Proceedings of UbiComp 2002: Ubiquitous Computing*, volume LNCS 2498, pages 361–70. Springer-Verlag, Berlin Heidelberg, 2002.
19. J. Mantyjarvi, J. Himberg, and T. Seppanen. Recognizing human motion with multiple acceleration sensors. In *Proceedings of the IEEE International Conference on Systems, Man, and Cybernetics*, pages 747–52. IEEE Press, 2001.
20. C. Randell and H. Muller. Context awareness by analysing accelerometer data. In B. MacIntyre and B. Iannucci, editors, *The Fourth International Symposium on Wearable Computers*, pages 175–176. IEEE Press, 2000.
21. A. Sugimoto, Y. Hara, T.W. Findley, and K. Yoncmoto. A useful method for measuring daily physical activity by a three-direction monitor. *Scandinavian Journal of Rehabilitation Medicine*, 29(1):37–42, 1997.
22. M. Uiterwaal, E.B. Glerum, H.J. Busser, and R.C. van Lummel. Ambulatory monitoring of physical activity in working situations, a validation study. *Journal of Medical Engineering & Technology.*, 22(4):168–72, 1998.
23. K. Van Laerhoven and O. Cakmakci. What shall we teach our pants? In *The Fourth International Symposium on Wearable Computers*, pages 77–83. IEEE Press, 2000.
24. K. Van Laerhoven, A. Schmidt, and H.-W. Gellersen. Multi-sensor context aware clothing. In *Proceedings of the 6th IEEE International Symposium on Wearable Computers*, pages 49–56. IEEE Press, 2002.
25. G. Welk and J. Differding. The utility of the Digi-Walker step counter to assess daily physical activity patterns. *Medicine & Science in Sports & Exercise*, 32(9):S481–S488, 2000.
26. I.H. Witten and E. Frank. *Data Mining: Practical Machine Learning Tools and Techniques with Java Implementations*. Morgan Kaufmann, 1999.

Recognizing Workshop Activity Using Body Worn Microphones and Accelerometers

Paul Lukowicz[1], Jamie A. Ward[1], Holger Junker[1], Mathias Stäger[1],
Gerhard Tröster[1], Amin Atrash[2], and Thad Starner[2]

[1] Wearable Computing Laboratory, ETH Zürich
8092 Zürich, Switzerland, www.wearable.ethz.ch
{lukowicz,ward}@ife.ee.ethz.ch
[2] College of Computing, Georgia Institute of Technology
Atlanta, Georgia 30332-0280
{amin,thad}@cc.gatech.edu

Abstract. The paper presents a technique to automatically track the progress of maintenance or assembly tasks using body worn sensors. The technique is based on a novel way of combining data from accelerometers with simple frequency matching sound classification. This includes the intensity analysis of signals from microphones at different body locations to correlate environmental sounds with user activity.

To evaluate our method we apply it to activities in a wood shop. On a simulated assembly task our system can successfully segment and identify most shop activities in a continuous data stream with zero false positives and 84.4% accuracy.

1   Introduction

Maintenance and assembly are among the most important applications of wearable computing to date; the use of such technology in tasks such as aircraft assembly [17], vehicle maintenance [4] and other on-site tasks [2,7] demonstrates a genuine utility of wearable systems.

The key characteristic of such applications is the need for the user to physically and perceptually focus on a complex real world task. Thus in general the user cannot devote much attention to interaction with the system. Further the use of the system should not restrict the operators physical freedom of action. As a consequence most conventional mobile computing paradigms are unsuitable for this application field. Instead wearable systems emphasizing physically unobtrusive form factor, hands free input, head mounted display output and low cognitive load interaction need to be used.

Our work aims to further reduce the cognitive load on the user while at the same time extending the range of services provided by the system. To this end we show how wearable systems can automatically follow the progress of a given maintenance or assembly task using a set of simple body worn sensors. With such *context* knowledge the wearable could pro-actively provide assistance without the need for any explicit action by the user. For example, a maintenance

A. Ferscha and F. Mattern (Eds.): PERVASIVE 2004, LNCS 3001, pp. 18–32, 2004.

support system could recognize which particular subtask is being performed and automatically display the relevant manual pages on the system's head-up display. The wearable could also record the sequence of operations that are being performed for later analysis, or could be used to warn the user if an important step has been missed.

1.1   Related Work

Many wearable systems explore *context* and proactiveness (e.g [1]) as means of reducing the cognitive load on the user. Much work has also been devoted to recognition methods, in particular the use of computer vision [20,24,25,16,15].

The application of proactive systems for assisting basic assembly tasks has been explored in [22], however this is built on the assumption of sensors integrated into the objects being assembled, not on the user doing the assembly.

Activity recognition based on body worn sensors, in particular acceleration sensors, has been studied by different research groups [11,14,23]. However all of the above work focused on recognizing comparatively simple activities (walking, running, and sitting). Sound based situation analysis has been investigated by Pelton *et al.* and in the wearables domain by Clarkson and Pentland [12, 5]. Intelligent hearing aids have also exploited sound analysis to improve their performance [3].

1.2   Paper Aims and Contributions

This paper is part of our work aiming to develop a reliable context recognition methodology based on simple sensors integrated in the user's outfit and in the user's artifacts (e.g. tools, appliances, or parts of the machinery) [10]). It presents a novel way of combining motion (acceleration) sensor based gesture recognition [8] with sound data from distributed microphones [18]. In particular we exploit intensity differences between a microphone on the wrist of the dominant hand and on the chest to identify relevant actions performed by the user's hand.

In the paper we focus on using the above method to track the progress of an assembly task. As described above such tasks can significantly benefit from activity recognition. At the same time they tend to be well structured and limited to a reasonable number of often repetitive actions. In addition, machines and tools typical to a workshop environment generate distinct sounds. Therefore these activities are well suited for a combination of gesture and sound–based recognition.

This paper describes our approach and the results produced in an experiment performed on an assembly task in a wood workshop. We demonstrate that simple sensors placed on the user's body can reliably select and recognize user actions during a workshop procedure.

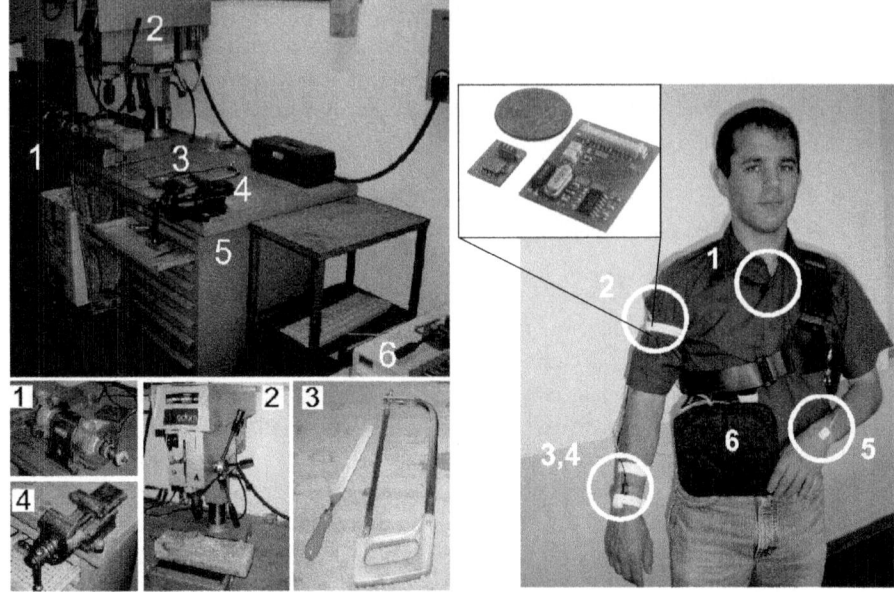

Fig. 1. The wood workshop (*left*) with (*1*) grinder, (*2*) drill, (*3*) file and saw, (*4*) vise, and (*5*) cabinet with drawers. The sensor type and placement (*right*): (*1,4*) microphone, (*2,3,5*) 3-axis acceleration sensors and (*6*) computer

2   Experimental Setup

Performing initial experiments on live assembly or maintenance tasks is inadvisable due to the cost and safety concerns and the ability to obtain repeatable measurements under experimental conditions. As a consequence we have decided to focus on an "artificial" task performed at the workbench of wood workshop of our lab (see Figure 1). The task consisted of assembling a simple object made of two pieces of wood and a piece of metal. The task required 8 processing steps using different tools; these were intermingled with actions typically exhibited in any real world assembly task, such as walking from one place to another or retrieving an item from a drawer.

2.1   Procedure

The assembly sequence consists of sawing a piece of wood, drilling a hole in it, grinding a piece of metal, attaching it to the piece of wood with a screw, hammering in a nail to connect the two pieces of wood, and then finishing the product by smoothing away rough edges with a file and a piece of sandpaper. The wood was fixed in the vise for sawing, filing, and smoothing (and removed whenever necessary). The test subject moved between areas in the workshop be-

Table 1. Steps of workshop assembly task

No	action
1	take the wood out of the drawer
2	put the wood into the vise
3	take out the saw
4	saw
5	put the saw into the drawer
6	take the wood out of the vise
7	drill
8	get the nail and the hammer
9	hammer
10	put away the hammer, get the driver and the screw
11	drive the screw in
12	put away the driver
13	pick up the metal
14	grind
15	put away the metal, pick up the wood
16	put the wood into the vise
17	take the file out of the drawer
18	file
19	put away the file, take the sandpaper
20	sand
21	take the wood out of the vise

tween steps. Also, whenever a tool or an object (nail screw, wood) was required, it was retrieved from its drawer in the cabinet and returned after use.

The exact sequence of actions is listed in Table 1. The task was to recognize all tool-based activities. Tool-based activities exclude drawer manipulation, user locomotion, and clapping (a calibration gesture). The experiment was repeated 10 times in the same sequence to collect data for training and testing. For practical reasons, the individual processing steps were only executed long enough to obtain an adequate sample of the activity. This policy did not require the complete execution of any one task (e.g. the wood was not completely sawn), allowing us to complete the experiment in a reasonable amount of time. However this protocol influenced only the duration of each activity and not the manner in which it was performed.

2.2   Data Collection System

The data was collected using the ETH PadNET sensor network [8] equipped with 3 axis accelerometer nodes and two Sony mono microphones connected to a body worn computer. The position of the sensors on the body is shown in Figure 1: an accelerometer node on both wrist and on the upper arm of the right hand, and a microphone on the chest and on the right wrist (the test subject was right handed).

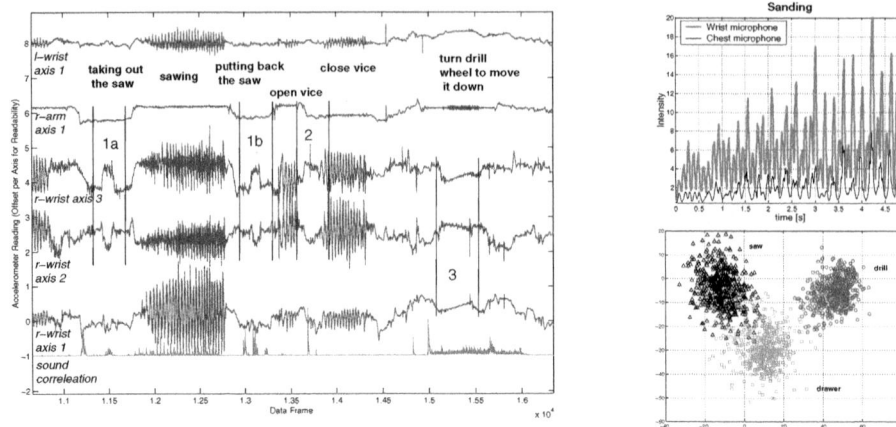

Fig. 2. Example accelerometer data from sawing and drilling (*left*); audio profile of sanding from wrist and chest microphones (*top right*); and clustering of activities in LDA space (*bottom right*)

As can be seen in Figure 1 each PadNET sensor node consist of two modules. The main module incorporates a MSP430149 low power 16-Bit mixed signal microprocessor (MPU) from Texas Instruments running at 6 MHz maximum clock speed. The current module version reads out up to three analog sensor signals including amplification and filtering and handles the communication between modules through dedicated I/O pins. The sensors themselves are hosted on an even smaller 'sensor-module' that can be either placed directly on the main module or connected through wires. In the experiment described in this paper sensor modules were based on a 3-axis accelerometer package consisting of two ADXL202E devices from Analog Devices. The analog signals from the sensor were lowpass filtered ($f_{cutoff} = 50Hz$) and digitized with 12Bit resolution using a sampling rate of 100Hz.

3   Recognition

3.1   Acceleration Data Analysis

Figure 2 (left) shows a segment of the acceleration data collected during the experiment. The segment includes sawing, removing the wood from the vise, and drilling. The user accesses the drawer two times and walks between the vise and the drill. Clear differences can be seen in the acceleration signals. For example, sawing clearly reflects a periodic motion. By contrast, the drawer access (marked as 1a and 1b in the figure) shows a low frequency "bump" in acceleration. This bump corresponds to the 90 degree turns of the wrist as the user releases the drawer handle, retrieves the object, and grasps the handle again to close the drawer.

Given the data, time series recognition techniques such as hidden Markov models (HMMs) [13] should allow the recognition of the relevant gestures. However, a closer analysis reveals two potential problems. First, not all relevant activities are strictly constrained to a particular sequence of motions. While the characteristic motions associated with sawing or hammering are distinct, there is high variation in drawer manipulation and grinding. Secondly, the activities are separated by sequences of user motions unrelated to the task (e.g the user scratching his head). Such motions may be confused with the relevant activities. We define a "noise" class to handle these unrelated gestures.

3.2   Sound Data Analysis

Considering that most gestures relevant for the assembly/maintenance scenario are associated with distinct sounds, sound analysis should help to address the problems described above. We distinguish between three different types of sound:

1. *Sounds made by a hand-tool:* - Such sounds are directly correlated with user hand motion. Examples are sawing, hammering, filing, and sanding. These actions are generally repetitive, quasi–stationary sounds (i.e. relatively constant over time - such that each time slice on a sample would produce an identical spectrum over a reasonable length of time). In addition these sounds are much louder than the background noise (dominant) and are likely to be much louder at the microphone on the user's hand than on his chest. For example, the intensity curve for sanding (see Figure 2 top right) reflects the periodic sanding motion with the minima corresponding to the changes in direction and the maxima coinciding with the maximum sanding speed in the middle of the motion. Since the user's hand is directly on the source of the sound the intensity difference is large. For other activities it is smaller, however in most cases still detectable.

2. *Semi-autonomous sounds:* These sounds are initiated by user's hand, possibly (but not necessarily) remaining close to the source for most of the sound duration. This class includes sound produced by a machine, such as the drill or grinder. Although ideal quasi-stationary sounds, sounds in this class may not necessarily be dominant and tend to have a less distinct intensity difference between the hand and the chest (for example, when a user moves their hand away from the machine during operation).

3. *Autonomous sounds:* These are sounds generated by activities not driven by the user's hands (e.g loud background noises or the user speaking).

Obviously the vast majority of relevant actions in assembly and maintenance are associated with handtool sounds and semi–autonomous sounds. In principle, these sounds should be easy to identify using intensity differences between the wrist and the chest microphone. In addition, if extracted appropriately, these sounds may be treated as quasi-stationary and can be reliably classified using simple spectrum pattern matching techniques.

The main problem with this approach is that many irrelevant actions are also likely to fall within the definition of hand-tool and semi–autonomous sound.

Such actions include scratching or putting down an object. Thus, like acceleration analysis, sound–based classification also has problem distinguishing relevant from irrelevant actions and will produce a number of false positives.

3.3   Recognition Methodology

Neither acceleration nor sound provide enough information for perfect extraction and classification of all relevant activities; however, we hypothesize that their sources of error are likely to be statistically distinct. Thus, we develop a technique based on the fusion of both methods. Our procedure consists of three steps:

1. Extraction of the relevant data segments using the intensity difference between the wrist and the chest microphone. We expect that this technique will segment the data stream into individual actions
2. Independent classification of the actions based on sound or acceleration. This step will yield imperfect recognition results by both the sound and acceleration subsystems.
3. Removal of false positives. While the sound and acceleration subsystems are each imperfect, when their classifications of a segment agree, the result may be more reliable (if the sources of error are statistically distinct).

4   Isolated Activity Recognition

As an initial experiment, we segment the activities in the data files by hand and test the accuracy of the sound and acceleration methods separately. For this experiment, the non-tool gestures, drawer and clapping, are treated as noise and as such are not considered here.

4.1   Accelerometer–Based Activity Recognition

Hidden Markov models (HMMs) are probabilistic models used to represent non-deterministic processes in partially observable domains and are defined over a set of states, transitions, and observations. Details of HMMs and the respective algorithms are beyond the scope of this paper but may be found in Rabiner's tutorial on the subject [13].

Hidden Markov models have been shown to be robust for representation and recognition of speech [9], handwriting [19], and gestures [21]. HMMs are capable of modeling important properties of gestures such as time variance (the same gesture can be repeated at varying speeds) and repetition (a gesture which contains a motion which can be repeated any number of times). They also handle noise due to sensors and imperfect training data by providing a probabilistic framework.

For gesture recognition, a model is trained for each of the gestures to be recognized. In our experiment, the set of gestures includes saw, drill, screw, hammer, sand, file and vise. Once the models are trained, a sequence of features can be passed to a recognizer which calculates the probability of each model given the observation sequence and returns the most likely gesture. For our experiments, the set of features consist of readings from the accelerometers positioned at the wrist and at the elbow. This provides 6 total continuous feature values - the x,y and z acceleration readings for both positions - which are then normalized to sum to one and collected at approximately 93 Hz.

We found that most of the workshop activities typically require only simple single Gaussian HMMs for modeling. For file, sand, saw, and screw, a 5 state model with 1 skip transition and 1 loop-back transition suffice because they consist of simple repetitive motions. Drill is better represented using a 7 state model, while grinding is again more complex, requiring a 9 state model. The vise is unique in that it has two separate motions, opening and closing. Thus a 9 state model is used with two appropriate loop-backs to correctly represent the gesture (See Figure 3). These models were selected through inspection of the data, an understanding of nature of the activities, and experience with HMMs.

4.2   HMM Isolation Results

For this project, a prototype of the Georgia Tech Gesture Recognition Toolkit was used to train the HMMs and for recognition. The Toolkit is an interface to the HTK toolkit [26] designed for training HMMs for speech recognition. HTK handles the algorithms for training and recognizing the Hidden Markov Models allowing us to focus primarily on properly modeling the data.

To test the performance of the HMMs in isolation, the shop accelerometer data was partitioned by hand into individual examples of gestures. Accuracy of the system was calculated by performing leave-one-out validation by iteratively reserving one sample for testing and training on the remaining samples for each sample. The HMMs were able to correctly classify 95.51% of the gestures over data collected from the shop experiments. The rates for individual gestures are given in Table 2.

4.3   Sound Recognition

Method
The basic sound classification scheme operates on individual frames of length t_w seconds. The approach follows a three step process: feature extraction, dimensionality reduction, and the actual classification.

The features used are the spectral components of each t_w obtained by Fast Fourier Transformation (FFT). This produces $N = \frac{f_s}{2} \cdot t_w$ dimensional feature vectors, where f_s is sample frequency. Rather than attempting to classify such large N-dimensional vectors directly, Linear Discriminant Analysis (LDA)[6] is employed to derive an optimal projection of the data into a smaller, M dimensional feature space (where M is the number of classes). In the "recognition phase", the LDA transformation is applied to the data frame under test to produce the corresponding $M - 1$ dimensional feature vector.

Using a labeled training-set, class means are calculated in the $M - 1$ dimensional space. Classification is performed simply by choosing the class mean which has the minimum Euclidean distance from the test feature vector (see Figure 2 bottom right).

Intensity Analysis

Making use of the fact that signal intensity is inversely proportional to the square of the distance from its source, the ratio of the two intensities I_{wrist}/I_{chest} is used as a measure of absolute distance of source from the user. Assuming the sound source is distance d from the wrist microphone and $d + \delta$ from the chest, the ratio of the intensities will be proportional to

$$\frac{I_{wrist}}{I_{chest}} \simeq \frac{(d + \delta)^2}{d^2} = \frac{d^2 + 2d\delta + \delta^2}{d^2} = 1 + \frac{2\delta}{d} + \frac{\delta^2}{d^2}$$

When both microphones are separated by at least δ, any sound produced at a distance d (where $d >> \delta$) from the user will bring this ratio close to one. Sounds produced near the chest microphone (e.g. the user speaking) will cause the ratio to approach zero whereas any sounds close to the wrist mic will make this ratio large.

Sound extraction is performed by sliding a window w_{ia} over the f_s Hz resampled audio data. On each iteration, the signal energy over w_{ia} for each channel is calculated. For these windows, the difference in ratio I_{wrist}/I_{chest} and its reciprocal are obtained, which are then compared to an empirically obtained threshold th_{ia}.

The difference $I_{wrist}/I_{chest} - I_{chest}/I_{wrist}$ provides a convenient metric for thresholding - zero indicates a far off (or exactly equidistant) sound; while above or below zero indicate a sound closer to the wrist or chest microphone respectively.

4.4   Results

In order to analyze the performance of the sound classification, individual examples of each class were hand partitioned from each of the 10 experiments. This provided at least 10 samples of every class - some classes had more samples on account of more frequent useage (e.g. vise). From these, two samples of each class were used for training while testing was performed on the rest.

Similar work[18] used FFT parameters of f_s=4.8kHz and t_w=50 ms (256 points), for this experiment t_w was increased to 100 ms. With these parameters LDA classification was applied to successive t_w frames within each of the class partitioned samples - returning a hard classification for each frame. Judging accuracy by the number of correctly matching frames over the total number of frames in each sample, an overall recognition rate of 90.18% was obtained. Individual class results are shown in the first column of Table 2. We then used intensity analysis to select those frames corresponding to where source intensity ratio difference surpassed a given threshold. With LDA classification applied only to these selected frames, the recognition improved slightly to a rate of 92.21% (second column of Table 2.)

To make a comparison with the isolated accelerometer results, a majority decision was taken over all individual frame results within each sample to produce an overall classification for that gesture. This technique resulted in 100% recognition over the sound test data in isolation.

Table 2. Isolated recognition accuracy (in %) for sound LDA, LDA with IA preselection, majority decision over IA+LDA, and for acceleration based HMM

Gesture	Sound			Acceleration
	LDA	IA+LDA	maj(IA+LDA)	HMM
Hammer	96.79	98.85	100	100
Saw	92.71	92.98	100	100
Filing	69.68	81.43	100	100
Drilling	99.59	99.35	100	100
Sanding	93.66	92.87	100	88.89
Grinding	97.77	97.75	100	88.89
Screwing	91.17	93.29	100	100
Vise	80.10	81.14	100	92.30
Overall	90.18	92.21	100	95.51

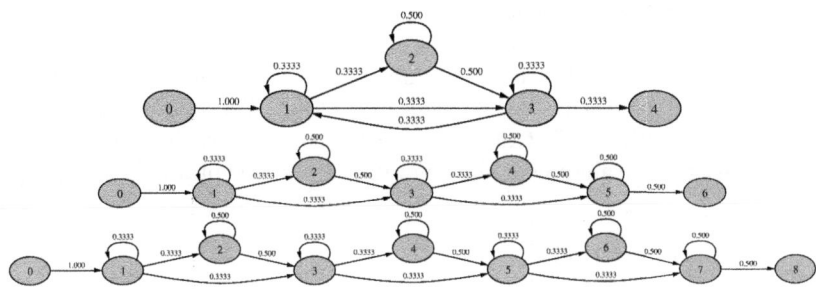

Fig. 3. HMMs topologies

5   Continuous Recognition

Recognition of gestures from a continuous stream of features is difficult. However, we can simplify the problem by partitioning the continuous stream into segments and attacking the problem as isolated recognition. This approach requires a method of determining a proper partitioning of the continuous stream. We take advantage of the intensity analysis described in the previous section as a technique for identifying appropriate segments for recognition.

Since neither LDA nor the HMM are perfect at recognition, and each is able to recognize a different set of gestures well due to working in different feature space, it is advantageous to compare their independent classifications of a segment. If the classification of the segment by the HMMs matches the classification of the segment by the LDA, the classification can be believed. Otherwise, the noise class can be assumed, or perhaps a decision appropriate to the task can be taken (such as requesting additional information from the user).

Thus, the recognition is performed in three main stages: 1) Extracting potentially interesting partitions from the continuous sequence, 2) Classifying these individually using the LDA and HMMs, and 3) Combining the results from these approaches.

5.1   LDA for Partitioning

For classification, partitioned data needs to be arranged in continuous sections corresponding to a single user activity. Such partitioning of the data is obtained in two steps: First, LDA classification is run on segments of data chosen by the IA. Those segments not chosen by intensity analysis are returned with classification zero. (In this experiment, classifications are returned at the same rate as accelerometer features); Secondly, these small window classifications are further processed by a larger (several seconds) majority decision window, which returns a single result for the entire window duration.

This partitioning mechanism helps reduce the complexity of continuous recognition. It will not give accurate bounds on the beginning and end of a gesture. Instead, the goal is to provide enough information to generate context at a general level, i.e., "The user is hammering" as opposed to "A hammering gesture occurred between sample 1500 and 2300." The system is tolerant of, and does not require, perfect alignment between the partitions and the actual gesture. The example alignment shown in Figure 4 is acceptable for our purposes.

5.2   Partitioning Results

Analysis of the data was performed to test the system's ability to reconstruct the sequence of gestures in the shop experiments based on the partitioning and recognition techniques described to this point. Figure 5 shows an example of the automated partitioning versus the actual events. The LDA classification of each partition is also shown. For this analysis of the system, the non-tool gestures, drawer and clapping, were considered as part of the noise class. After applying the partition scheme, a typical shop experiment resulted in 25-30 different partitions.

5.3   HMM Classification

Once the partitions are created by the LDA method, they are passed to set of HMMs for further classification. For this experiment, the HMMs are trained on individual gestures from the shop experiments using 6 accelerometer features from the wrist and elbow. Ideally, the HMMs will return a single gesture classification for each segment. However, the segment sometimes includes the beginning or end of the next or previous gesture respectively, causing the HMMs to return a sequence of gestures. In such cases, the gesture which makes up the majority of the segment is used as the classification. For example the segment labeled "B" in Figure 4 may return the sequence "hammer vise" and would then be assigned as the single gesture "vise."

5.4   Combining LDA and HMM Classification

For each partitioned segment, the classification of the LDA and HMM methods were compared. If the classifications matched, that classification was assigned the segment. Otherwise, the noise class was returned.

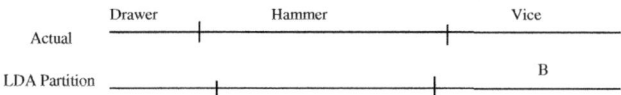

Fig. 4. Detailed example of LDA partitioning

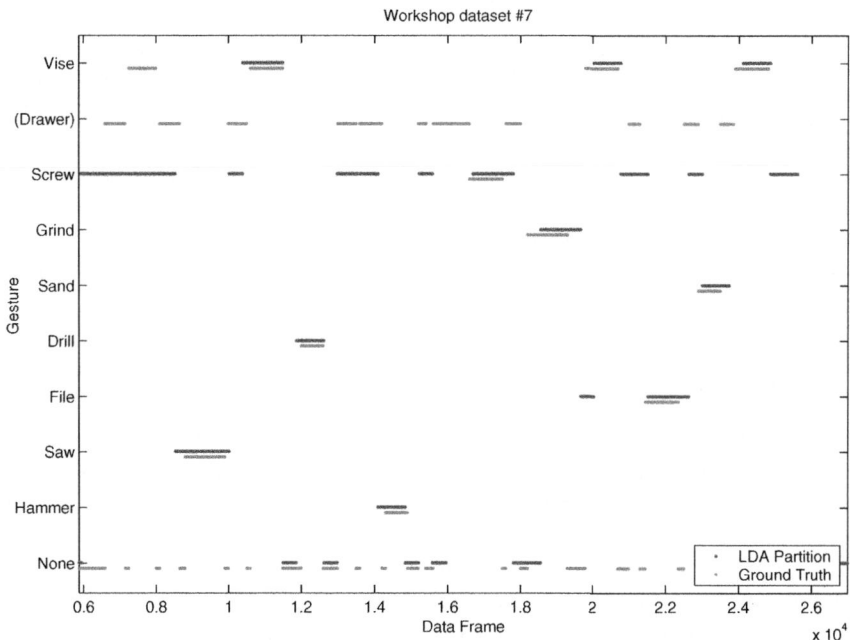

Fig. 5. LDA partitions versus ground truth on a typical continuous dataset

Table 3. Continuous recognition accuracy per gesture (Correct | Insertions | Deletions | Substitutions | Accuracy) and probability of gesture given classification P(G|Class)

Gesture	HMM					LDA					HMM + LDA					P(G|Class)
	C	I	D	S	Acc	C	I	D	S	Acc	C	I	D	S	Acc	
Hammer	8	2	0	1	66.7	9	1	0	0	88.9	8	0	1	0	88.9	1.00
Saw	9	0	0	0	100	9	1	0	0	88.9	9	0	0	0	100	1.00
Filing	10	0	0	0	100	9	7	0	1	23.2	9	0	1	0	90	1.00
Drilling	9	7	0	0	22.2	9	1	0	0	88.9	9	0	0	0	100	1.00
Sanding	8	0	0	1	77.8	9	8	0	0	11.1	8	0	1	0	88.9	1.00
Grinding	11	13	0	0	-18.2	9	0	0	2	81.8	9	0	2	0	81.8	1.00
Screw	5	1	0	4	44.4	9	75	0	0	-733.3	4	0	5	0	44.4	1.00
Vise	42	0	0	1	97.7	34	1	2	7	76.6	36	0	7	0	83.7	1.00
Overall	102	23	0	7	72.5	97	94	2	10	2.8	92	0	17	0	84.4	1.00

Table 3 shows the number of correct classifications (C), insertions (I), deletions (D), and substitutions(S) for the HMMs, the LDA, and the combination. Insertions are defined as noise gestures identified as a tool gesture. Deletions are tool gestures recognized as noise gestures. A substitution for a gesture occurs when that gesture is incorrectly identified as a different gesture. In addition, the accuracy of the system is calculated based on the following metric:

$$\%Accuracy = \frac{Correct - Insertions}{TotalSamples}$$

The final column reports the probability of a gesture having occurred given that the system reported that gesture.

Clearly, the HMMs and LDA each perform better than the other on various gestures and tended to err in favor of a particular gesture. When incorrect, LDA tended to report the "screw" gesture. Similarly, the HMMs tended to report "grinding" or "drilling." Comparing the classification significantly helps address this problem and reduce the number of false positives, thus increasing the performance of the system as a whole. The data shows that the comparison method performed better than the HMMs and the LDA in many cases and improved the accuracy of the system.

6   Discussion

Although the accuracy of the system in general is not perfect, it is important to note that the combined HMM + LDA method results in no insertions or substitutions. This result implies that when the system returns a gesture, that gesture did occur. While the system still misses some gestures, the fact that it does not return false positives allows a user interface designer to be more confident in his use of positive context.

Of course for many applications deletions are just as undesirable as false positives. In a safety monitoring scenario for example, any deletions of alarm or warning events would naturally be unnaceptable. In such cases it would be better for the system to return some warning, however erroneous, rather than none at all. On the other hand, if one sensor is known to produce many false positives in particular circumstances, whereas another is known to be extremely reliable for the same, then some means of damping the influence of the first in favour of the second sensor would be desirable.

The simple fusion scheme described in this paper could be modified to accomodate these issues by weighting sensor inputs based on knowledge of their reliability in given circumstances. Such weighting, together with decision likelihood information from individual classifiers, would allow a more intelligent fusion scheme to be developed. This will be the focus of future work.

7   Conclusion

We have shown a system capable of segmenting and recognizing typical user gestures in a workshop environment. The system uses wrist and chest worn microphones and accelerometers, leveraging the feature attributes of each modality to improve the system's performance. For the limited set analyzed, the system demonstrated perfect performance in isolated gesture testing and a zero false positive rate in the continuous case. In the future, we hope to apply these promising techniques, together with more advanced methods for sensor fusion, to the problem of recognizing everyday gestures in more general scenarios.

References

1. D. Abowd, A. K. Dey, R. Orr, and J. Brotherton. Context-awareness in wearable and ubiquitous computing. *Virtual Reality*, 3(3):200–211, 1998.
2. Len Bass, Dan Siewiorek, Asim Smailagic, and John Stivoric. On site wearable computer system. In *Proceedings of ACM CHI'95 Conference on Human Factors in Computing Systems*, volume 2 of *Interactive Experience*, pages 83–84, 1995.
3. Michael C. Büchler. *Algorithms for Sound Classification in Hearing Instruments*. PhD thesis, ETH Zurich, 2002.
4. C. Buergy, Jr. J. H. Garrett, M. Klausner, J. Anlauf, and G. Nobis. Speech-controlled wearable computers for automotive shop workers. In *Proceedings of SAE 2001 World Congress*, number 2001-01-0606 in SAE Technical Paper Series, march 2001.
5. B. Clarkson, N. Sawhney, and A. Pentland. Auditory context awareness in wearable computing. In *Workshop on Perceptual User Interfaces*, November 1998.
6. R. Duda, P. Hart, and D. Stork. *Pattern Classification, Second Edition*. Wiley, 2001.
7. J. H. Garrett and A. Smailagic. Wearable computers for field inspectors: Delivering data and knowledge-based support in the field. *Lecture Notes in Computer Science*, 1454:146–164, 1998.
8. N. Kern, B. Schiele, H. Junker, P. Lukowicz, and G. Tröster. Wearable sensing to annotate meeting recordings. In *6th Int'l Symposium on Wearable Computers*, pages 186–193, October 2002.

9. F. Kubala, A. Anastasakos, J. Makhoul, L. Nguyen, R. Schwartz, and G. Zavaliagkos. Comparative experiments on large vocabulary speech recognition. In *ICASSP*, Adelaide, Australia, 1994.
10. Paul Lukowicz, Holger Junker, Mathias Staeger, Thomas von Bueren, and Gerhard Troester. WearNET: A distributed multi-sensor system for context aware wearables. In G. Borriello and L.E. Holmquist, editors, *UbiComp 2002: Proceedings of the 4th International Conference on Ubiquitous Computing*, pages 361–370. Springer: Lecture Notes in Computer Science, September 2002.
11. J. Mantyjarvi, J. Himberg, and T. Seppanen. Recognizing human motion with multiple acceleration sensors. In *2001 IEEE International Conference on Systems, Man and Cybernetics*, volume 3494, pages 747–752, 2001.
12. V. Peltonen, J. Tuomi, A. Klapuri, J. Huopaniemi, and T. Sorsa. Computational auditory scene recognition. In *IEEE Int'l Conf. on Acoustics, Speech, and Signal Processing*, volume 2, pages 1941–1944, May 2002.
13. L. R. Rabiner and B. H. Juang. An introduction to hidden Markov models. *IEEE ASSP Magazine*, pages 4–16, January 1986.
14. C. Randell and H. Muller. Context awareness by analysing accelerometer data. In *Digest of Papers. Fourth International Symposium on Wearable Computers.*, pages 175–176, 2000.
15. J. M. Rehg and T. Kanade. DigitEyes: vision-based human hand tracking. School of Computer Science Technical Report CMU-CS-93-220, Carnegie Mellon University, December 1993.
16. J. Schlenzig, E. Hunter, and R. Jain. Recursive identification of gesture inputs using hidden Markov models. *Proc. Second Annual Conference on Applications of Computer Vision*, pages 187–194, December 1994.
17. D. Sims. New realities in aircraft design and manufacture. *Computer Graphics and Applications*, 14(2), March 1994.
18. Mathias Stäger, Paul Lukowicz, Niroshan Perera, Thomas von Büren, Gerhard Tröster, and Thad Starner. Soundbutton: Design of a low power wearable audio classification system. 7th Int'l Symposium on Wearable Computers, 2003.
19. T. Starner, J. Makhoul, R. Schwartz, and G. Chou. On-line cursive handwriting recognition using speech recognition methods. In *ICASSP*, pages 125–128, 1994.
20. T. Starner, B. Schiele, and A. Pentland. Visual contextual awareness in wearable computing. In *IEEE Intl. Symp. on Wearable Computers*, pages 50–57, Pittsburgh, PA, 1998.
21. T. Starner, J. Weaver, and A. Pentland. Real-time American Sign Language recognition using desk and wearable computer-based video. *IEEE Trans. Patt. Analy. and Mach. Intell.*, 20(12), December 1998.
22. Bernt Schiele Stavros Antifakos, Florian Michahelles. Proactive instructions for furniture assembly. In *4th Intl. Symp. on Ubiquitous Computing. UbiComp 2002.*, page 351, Göteborg, Sweden, 2002.
23. K. Van-Laerhoven and O. Cakmakci. What shall we teach our pants? In *Digest of Papers. Fourth International Symposium on Wearable Computers.*, pages 77–83, 2000.
24. C. Vogler and D. Metaxas. ASL recognition based on a coupling between HMMs and 3D motion analysis. In *ICCV*, Bombay, 1998.
25. A. D. Wilson and A. F. Bobick. Learning visual behavior for gesture analysis. In *Proc. IEEE Int'l. Symp. on Comp. Vis.*, Coral Gables, Florida, November 1995.
26. S. Young. *HTK: Hidden Markov Model Toolkit V1.5*. Cambridge Univ. Eng. Dept. Speech Group and Entropic Research Lab. Inc., Washington DC, 1993.

"Are You with Me?" – Using Accelerometers to Determine If Two Devices Are Carried by the Same Person

Jonathan Lester[1], Blake Hannaford[1], and Gaetano Borriello[2,3]

[1] Department of Electrical Engineering, University of Washington
[2] Department of Computer Science and Engineering, University of Washington
[3] Intel Research Seattle
Seattle, WA, USA
{jlester@ee,blake@ee,gaetano@cs}.washington.edu

Abstract. As the proliferation of pervasive and ubiquitous computing devices continues, users will carry more devices. Without the ability for these devices to unobtrusively interact with one another, the user's attention will be spent on co-ordinating, rather than using, these devices. We present a method to determine if two devices are carried by the same person, by analyzing walking data recorded by low-cost MEMS accelerometers using the coherence function, a measure of linear correlation in the frequency domain. We also show that these low-cost sensors perform similarly to more expensive accelerometers for the frequency range of human motion, 0 to 10Hz. We also present results from a large test group illustrating the algorithm's robustness and its ability to with-stand real world time delays, crucial for wireless technologies like Bluetooth and 802.11. We present results that show that our technique is 100% accurate using a sliding window of 8 seconds of data when the devices are carried in the same location on the body, is tolerant to inter-device communication latencies, and requires little communication bandwidth. In addition we present results for when devices are carried on different parts of the body.

1 Introduction

For the past 30 years, the dominant model for using our computing devices has been interactive. This approach puts the human in a feedback loop together with the computer. A user generates input and the computer responds through an output device, this output is then observed by the user who reacts with new input. When the ratio of humans to computing devices was close to 1:1, this was a reasonable approach. Our attention was commanded by one device at a time, our desktop, laptop, or handheld. This was appropriate as our tasks often involved manipulating information on the computer's screen in word processing, drawing, etc.

Today, the conditions of human-computer interaction are rapidly changing. We have an ever-increasing number of devices. Moreover, they are becoming deeply embedded into objects, such as automobiles. Many of these devices have a powerful CPU inside of them, however, we do not think of them as computing devices.

There are two main implications of this explosion in the number of computing devices. First, the human user can no longer be in the loop of every interaction with and

A. Ferscha and F. Mattern (Eds.): PERVASIVE 2004, LNCS 3001, pp. 33–50, 2004.
© Springer-Verlag Berlin Heidelberg 2004

between these devices; there are just too many, the interactive model simply does not scale. Devices must share information appropriately or they will end up demanding even more of our time. To further complicate matters, each device will necessarily have a different specialized user interface due to their different functions and form-factor. Second, as these new devices are embedded in other objects, we often are not even aware there are computing devices present, because our focus is on our task, not on the devices.

Invisibility is an increasingly important aspect of user interaction, with the principal tenet being "do not distract the user from the task at hand". An example of this is package delivery, which now includes tablet-like computers to collect signatures, RFID tracking of packages, and centralized databases to provide web services to customers, such as the current location of their parcel. Delivery truck drivers, cargo handlers, and recipients do not want a user interface to slow down a package in reaching its destination. They prefer if the devices gather input, explicitly or implicitly, and communicate the data amongst themselves. There is no reason for users to take an interactive role with all the steps, nor do they want to. Another example draws on devices becoming so cheap that they are viewed as a community resource. In hospitals, nurses and physicians carry clipboards, charts, and folders that could provide more timely information if they were electronic devices connected to the hospital's infrastructure. Many individuals would use these devices as they use their paper versions now. One way to enhance the user interaction with these devices would be for the devices themselves to recognize whom they were being carried by.

Motivated by these examples, we are investigating methods for devices to determine automatically when they should interact or communicate with each other. Our goal is to enable devices to answer questions such as:

- Is the same person carrying two devices? With what certainty?
- Are two devices in the same room? For how long?
- Are two devices near each other? How near?
- What devices did I have with me when I came in? When I went out?

Different applications will want answers to a different set of these and other questions. We are developing a toolkit of technologies and methods that can be used by interaction designers to create systems with a high degree of invisible interactions between many devices. This paper presents our work on developing methods for answering the first question.

We assume a world where a user will carry a changing collection of devices throughout a day. These might include a cell phone, a laptop, a tablet, and a handheld. In addition, the collection may include more specialized devices such as RFID or barcode scanners, GPS receivers, wrist-watch user interfaces, eyeglass-mounted displays, headphones, etc. These devices may be tossed into a pocket, strapped to clothing, worn on a part of the body, or placed in a backpack or handbag. We posit that it will be an insignificant addition to the cost of these devices if they include a 3-axis accelerometer. We also expect these devices to have a means of communicating with each other through wireless links, such as Bluetooth or 802.11. Recent work in wireless sensor networks is demonstrating that the communicating nodes may become as small as "smart dust"[1] and function with a high degree of power efficiency.

Here we consider the problem of how easily and reliably two such devices can autonomously determine whether or not they "belong" to the same user by comparing their acceleration over time. If the acceleration profiles of the two devices are similar enough, they should be able to conclude that the same person is currently carrying them. In practical situations, one of these devices may be a personal one (e.g., a wristwatch or pager on a belt that is "always" with the same person) while the other is a device picked-up and used for a period of time. The question we posed is: "How reliably can accelerometer data be used to make the determination that two devices are on the same person?"

2 Related Work

Many techniques exist that could possibly be used to answer our question. We could use capacitive-coupling techniques to determine if two devices are touching the same person [2], however, this requires direct physical contact with both devices and is highly dependent on body geometry and device placement. A second approach is to use radio signal strength [3] to determine proximity of two devices. However, RF signal strength is not a reliable measure of distance and is also highly dependent on body orientation and placement of devices. Furthermore, these RF signals could be received by nearby unauthorized devices on another person or in the environment.

The approach we investigate in this paper is to directly measure the acceleration forces on two devices and then compare them over a sliding time window. There has been much previous work in using accelerometers for gesture recognition and device association. We describe the contributions of three different pieces of related work: two from the ubiquitous computing research community and one from bioengineering instrumentation.

Gesture recognition using accelerometers has been used to develop a wand to remote control devices in a smart space [4] and a glove that uses sensing on all the fingers to create an "air keyboard" for text input [5]. This work is primarily concerned with using accelerometer data as part of the process in computing the position of an object, in these cases, a plastic tube or finger segments, respectively. By observing the variations in position over time, gestures can be recognized.

Device association is the process by which two devices decide whether they should communicate with each other in some way. Work at TeCO used accelerometers to create smart objects (Smart-Its) that could detect when they were being shaken together [6]. The idea was to associate two devices by placing them together and shaking the ensemble. Similar accelerations on the two devices would allow the connection to be established. The assumption is that it would be unlikely that two devices would experience the same accelerations unintentionally. Hinckley has developed a similar technique that uses bumping rather than shaking [7]. In both of these cases, the analysis of the accelerometer data is in the time domain, which can be sensitive to latencies in communication between the devices. Both techniques also have similarities in that the decision is strictly binary, instead of computing a probability that the two devices are being intentionally associated. The principal difference between this work and ours is more fundamental, while these two contributions exploit explicit user-initiated interactions (shaking or bumping) our focus is on making the determination implicitly and independent of the user's attention.

The work that is closest to ours, and provided much of our inspiration, used accelerometers to determine if the trembling experienced by a patient with Parkinson's Disease was caused by a single area of the brain or possibly multiple areas of the brain [8]. Physicians developed accelerometer sensors that were strapped to patients' limbs, data was collected, and an off-line analysis determined if the shaking was correlated in the frequency domain. Shaking in the same limb was found to be highly correlated; however, shaking across limbs was found to be uncorrelated. The key observation in this work is that Parkinson's related shaking is likely due to multiple sources in the brain that may be coupled to each other. We use a very similar approach, but with an on-line algorithm which can be running continuously within the devices being carried rather than strapped to the body.

Researchers attempting to identify activities in real-time have tried to identify activities such as: standing, sitting, walking, lying down, climbing stairs, etc. using more structured placement of multiple sensors and analysis methodologies like neural networks and Markov models. See [9] for a detailed overview of this work.

3 Our Approach

In order to provide a useful detection tool that doesn't require any user interaction the input to our system must come from an existing, natural action. In this paper we focus on the activity of human walking.

Although there are a number of different actions that a person regularly performs, walking provides a useful input because of its periodic nature. Human locomotion is regulated by the mechanical characteristics of the human body [10] as much as conscious control over our limbs. This regular, repeated activity lends itself to an analysis in the frequency domain, which helps reduce the effect of problems like communication latencies, device dependent thresholds, or the need for complex and computationally expensive analysis models.

We have two aims in this paper. First, we want to assess the quality of acceleration measurements obtained from low cost accelerometers, to ensure that they are appropriate for this application and that their measurements have a physical basis. Second, we want to determine whether there is sufficient information in the accelerations of two devices to determine whether they are being carried together.

4 Methods

Three different 3-axis MEMS accelerometers were used for our experiments, two were low cost commercial accelerometers from Analog Devices [5, 6] and STMicroelectronics, and the third was a calibrated accelerometer from Crossbow Technologies. Table 1 lists the accelerometers along with some specifications for each accelerometer.

Table 1. Selected specifications for three accelerometers

Manufacturer	Model	Axis	Sensitivity	Bandwidth[1]
Analog Devices	ADXL202E	2-axis[2]	± 2G	≈100Hz
STMicroelectronics	LIS3L02	3-axis	± 2G/6G[3]	≈100Hz
Crossbow Technologies	CXL02LF3	3-axis	± 2G	50Hz

Since the accelerometers used in these experiments provide 3-axis outputs (X, Y, and Z) data can be processed on a given set of axes, or taken as a whole. We assume a random, and possibly continuously changing, orientation and thus we take the magnitude of the force vector by combining the measurements from all 3 axes using Eqn. 1 to derive a net acceleration independent of orientation. Each accelerometer is capable of measuring the 1G gravity field present on the Earth, and although this information can be used to estimate the direction of the gravity vector, we subtract this offset from our data to reduce the DC offset in the Fast Fourier Transform (FFT) frequency spectra.

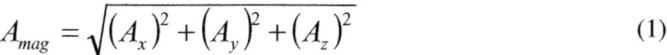

$$A_{mag} = \sqrt{\left(A_x\right)^2 + \left(A_y\right)^2 + \left(A_z\right)^2} \qquad (1)$$

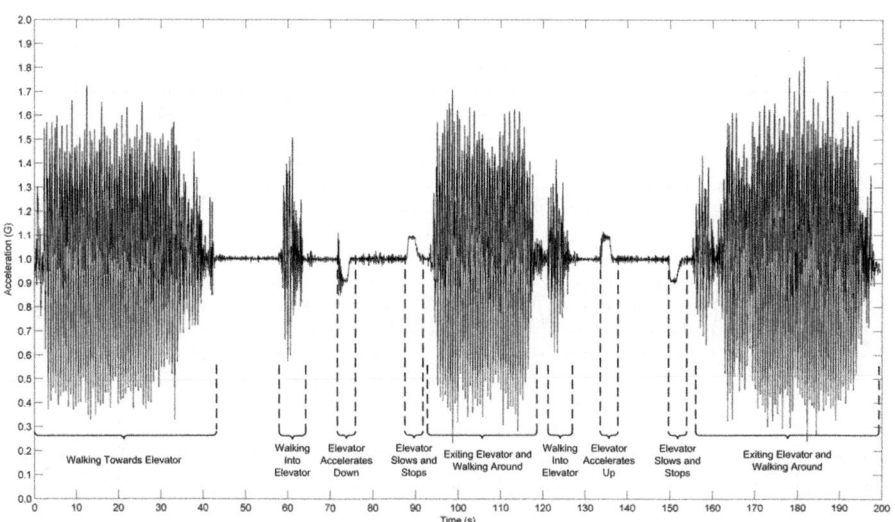

Fig. 1. Example of filtered magnitude data from accelerometer, similar to [11]. Data shows subject walking, riding down an elevator, walking, riding up an elevator, and walking

Figure 1 shows the magnitude of the output of the accelerometers for a subject walking and riding an elevator. For this experiment, as well as the majority of the experiments in this paper, the subject carried the accelerometer in a fanny pack worn around the waist. The use of the fanny pack ensured that the placement of the sensors

---

[1] Bandwidth as configured in this experiment

[2] Two ADXL202E's are mounted at 90° to provide 3-axis (plus one redundant axis)

[3] The LIS3L02 has a pin selectable sensitivity between ±2G and ±6G

was consistent between different test subjects. However, we also show results for carrying the accelerometers elsewhere on the body.

4.1   Accelerometer Characterization and Experimental Setup

The purpose of using a precise calibrated accelerometer is to characterize the performance of the low cost accelerometers; to ensure that the data provided is an accurate representation of the physical world. Without this verification, any experimental results could be based on incorrect data or rely upon artifacts inherent to the sensors and not the physical environment.

The physical mechanics and low-level neural circuits of the human body control the way humans walk, more than the conscious control of limbs. This automatic control makes walking a regular periodic activity, which has been studied extensively in the biomechanics community. It is widely accepted that the useful frequency spectra of human motion lies within the range of 0 – 10Hz [10]. Using this range as our region of interest we focus our study of accelerometers on the 0 – 10Hz range.

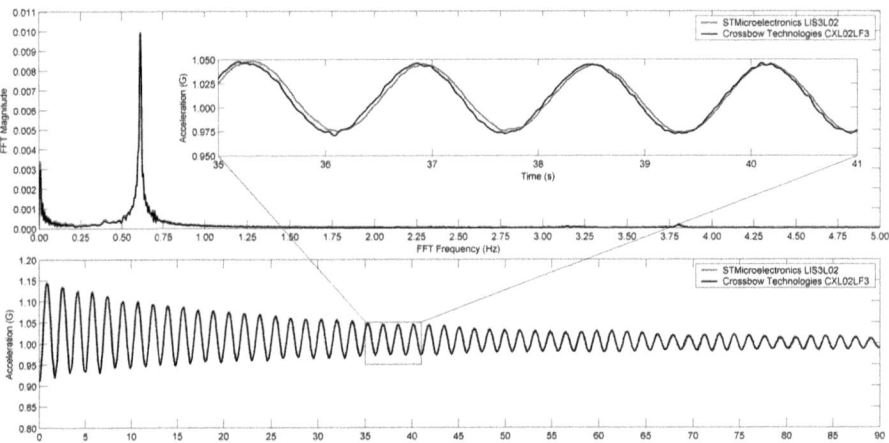

Fig. 2. Acceleration magnitudes and their corresponding FFT spectra collected during a swinging pendulum test showing that both signals are very similar in the time and frequency domains

To characterize the accelerometers, several experiments were conducted using different combinations of accelerometers. Data collection used a custom interface board, (approx. 9cm x 5cm) designed to sample the data at 600Hz and record data through a serial connection to an iPaq, and the AD128 commercial data collector from Crossbow Technologies. Our results showed that for low frequencies, 0 – 10Hz, the ADXL202E and LIS3L02 (the smaller and cheaper accelerometers) performed similarly to the CXL02LF3 (the larger and more expensive, but more robust, accelerometer), producing very consistent magnitude and FFT spectra. These results allow us to reliably conclude that low cost accelerometers are appropriate for collecting data in this frequency range and that they provide reliable data. Figure 2 shows an experiment performed with two accelerometers placed on a swinging pendulum, both traces are nearly identical in the time and frequency domains.

4.2   Coherence

Given a time series of acceleration magnitude signals like those found in Figure 1, it was necessary to find a reliable method for examining the data. Ben-Pazi *et al.*[8] used the coherence function to analyze the origin of rest tremors in patients with Parkinson's. By examining their use of coherence with biological accelerometer data, we reasoned that such methods could be expanded to other signals like the accelerations experienced by the body when walking.

Coherence, $\gamma_{xy}(f)$, is a normalized function of frequency derived from the cross-spectrum of two signals, S_{xy}, divided by the power spectrum of each signal, S_{xx} and S_{yy} [12, 13]. Coherence measures the extent to which two signals are linearly related at each frequency, with 1 indicating that two signals are highly correlated at a given frequency and 0 indicating that two signals are uncorrelated at that frequency [8, 12, 14]. Although coherence is a complex quantity, it is often approximated using the magnitude squared coherence (MSC), C_{xy}, shown in Eqn. 2. As all coherences used in this paper are real valued we will refer to the MSC estimator simply as the coherence.

$$C_{xy}(f) = \left|\gamma_{xy}(f)\right|^2 = \frac{\left|S_{xy}(f)\right|^2}{S_{xx}(f)S_{yy}(f)} \tag{2}$$

If the two signals used to compute the coherence are identical, then the coherence gives a unity result for all frequencies. Similarly if two signals describe completely uncorrelated random processes, the coherence will be zero for all frequencies. For example, two separate audio signals, recorded from a single opera singer, would have a high coherence (near 1) for most frequencies, whereas audio signals from two different opera singers performing completely different scores would have low coherence (near 0). And of course, two similar audio signals would have coherences spread throughout the range.

Eqn. 2 would always result in a unity magnitude for all frequencies (though the imaginary component may not be 1) if a single window were used to estimate the spectral density. A commonly used multiple windowing method is called weighted overlapped segment averaging (WOSA) and involves splitting two signals, X and Y, into equal length windowed segments. The FFT of these segments is taken and their results are averaged together to estimate the spectral density. As the name suggests, segments can be overlapped to reduce the variance of the spectral estimate (an overlap of 50% is fairly common), however, overlapping is more computationally expensive. As our goal is to use these algorithms in environments with limited computing power, we limit ourselves to two windowed segments, no overlapping segments, and a FFT size equal to the size of the windowed segments [12, 15, 16].

Different window sizes/types, as well as different sized FFTs, can be used to obtain coherences with different attributes. For simplicity, a common Hanning window is used in our calculations. The number of windows used in a coherence calculation determines the significance level of the coherence output: the more windows used, the lower the coherence value has to be to signify a strong coherence. In general, more windows provide smoother coherences with less variability but require more computations.

Due to the fact that the coherence is a function of frequency, it was necessary to determine a method of computing a scaled measure of similarity based on coherence. A basic measure is to integrate the area under the coherence curve for a given frequency range. By leveraging the fact that physical human motion rests below the 10Hz range, we used a normalized integration over the range from 0 – 10Hz to get a 0 to 1 measure of the coherence (Eqn. 3). This result is expressed as P_{xy}, and is the measure of similarity between our two acceleration signals.

$$P_{xy} = 0.1 \int_0^{10} C_{xy}(f) \ df \tag{3}$$

Although more complex methods are possible, the results presented in this paper only use this basic normalized integration of the coherence from 0 – 10Hz. This allows us to reliably quantify the coherence output using relatively fast and computationally inexpensive methods.

5   Experimental Data and Results

Two experiments are discussed in this paper. In the first experiment, the single-person experiments, six test subjects walked normally for ≈30 meters wearing two accelerometers in a fanny pack around their waist. Each subject walked this distance eight times; twice with four different combinations of accelerometers and data collectors (listed in Table 2), giving a total of 48 pairs of recordings for the first experiment.

Table 2. Walking trials and equipment for each subject

Walking Trial	First Sensor	Second Sensor	Data Collector
1, 2	ADXL202E	CXL02LF3	AD128
3, 4	LIS3L02	CXL02LF3	AD128
5, 6	ADXL202E	CXL02LF3	Custom Board
7, 8	LIS3L02	CXL02LF3	Custom Board

In the second experiment, the paired walking experiments, six subjects walked a distance of ≈61 meters in pairs with another test subject. Again, each of the subjects wore two accelerometers in a fanny pack around their waist and data was collected using a custom interface board and an iPaq. During this experiment each subject walked with every other subject for a total of 15 pairs. Each pair of subjects walked the 61 meter distance four times; the first two times they walked casually with the other person, and the last two times they walked purposely in-step with the other person. There were a total of 60 pairs of recordings for the second experiment.

The data from these two experiments was then manually trimmed so that only the walking data from each pair of recordings was selected. It is possible to use this method on a continuous stream of data by, for example, performing the calculations on 1-second wide windows of data and using the combined result over the past 8 seconds. However, it is necessary to divide the data up so we can perform a useful analysis on the entire data set because several walking trials are recorded back-to-back with short pauses separating them.

5.1 Single-Person Walking Experiment

Figure 3 shows a sample of the data collected from the first experiment. Fifteen seconds of acceleration magnitude data from two accelerometers (bottom plot) were used to calculate the FFT spectra and the coherence (top plot). The two signals compared in Figure 3 were from data recorded on the same individual, during the same walking trial. The high coherence shown in Figure 3 indicates that the two signals are highly correlated at most of the frequencies. Notice also that the two FFT spectra appear similar as the two signals recorded the same walking pattern.

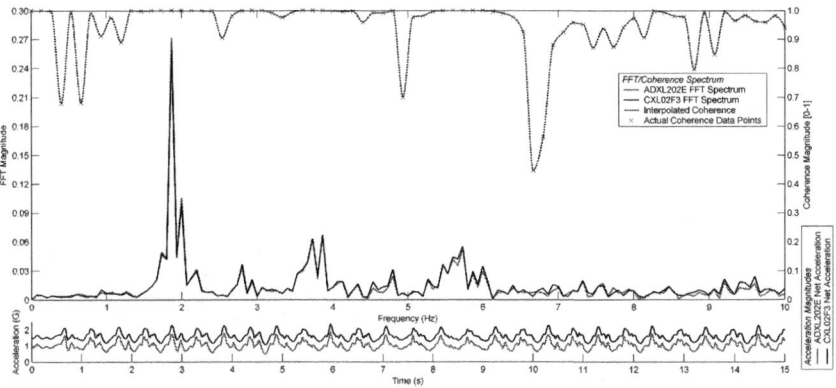

Fig. 3. Coherence and FFT spectra (top) of matching accelerometers on the same person during a 15 second walking segment – (bottom) acceleration magnitudes as a function of time. Coherence is nearly 1 for the 0 – 10Hz range, indicating two highly correlated signals

Figure 4 shows a similar figure, however, this time the two signals compared are not from the same individual. The data from the CXL02LF3 sensor in Figure 4 is the same 15-second segment shown in Figure 3 while the ADXL202E data is from a different person's walking segment, which was chosen because of its similar time plot. The wildly varying coherence shown in Figure 4 indicates that the two signals are not highly correlated, despite appearing similar in the time plots. The FFT spectra shown in Figure 4 are noticeably different as well.

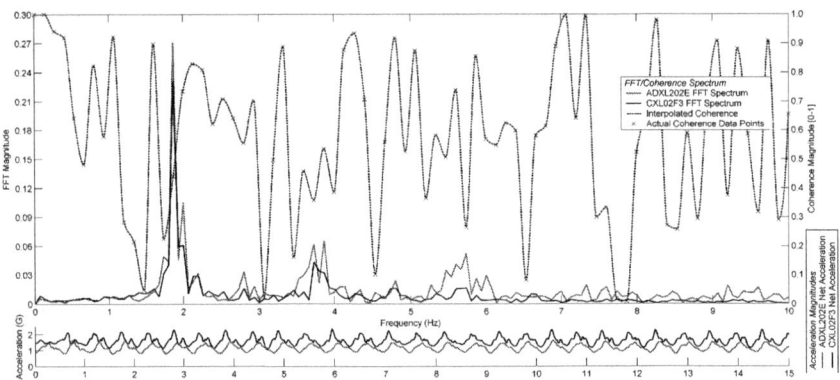

Fig. 4. Coherence and FFT spectra (top) of acceleration measured on different subjects during a 15 second walking segment – (bottom) acceleration magnitudes as a function of time. Coherence is not 1 for most of the 0 – 10Hz range, indicating two uncorrelated signals

To analyze the data in the first experiment we took the first 8 seconds of walking data from each trial and computed the coherence of each possible pairing of recordings. This created a 48 x 48 matrix (not shown) of coherence results. The diagonals of this matrix correspond to the (matching) coherences between pairs of sensors worn on the same person during the same trial, and the off diagonals correspond to the (non-matching) coherences between pairs of sensors worn on different subjects during different trials. Viewing the 0 to 1 range of the coherence integral as a percentage from 0 – 100%, we found that the matching diagonal coherences had a mean of 90.2% with a standard deviation of 5.2%. By contrast the non-matching, off diagonal, coherences had a mean of 52.3% ± 6.1% std.

5.2 Paired Walking Experiment

To analyze the data in the second paired walking experiment, we again took the first 8 seconds of walking data and computed the coherence of each possible pairing of recordings. Unlike the first experiment, we now have four sets of accelerometer data, instead of two, because each of the two people walking together carries two accelerometers. Two accelerometers correspond to the first person in the pair, test subject A, and the other two correspond to the second person, test subject B[4]. We can therefore create two 60x60 matrices of coherence results, one corresponding to the two accelerometers on test subject A and the other to test subject B. Figure 5 shows the 60x60 matrix created using only the data from test subject A (test subject B's data creates a similar matrix and is not shown).

---

[4] Test subject A and test subject B refer to the first and second person in each trial pair and not to any specific test subjects in the trials.

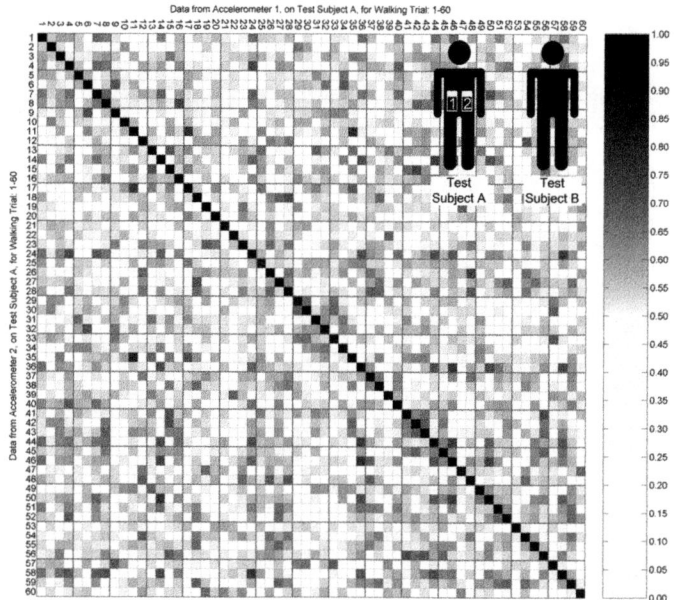

Fig. 5. Coherence values between all possible pairs of accelerometer 1 and accelerometer 2 worn on test subject A. In this matrix, coherences are calculated only with the first 8 seconds of walking data. The dark diagonal indicates high coherence between sensors worn on the same person at the same time

Again, the diagonals correspond to the coherence of matching pairs of accelerometers worn on the same person, test subject A, at the same time. As we would expect, the diagonals show a high coherence with a mean of 95.4% ± 3.1% std. The off diagonals show a low coherence with a mean 52.9% ± 6.6% as they correspond to sensor data from different people recorded at different times.

Another useful matrix to examine would be the coherence between an accelerometer on test subject A and another accelerometer on test subject B. If coherence were affected by the two people walking together casually, or in step, we would expect to see high coherences on the diagonal of this matrix. Again, since each subject was carrying two accelerometers we could construct more than one such matrix, but we only show one of these matrices in Figure 6. To create this matrix we took the first accelerometer data from subject A and the first accelerometer data from subject B in each pair and computed the coherence of each possible pairing of recordings (other combinations of accelerometers on different test subjects produce similar results). The diagonals, in Figure 6, correspond to the coherences of two accelerometers worn on different individuals, who walked together at the same time. The off-diagonals represent the coherences of two accelerometers worn on different individuals, recorded at different times. Overall this matrix shows a low coherence, with the diagonals having a mean of 56.9% ± 8.4%, and the off-diagonals having a mean of 53.1% ± 6.4% (the entire matrix has a mean of 53.1% ± 6.8%).

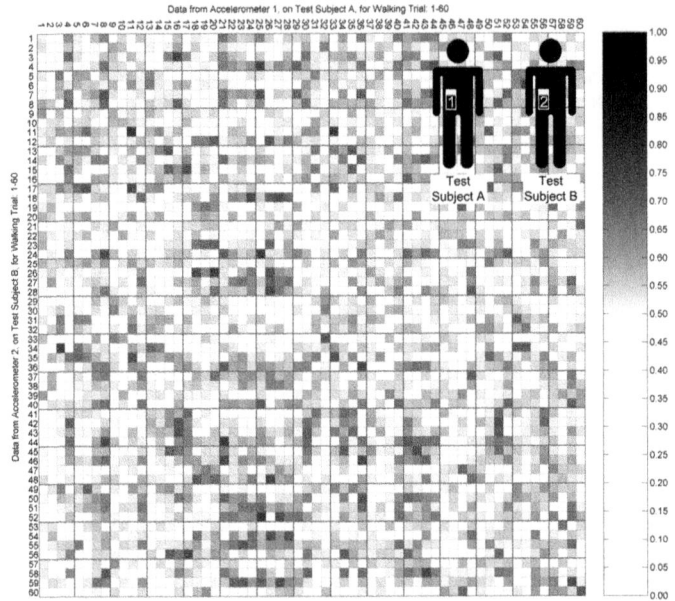

Fig. 6. Coherence values between all possible pairs of accelerometer 1, worn on test subject A, and accelerometer 2, worn on test subject B. In this matrix, coherences are calculated only with the first 8 seconds of data from the walking trials. The light shading throughout the matrix indicates low coherence between accelerometers on different people. The second two rows/cols of each 4x4 box are the in-step trials in which subjects intentionally tried to walk synchronously

In Figure 6, the 3rd and 4th row/column of each of the 4x4 boxes in the matrix represent data from when the subjects were walking together in step. The in-step data has a mean of 61.8% ± 7.6% while the casual data has a mean of 52.0% ± 5.9%.

6   Practical Considerations

In the results presented so far, only the first 8 seconds of walking data were used to generate coherence values. However, different sized segments of data can be used to calculate the coherence, with longer segments of walking data providing more information and shorter segments providing less information. The obvious question is how large do the segments need to be so as to get 'enough' information.

To address this question, we re-analyzed the data from our first, single-person, experiment by splitting the entire data set in segments of 1 – 15 seconds (in 1-second increments). We would expect there to be enough information in a segment if coherence is able to successfully pick the corresponding sensor pairs from the same segment. Since the walking data from the first experiment was typically only about 20 seconds in length, this gives us 48 pairs, for 15 seconds, all the way up to 1,212 pairs, for 1 second segments. For each segment we computed all the possible coherence pairs and recorded the pair with the largest coherence. If the largest coherence belonged to the same pair as our chosen signal then the algorithm correctly determined the best pairing. We then repeated the process using the next segment until all segments had been processed. Figure 7, shows the results of this analysis in the trace labeled 'No Delay', which shows that if 8-second, or larger, segments are used then the largest coherence is between the correct matching pairs 100% of the time.

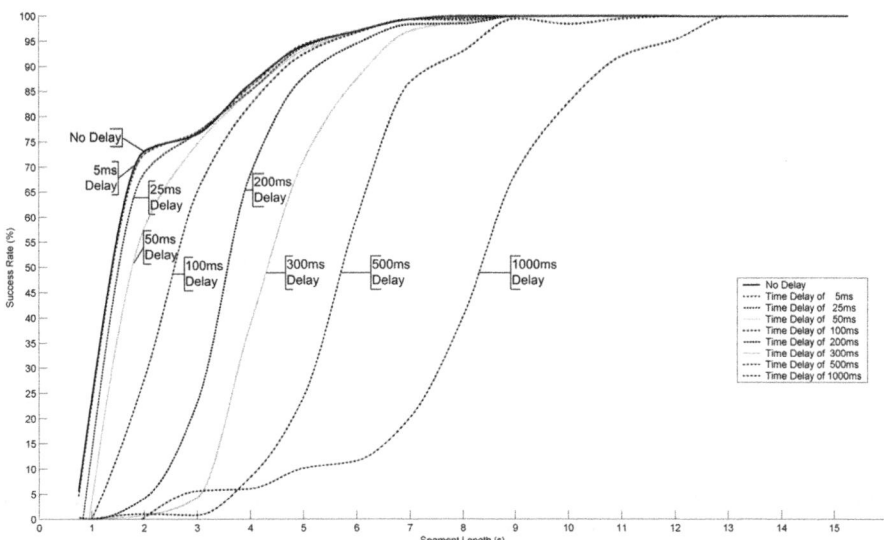

Fig. 7. Interpolated success rates showing the effects of time delay on the coherence results, using different length of segments. For segment lengths greater than eight seconds the success rate remains >95% even with a 500ms time delay

Another practical consideration in deploying this algorithm is understanding its sensitivity to communication latencies between the two devices. Although data transfers would be done in large chunks, i.e. 1 or 8 second sliding windows, it is desirable if the system is relatively insensitive to communication delays. Wireless synchronization methods do exist; however, they can require more power hungry radio communication. Given that our analysis uses magnitudes in the frequency domain, we would expect some insensitivity to phase shifts in the data. Figure 7 also shows how our results change as we vary the latency between the two devices. We simulated latencies ranging from 0 to 1000ms. Each curve shows the accuracy for different segment lengths at these different latencies. The results are only slightly degraded up to 500ms with more substantial differences above that latency. We claim that this is more than adequate for two devices likely to be in short-range radio contact. For example, Bluetooth packet latency is on the order of 50ms in the worst case.

Another practical consideration is the amount of data to be transferred to calculate the coherence. Because our frequency range lies in the 0 – 10Hz range, we can filter the signal and sample at 20Hz. At this sampling rate it is only necessary to transmit 320 Bytes to the other device to send an 8-second acceleration magnitude (8 seconds x 20Hz x 2 bytes), which could fit in a single Bluetooth packet. This would enable devices with little computational power or battery life to have a more powerful device perform the calculation. However, if it were necessary for the sampling rate to remain at the 600Hz rate we currently use, we could simply transmit the 0 – 10Hz FFT coefficients necessary for the coherence calculation. Figure 8 shows the situation where accelerometers are still sampled at 600Hz and the devices do a wireless handshake and then transmit the FFT data.

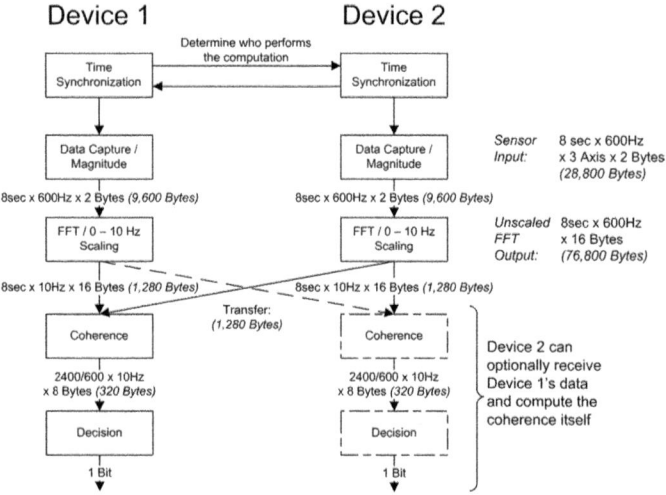

Fig. 8. Hypothetical data exchange as it would take place on embedded devices. Results in Figures 2 – 7 use higher resolution computations. Labels on the arrows indicate the data that must pass from one step to the next. Eight-second segments are used in this example, resulting in 4,800 magnitude data points and two 2,400-point FFTs

The two devices can coordinate which of them will make the determination. The data need only be sent to one device, most likely the one with the processor capacity required, or the most battery power, etc. Alternatively, or possibly for security reasons, both devices can make the determination that they are likely on the same person. Of course, a completely secure implementation would require a more complex coordination process.

7   Discussion

As the activities of interest in our experiments are human motion, we are able to focus on the frequency range of 0 – 10Hz. Our initial experiments using a pendulum provided a well-controlled environment for closely measuring the behavior of our accelerometers against significantly more costly calibrated accelerometers like the Crossbow Technologies CXL02LF3. The pendulum experiment clearly showed that the low-cost accelerometers produced similar acceleration magnitude data and very similar FFT spectra for the sampling rates we required.

Our first, single-person, experiments with different accelerometer pairs and data collectors confirmed that the accelerometers performed similarly in the frequency range of interest. Using these results we believe it safe to conclude that these inexpensive accelerometers are appropriate and well suited to the tasks of recording highly periodic human motion. We also believe that they provided measurements which were consistent with the physical world and were not merely the artifacts associated with a particular accelerometer.

Our first experiment also showed us that the coherence integral does provide a reasonable measure of the correlation of two acceleration magnitudes. We were able to show that 8-second segments, recorded from the same person at the same time, showed mean coherences of 90.2% and that 8-second segments, from different people recorded at different times, showed mean coherences of 52.3%. The second experiment, the paired walking experiments, further expanded these results showing that accelerometers on the same person have mean coherences of 95.4%, with accelerometers on different people having a mean of 52.9%. Accelerometers on two different individuals have mean coherences of 56.9% and 53.1%, even with people purposely attempting to confuse the results by walking in-step. This large separation is also clearly evident in Figures 5 and 6, showing that we can indeed determine when sensors were worn on the same individual. There were no false positive results from the sensor readings from the same person (recorded during different walks), people walking casual together, and even people purposely trying to walk in-step together.

8   Conclusions and Future Work

We set out with two goals for this paper: (1) showing that we can accurately determine when two accelerometers are being carried by the same person using and (2) showing that inexpensive accelerometers are adequate for this task. We believe our experimental data strongly supports positive conclusions in both cases. We have also answered our original question: "How reliably can accelerometer data be used to make the determination that two devices are on the same person?" by showing that we can reliably determine if two devices are being carried by the same person using just 8 seconds of walking data. We can even differentiate between devices carried by two people walking together.

As we are concerned only with the physics of human motion we are only interested in frequencies up to 10Hz. This is of great advantage for small, portable, and low-power devices as the accelerometers only need to be sampled at 20Hz to meet the Nyquist criterion. Low sampling rates will also help facilitate the use of wireless communication to offload computationally expensive portions of this algorithm to devices with more powerful processors and/or batteries.

Our results for determining whether two devices are carried by the same person are currently limited to people that are walking – either independently or side-by-side and even in-step. We use the values of frequency coherence between 0 and 10Hz to provide us with a rough likelihood that the two devices were being carried by the same person by integrating the area under the coherence curve. The results are not only very positive but they are also very robust. If we use an 8-second window for the comparison, our results are 100% accurate (no false positives, no false negatives). We feel that the most impressive result is that segments which should have high coherences have mean coherences in the 90 – 95% range with standard deviations on the order of 5–6% and segments that should have low coherences have means in the range of ≈52% with standard deviations of ≈6%. This substantial separation illustrates our algorithm's robustness and even holds up when people attempt to purposely walk in-step to fool our algorithm.

If we narrow our data window, the results necessarily degrade, but only moderately. For example, with a 5 second window, we still have ≈95% accuracy, and even at 2 seconds the accuracy is still in the ≈75% range. This is encouraging as it is likely that other sensors can be used in conjunction with the accelerometers (e.g., microphones) to provide additional data to make the determination.

The major limitations of our current approach are that we require the user(s) to walk and that we only consider devices in a fanny pack worn around the waist. We have yet to extensively study the cases of other everyday motions. For example, sitting and typing at a desk will likely require more analysis in the time-domain as there is less periodicity in the person's movement. Again, we see the technique presented here as only one of many analysis tools available to the interaction designer. For example, this method could be enhanced with additional sensor data, simple time domain analysis, etc.

We also conducted preliminary experiments with devices at other locations on the body, including accelerometers on the wrist, placed in one or both pockets, in a backpack, and in a fanny pack. Figure 9 shows some preliminary experimental results with pairs of sensors worn on the same person in a pocket/fanny pack, pocket/backpack, and pocket/wrist. Table 3 lists the statistics gathered from this preliminary experiment. While the results from different locations for the accelerometers are not quite as good as when both accelerometers are in the same fanny pack, they still show promise. Coherence clearly is an important tool for solving the device association problem. However, it is also clear that coherence will need to be enhanced with other techniques appropriate to the usage model.

Table 3. Stastics from coherence analysis of three different preliminary experiments

Experiment	Diagonal	Off-Diagonal	Average Success Rate
Pocket / Fanny Pack	76.7% ± 4.6%	55.3% ± 8.8%	87.5%
Pocket / Backpack	77.1% ± 6.0%	54.9% ± 8.8%	81.3%
Pocket / Wrist	75.7% ± 6.5%	51.9% ± 6.9%	70%

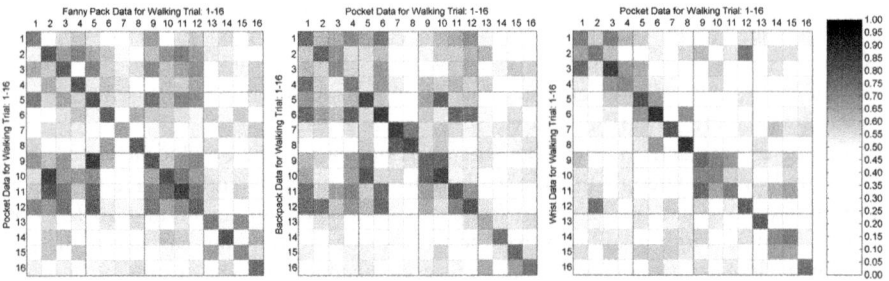

Fig. 9. Data from three different experiments using sensors worn on different parts of the body, Pocket/Fanny Pack, Backpack/Pocket, and Wrist/Pocket

Our future plans are to incorporate these methods and others into a general purpose sensor platform that can be easily integrated with a variety of devices such as cellphones, key-chains, digital cameras, and wrist-watches, as well as PDAs, tablets, and laptops. We are currently designing a new sensor board with accelerometers, temperature, barometric pressure, microphones, light, and inertial sensors that will fit into a package roughly 3 x 3 cm. We will be using an embedded ARM7 processor that will be able to handle the computations easily along with Bluetooth RF communication.

When this new sensor platform is complete, we will also conduct user studies for typical use in two scenarios: borrowable cameras that can keep track of what other devices they were being carried with when each picture was taken, and automatic synchronization of data between PDAs, laptops, and wrist-watches.

Acknowledgements. We would like to thank David Mizell for starting this project. Adam Rea and Ken Smith at Intel Research Seattle provided valuable support in getting our instrumentation in place. We would also like to thank Crossbow Technologies and STMicroelectronics for the use of their accelerometers. And most of all, we would like to thank our anonymous test subjects for their patience and willingness to participate in our experiments. Jonathan Lester was supported by a gift from Intel Research Seattle.

References

1. Warneke, B., Last, M., Leibowitz, B., Pister, K.S.J.: Smart Dust: Communicating with a Cubic-Millimeter Computer. In: IEEE Computer Magazine, (Jan. 2001) 44–51
2. Partridge, K., Dahlquist, B., Veiseh, A., Cain, A., Foreman, A., Goldberg, J., and Borriello, G.: Empirical Measurements of Intrabody Communication Performance under Varied Physical Configurations. In: Symposium on User Interface Software and Technology, Orlando, Florida (UIST 2001) 183–190
3. Brunette, W., Hartung, C., Nordstrom, B., and Borriello, G.: Proximity Interactions between Wireless Sensors and their Application. In: Second ACM International Workshop on Wireless Sensor Networks and Applications, San Diego, California (WSNA 2003) 30–37
4. Wilson, A., and Shafer, S.: XWand: UI for Intelligent Spaces. In: Conference on Human Factors and Computing Systems, Ft. Lauderdale, Florida (CHI 2003) 545–552
5. Perng J., Fisher B., Hollar S., and Pister K. S. J.: Acceleration Sensing Glove. In: 3rd International Symposium on Wearable Computers, San Francisco, California (1999) 178–181
6. Holmquist, L., Mattern, F., Schiele, B., Alahuhta, P., Beigl, M., Gellersen, H.: Smart-Its Friends: A Technique for Users to Easily Establish Connections between Smart Artefacts. In: Proceedings of Ubicomp (2001)
7. Hinckley, K.: Synchronous Gestures for Multiple Persons and Computers. In: Symposium on User Interface Software and Technology, Vancouver, British Columbia (UIST 2003) 149–158
8. Ben-Pazi, H., Bergman H., Goldberg J. A., Giladi N., Hansel D., Reches A., and Simon E. S.: Synchrony of Rest Tremor in Multiple Limbs in Parkinson's Disease: Evidence for Multiple Oscillators. In: Journal of Neural Transmission, Vol. 108 (3) (2001) 287–296
9. Bao, L.: Physical Activity Recognition from Acceleration Data under Semi-Naturalistic Conditions. Master's thesis, Massachusetts Institute of Technology, Dept. of Electrical Engineering and Computer Science (2003)
10. Winter, D.: Biomechanics and Motor Control of Human Movement (2nd ed.). New York: Wiley (1990)
11. Lukowicz P., Junker H., Stäger M., von Büren T., and Tröster G.: WearNet: A Distributed Multi-Sensor System for Context Aware Wearables. In: Proc. of the 4th International Conference on Ubiquitous Computing, Göteborg, Sweden (2002) 361–370

12. Carter, C.: Tutorial Overview of Coherence and Time Delay Estimation. In: Coherence and Time Delay Estimation – An Applied Tutorial for Research, Development, Test, and Evaluation Engineers. IEEE Press (1993) 1–23
13. Carter, C.: Coherence and Time Delay Estimation. In: Proc. of the IEEE, Vol. 75. (1987) 236–255
14. Shulz, M., and Stattegger, K.: Spectrum: Spectral Analysis of Unevenly Spaced Paleoclimatic Time Series. In: Computers & Geosciences, Vol. 23, No. 9 (1997) 929–945
15. Carter, C., Knap, C., and Nuttall, A,: Estimation of the Magnitude-Squared Coherence Function Via Overlapped Fast Fourier Transform Processing. In: Coherence and Time Delay Estimation – An Applied Tutorial for Research, Development, Test, and Evaluation Engineers. IEEE Press (1993) 49–56
16. Welch, P.: The Use of Fast Fourier Transform for the Estimation of Power Spectra: A Method Based on Time Averaging Over Short, Modified Periodograms. In: IEEE Transactions on Audio and Electroacoustics, Vol. 15, No. 2 (1967) 70–73

Context Cube: Flexible and Effective Manipulation of Sensed Context Data

Lonnie Harvel[1,2], Ling Liu[2], Gregory D. Abowd[2], Yu-Xi Lim[1],
Chris Scheibe[1,2], and Chris Chatham[1,2]

[1]School of Electrical and Computer Engineering
[2]College of Computing & GVU Center
Georgia Institute of Technology
Atlanta, Georgia 30332-0280, USA
{ldh,lingliu,abowd}@cc.gatech.eduHYPERLINK

Abstract. In an effort to support the development of context-aware applications that use archived sensor data, we introduce the concept of the Context Cube based on techniques of data warehousing and data mining. Our implementation of the Context Cube provides a system that supports a multi-dimensional model of context data and with tools for accessing, interpreting and aggregating that data by using concept relationships defined within the real context of the application. We define the Context Cube information model, demonstrate functional applications that we have developed with the system, and explore possible applications that may be developed more easily using our tool.

1 Introduction

Humans are good at making use of implicit context information when interacting with other people and the surrounding environment, whereas computers usually require the interpretation of the context information to be explicitly provided in order to use it appropriately. One of the challenges for ubiquitous computing is to bridge the gap between captured context data and the different levels of interpretation of captured context information needed by applications. Working with archives of context data provides opportunities for understanding user behavior by analysis and assessment of that behavior over time [17, 18, 19]. In this paper, we focus on accessing and manipulating an archive of sensed context data to support applications in context-aware and everyday computing, using concept hierarchies to relate context data to real-world definitions.

Currently, most context-aware applications focus on the immediate state. In contrast, there are many applications, envisioned as supportive of our everyday lives, which require more complicated context that we cannot sense directly. For example, providing a summary of a person's daily movement, determining when and how often a family gathers for meal time, assessing an elder relatives level of social interaction, or understanding space usage within a building in order to optimize HVAC resources all require processing of the lower-level context data and the interpretation or analysis of that data over time.

A. Ferscha and F. Mattern (Eds.): PERVASIVE 2004, LNCS 3001, pp. 51-68, 2004.

There are tools in the field of data warehousing and mining that we can use to manipulate low-level context data into higher-level information. However, to apply these techniques: context must be effectively modeled, structured to allow for interpretation and aggregation, and made easily available to applications and mining engines. In this paper, we describe the Context Cube, a system that supports a multi-dimensional model of context data and provides tools for accessing, interpreting and aggregating that data.

1.1 Motivating Example

Of the possible application domains that would benefit from accessing historical context, much of this work focuses on one of our larger research agendas: aiding senior adults who remain in their homes. Many adult children live far away from their aging parents. One of the biggest concerns for these children is the everyday well being of the parent. In fact, many times the decision to move a parent out of their own home and into some assisted living facility is the desire to maintain the peace of mind of the adult children. Mynatt *et al.* proposed the Digital Family Portrait (DFP) as an enhanced information appliance in the home of the adult child [1]. The DFP portrays an individual, the parent, whom we assume lives in a sensor-rich home that tracks certain critical parameters of everyday life.

Fig. 1. The primary Digital Family Portrait display, the image is a substitute

The center of the portrait is a picture of the parent, and the border is a visualization of the activity of the parent over the last 28 days (see Figure 1). In the prototype proposed by Mynatt et al., the visualization consisted of butterfly icons to reflect physical activity of the parent. A small (medium, large) icon represents little (medium, much) activity for that parent for that day. The number of detected room transitions for the parent on a given day determined the size of the icon. This version of the Digital Family Portrait demonstrates a simple example of analyzing contextual data, in this case room-level position data, over some period of time to produce an indicator of trends. In this case, the visualization leaves up to the adult child any decision on whether a significant trend has emerged. More automated techniques might automatically signal when one day's movement trend is out of the ordinary.

The Digital Family Portrait serves as a useful motivation for storing, manipulating and visualizing a history of contextual data. There are many other examples, particularly in the case of monitoring the well being of an aging individual, in which trends of contextual data, specifically where an individual is and what they are doing, are important indicators of potential problems. Once we have gone to the trouble of instrumenting a space to detect human behaviors, we need to provide better ways to access and manipulate large collections of data. This paper proposes a technique for simplifying the development of this important class of context-monitoring applications.

1.2 Overview

We introduce the concept of the Context Cube Model in Section 2. Section 2.1 describes the structure for modeling the relationships between context data and with associated context derived from world knowledge. Section 2.2 describes how we use datacube operations with this model, and section 2.3 describes our actual implementation of this system. In Section 3 we present four applications that have been built on context cubes. The first application is a new version of the Digital Family Portrait that is working on live data from an instrumented home. The second application shows the co-location history of two individuals and demonstrates how new applications can share existing context cubes and associated context. The third application is a visualization of activity levels in the different areas of a home. It is also built from the same context cube structures as the previous two, but focuses on the location aspect instead of identity. Unlike the previous applications, the fourth example is a hypothetical design, given in greater detail, to demonstrate how a set of context cubes and associated context would be developed. It also shows how existing context cubes, in this case the ones from the previous application examples, can be augmented with new cubes to provide for new applications. We give a brief overview of the related work in the context-aware computing and database communities in Section 4, and our concluding discussion in Section 5.

2 Bridging the Context Divide – The Context Cube Model

As the amount of stored context grows, we are reaching the point where we are data rich, but information poor. There is growing effort in the context-aware community to extract interesting knowledge and information (rules, regularities, constraints, patterns) from large collections of context data. Without this level of analysis, it will be hard to build context-aware applications that need sophisticated context knowledge and historical context information. Some forms of this data analysis are difficult, even impossible, to compute with standard SQL constructs. In particular, aggregations over computed categories, roll-up and drill-down of aggregation hierarchies, and symmetric aggregation require a more powerful data representation. To meet this need, we have adapted a technique from data warehousing, namely the data cube, and provided a system for creating and managing the resulting context cubes. In this section, we present a brief overview of how this adaptation applies to context-aware applications. For more detail on data cubes in general, we refer the reader to the work of Gray et al. [2].

2.1 Context Dimensions and the Context Cube

Our idea behind using a warehouse is to represent relationships within the actual structure of the data. The benefits are threefold: easier representation and processing of queries [3], the inclusion of expanded definitions and relationships, and the creation of new context constructed from analysis of the existing data. We represent the relationships between context using dimensions. As an example, take location, a single dimension of context. Many applications, most notably tour guide systems, have been built with a simple knowledge of immediate location. However, other applications, like reminder services, need both location and time. We now have a two dimensional relationship as shown in Figure 2. If there is a need, as in the Digital Family Portrait, to perform an analysis or calculation over the data contained within this relationship, a materialized representation may simplify queries and improve performance. After adding a third dimension, identity, we are able to represent the three dimensional relationship (location, time, identity) as well as the two dimensional relationships of (location, time), (location identity), and (time, identity).

Thus, a *Context Cube* is a model for representing and manipulating multidimensional context, based on the principles of data cubes from the traditional database and data mining community [2, 3]. Each dimension of the cube represents a context dimension of interest. A relation of N attributes produces an N-dimensional context cube. To structure these context relationships and capture the quantitative measures of these relationships, we construct a *context fact table*. This table contains both the factual context data on each of the dimensions, and the *derived context* that are of importance to the domain analysts and are obtained by analysis and summarization over the captured context relationships. The derived context may be viewed as residing in the boxes of the context cube. We access the derived context by using the dimensions as filter conditions. Figure 3 shows the design of a context fact table as a star schema. The fact table represents the context that we will gather to form the dimensions (Location, Time, Identity) as well as the derived context "count", a measure of room transitions.

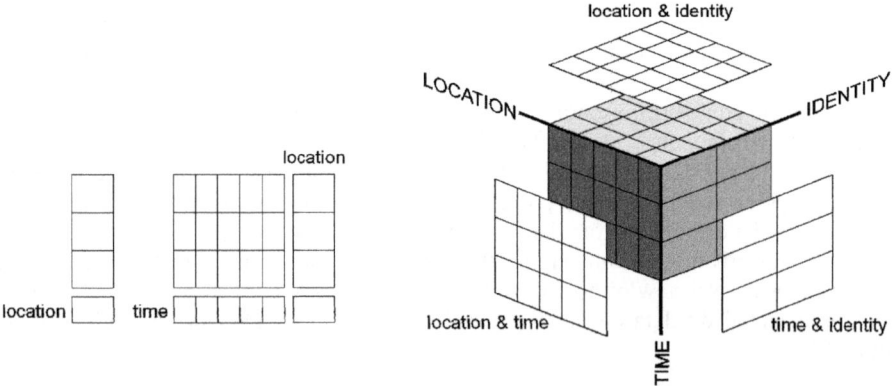

Fig. 2. Examples of data cubes. From left to right, we depict an array of locations, a matrix showing all the possible relationships between location and time, and finally a cube showing all the possible relationships between location, time and identity

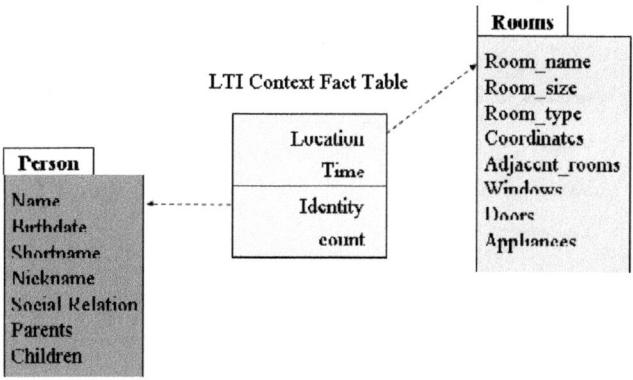

Fig. 3. An example star schema showing a context fact table for a Location, Time, Identity cube with context attribute tables for Person associated with the Identity dimension and Rooms associated with the Location Dimension.

The star schema also shows the connection of the context fact table to the *context attribute tables*. The context attribute tables allow for an expanded definition of an individual point of context data. They can be used by domain specialists who are interested in using higher levels of context to interpret and reason about the captured context fact data. One advantage of the star schema approach to context modeling is the reusability of context knowledge both within a domain and across domains. In our example design, used to form a simple context cube to support the Digital Family Portrait, we connect the Location dimension to the Rooms context attribute table, and the Identity dimension to the Person context attribute table. We can use the context attribute tables to provide alternative interpretations and associated information. Another important purpose of the context attribute tables is to enable the use of the *context dimension hierarchy* to introduce different levels of abstractions to the captured context data and context relationships.

A context dimension hierarchy is a concept hierarchy on the given context dimension. It is a mechanism that enables higher-level abstractions and context interpretations to the captured context data. Context dimension hierarchies allow us to manipulate a cube, along one of the context dimensions, using a combination of the data in the context fact table and the associated context in the context attribute table. Figure 4 shows two possible hierarchies constructed from the Location dimension and the Rooms attributes. In each of the examples, the leaf nodes are room names from the [removed for anonymity]. The first hierarchy classifies the locations by room type. The second hierarchy classifies the locations by region. If we only wanted information about Bedrooms, we could use the first hierarchy to make that selection, if however, we wanted information about Bedrooms on Floor 1, we would need to use both hierarchies to make the selection.

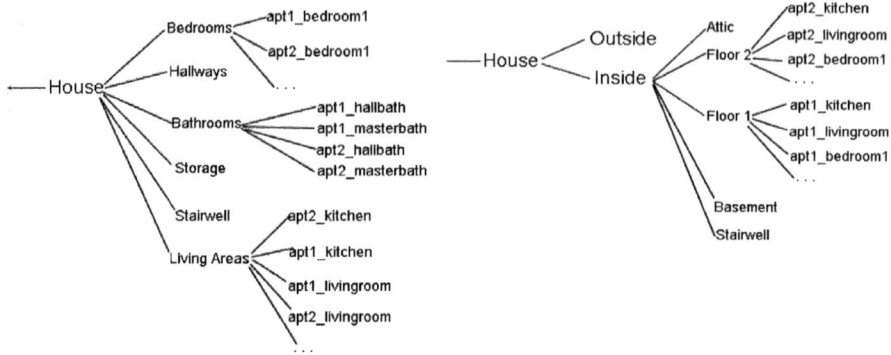

Fig. 4. Two context dimension hierarchies, the first depicts a classification based on room types and the second is based on regions

2.2 Context Cube Operations

We have shown how context cubes can be used for answering context queries at different levels of context abstraction. In this section, we present the four context operators used to manipulate the context cube: *slice, dice, roll-up,* and *drill-down.* These operators enrich the context cube model to support a much broader collection of context queries with efficient implementation methods. As evidenced in Gray's et al. seminal work on data cubes, cubes and cube operators provide an easier and more effective way to analyze multidimensional data sets. For example, using calls to cubes, a few lines of code can accomplish what would otherwise require a convoluted query with dozens of lines [2]. The result of a cube operation is another cube allowing us to perform operations in combination. To provide an intuitive understanding of these four cube operators, we provide an informal description of each operation with some illustrative examples.

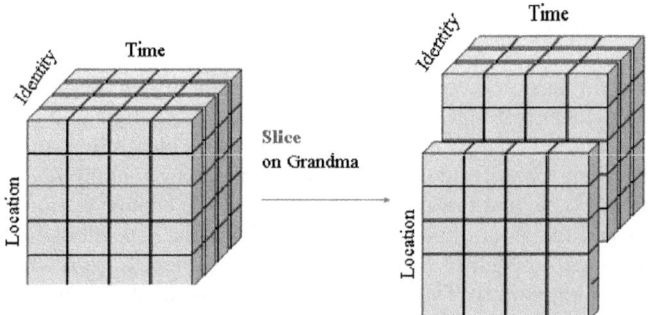

Fig. 5. A slice on Grandma returns Location and Days information about Grandma alone with count showing the number of times Grandma was in a given location on a given day

A *slice* is a selection on one dimension of an N dimension cube. For example, in the LTI (Location, Time, Identity) cube, used in the Digital Family Portrait, a slice could be on one user, or a group of users, selecting all conditions from the other two dimensions. We can express a slice using a representation from the context fact table, the context attribute table, or from the context dimension hierarchy. Figure 5 shows a slice on a single user, Grandma, over the three-dimensional context cube LTI. We can view a slice as an operator that cuts the cube by slicing it over the identity of Grandma.

The *dice* operator is a selection to all dimensions of an N dimension cube. Again using the LTI cube, we can dice for Grandma, in the Kitchen, Today. A dice is an easy way of doing a sequence of slices, Figure 6.

Roll-up generates a new cube by applying an aggregate function on one dimension. The roll-up operator uses the structure defined in a context dimension hierarchy and information in a context attribute table to determine how the roll-up should be performed. The values of the derived context are updated, when necessary, to reflect the change in resolution or scale of the filter conditions applied to the dimension. Again using the LTI context cube, a roll-up on Identity might generate a cube that uses the type of Social Relationship such as Family, Friend or Business instead of individual names. Another roll-up on identity might generate a cube that uses age group or faculty/grads/undergrads as the identity dimension unit instead of person names. A rollup on Location might generate a cube that uses Room Type as the scale instead of Room Name.

Drilldown is the reverse operation of rollup. It generates a context cube with finer granularity on one of the n dimensions. A drilldown on Years generates a cube with time in Months. Drilldown on the location region Bedroom would generate a cube with location in each of the bedrooms in the house. Drilldown on identity in a senior age group (suppose it is age from 65 above) would generate a cube with seniors above 65.

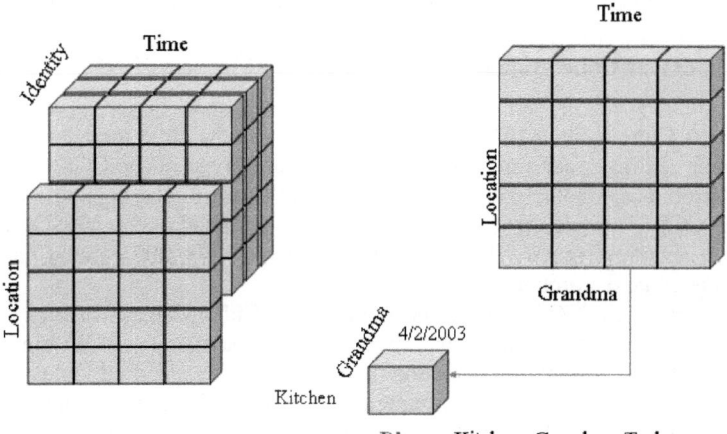

Dice on Kitchen, Grandma, Today

Fig. 6. A dice on Kitchen, Grandma, Today results in a cube that matches only those filter conditions with a count showing the number of times Grandma was in the Kitchen on that day

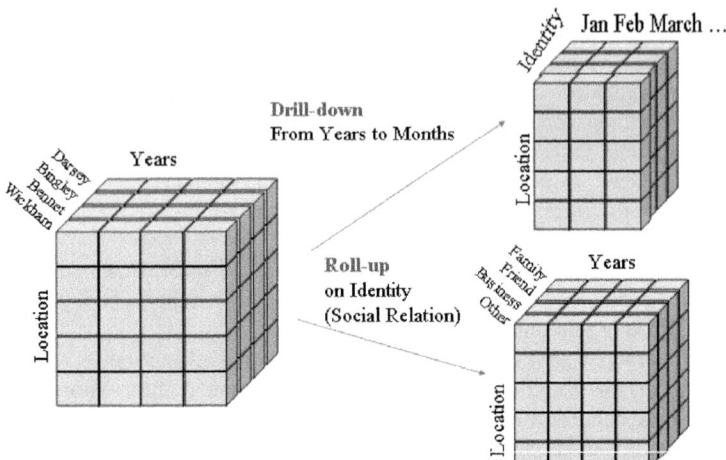

Fig. 7. A roll-up on Identity based on Social Relation, a drill-down on time going from Years to Months

Another operation over a cube, though not used for manipulations, is the ability to construct context dimension aggregate lattices. Given a context dimension table and a corresponding cube, one can generate the aggregate lattice that contains all possible combinations of aggregates one can apply to the context dimension table data. It would require a great deal of space to store a set of context cubes that represented the whole context dimension aggregate lattice for a given cube and set of context dimension hierarchies. One strategy for deciding what to construct and maintain is to use a greedy algorithm that retains the context cubes most commonly referenced. For our system, we have designed a registration-based system that allows applications to inform the Context Cube system of its data needs and an appropriate context cube will then be generated and maintained. The details of the implementation follow.

2.3 The Context Cube System

The Context Cube System (CCS) implements the context cube model and provides management and interface functionality. The first prototype of the CCS consists of a context warehouse, a context cube constructor, and context cube manipulation operations. The CCS is a package of Java applications constructed over a MySQL database running on a Solaris server. There are currently over 1,000,000 context data points stored in the context warehouse.

At the heart of the CCS is the Cube Manager. A Cube Manager creates a context cube, maintains the cube, handles data requests and cube operations. Within the Cube Manager, a class call sets up a data store and accesses the specified tables and fields to get the raw data from the context warehouse. Also in the Cube Manager, you can specify at what base scale/resolution you want the system to construct the context cube. We cannot perform a drill-down operation on a cube below the base granularity of the data. Specifying DAY as the default scale sets the base data resolution to Day along the Time dimension. A similar definition can be made on each of the dimensions of a context cube as long as the context dimension hierarchy information exists.

Once you bring in the raw data, you call the Cube class to process the raw data into a new cube. There are I/O classes to write out a cube, as a context fact table, to the database and for accessing an existing cube. The Cube class also provides the structure for N-dimensional arrays. A Cube Manager only requires a small amount of programming; the CCS does the rest of the work.

The average size of a Cube Manager for our current context cubes is about 25 lines of code. In future work, we hope to provide a GUI interface for constructing Cube Managers as well as providing for a construction language to allow for application control. Application interaction with a context cube is done either with ODBC/JDBC calls or through a PHP and XML interface. We use the latter method to provide data to our Flash based interactive displays, like the Digital Family Portrait.

Above the Cube Manager is the Cube Register. This maintains the information about all the context cubes in the system. It knows how to construct the cubes, where the cube data originates, and the status of the cubes. The Cube Register also serves to maintain cubes that update periodically (like every night) as well as launching, or relaunching, Cube Managers as needed. The Cube Register contains the data table dependency information needed by dynamic cubes that update based on *context triggers*. A context trigger signals a Cube Manager that an event has occurred in the context environment that necessitates an update to its context cube. We constructed our context triggers as Context Toolkit widgets [8]. The Cube Register also maintains a qualitative description of the cubes.

From this registry, application developers can determine what kinds of cubes and context dimensions are already available. Information on available context dimension hierarchies is kept in the Dimension Register. The Dimension Register functions like the Cube Register, but knows about available context dimension hierarchies in the system.

The time to construct a cube is related primarily to the size of the cube and secondly to the amount of data in the data source tables. However, as Figure 8 shows, the time it takes to compute a cube grows slower than the size of the cube. Making the initial connection to the database is a fundamental time cost for all cubes. In designing dynamic cubes updated by context triggers, we must consider the time necessary to construct the cube. In most cases, only small cubes are appropriate. Though larger cubes can be set to update with context triggers, it will probably be more viable to have them update at regular, predetermined intervals.

3   Building Applications Using the Context Cube System

In this section, we provide four examples of the context cube in use. The first three examples demonstrate how the cube supports data related, query-based context applications by performing cube operations on a single context cube. The final example is hypothetical and shows a more complex use of context dimension hierarchies, and demonstrates a technique for aggregating cubes.

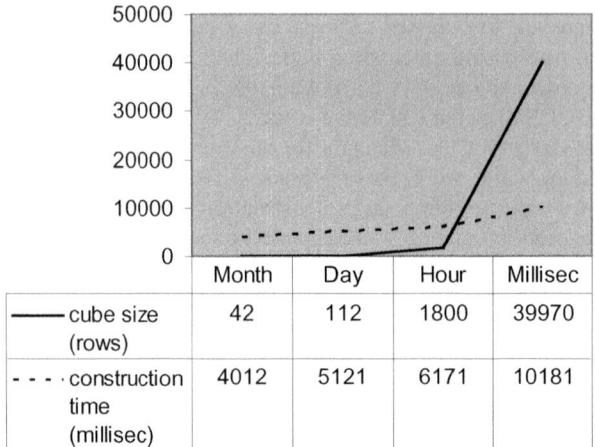

	Month	Day	Hour	Millisec
——— cube size (rows)	42	112	1800	39970
- - - construction time (millisec)	4012	5121	6171	10181

Fig. 8. Construction time of context cubes, in milliseconds, compared to the size of the cube. Month, Day, Hour, and Millisec refer to the base resolution of the data in the Time dimension. All cubes are for 28 days of data.

3.1 Supporting the Digital Family Portrait

The Digital Family Portrait, described earlier, provides a visual representation of an elderly individual's daily activity level. It also provides information on temperature and other environmental data. Direct reporting of the single-dimensional context can provide the environmental data. The LTI cube provides rapid querying and basic analysis of the multi-dimensional context data.

In the cube operations shown earlier in Figure 6, the first slice is on Grandma, interpreted by the Person context attribute table. This creates a Grandma cube containing Time and Location data. The slice along the Time dimen-

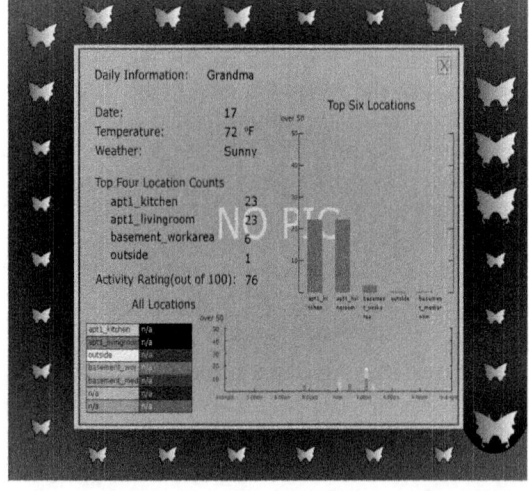

Fig. 9. A drill-down on an individual day

sion is a simple date, provided by the application. This can be constructed by a single dice operation. Currently, the Digital Family Portrait defines activity as moving from hall-to-room or room-to-room. The context dimension hierarchy associated with the

Location dimension provides the hall and room generalization. Using this information, the correct position changes can be easily derived and stored as the value of the remaining cell. This calculation is stored as the derived context "count", and the context cube maintains the value.

Instead of slice on the time dimension as the second step, we could have diced the cube with a range of dates to include the past 28 days. The result would be a cube that holds the defined activity value for those days. This is the information needed to support the DFP, so it will be able to request that information via a simple query. This is an example of how we can glean information from basic position data, by using warehousing techniques and the Context Cube.

Other manipulations of the LTI cube would produce the information for an example of family gatherings. In that case, we would need a context dimension hierarchy that encoded information about relationships so that we could distinguish between family and other visitors to the home. We would use the same raw data, but this application would request a different set of operations to create the context cube it needed. It could also use the same context dimension hierarchies as those created to support the DFP.

3.2 Visualizing Co-location

This application is similar to the DFP but instead of tracking the activity level of one individual, it shows a representation of two people's co-location history. The figure shows a drill-down on a single day. Colored sections in the schedule show the individual's presence in the home. Times of co-location are highlighted in the space between the schedules. The distances between the figures in the border reflect the level of co-location for that day. Days with only one person's data have one figure; days with no data for either individual are grayed out.

This is again derived from the LTI cube. We perform a slice on two individuals instead of just one. Next, we perform a roll-up on the Time dimension to provide Location information at the MINUTE resolution. Finally, we need to perform a roll-up on Location. In our original implementation, we provided information at the room level. However, the results were counterintuitive. In this Home, the kitchen and living room are open and adjacent

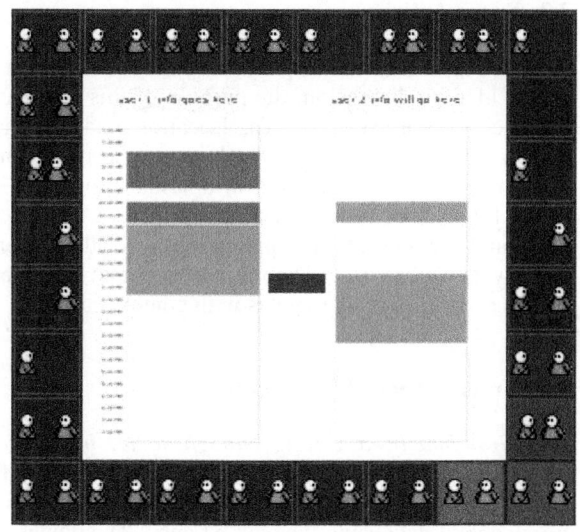

Fig. 10. A drilldown in the co-location application.

spaces. Two individuals, located in the kitchen and living room, would be able to see

Fig. 11. A visualization of space activity for the month of January 2003

and talk with each other easily, but our system would not acknowledge that as co-located. We had to construct a new interpretation of the Location roll-up to account for this inaccuracy in our results.

3.3 Space Activity

We have also built this application over the LTI cube. However, instead of focusing on the Identity dimension, the purpose of this application is to represent activity in a specific location over time. The Location data is at the base room level and we have done a roll-up on Identity to ALL, a way to represent a total aggregation of that dimension. The application shows a visualization of the activity levels in individual rooms for selected time periods.

Future versions of the application will allow users to manipulate Location and Identity as well as time. The purpose is to show how easy it is to manipulate interpretations and generate aggregates with context cubes.

3.4 Tracking Social Interaction

Has Grandma been getting enough social interaction lately? In this hypothetical example, we are looking at the level of interaction for an elderly individual living alone. This is an example of the kind of qualitative information that can be gleaned from context data using the context cube. The power of the cube is to provide organization and structure to a growing archive of information. By classifying and grouping data points, based on concepts derived from the real world, we make it easier for applications to analyze that data quickly.

The first step is to determine what constitutes a social interaction. For the purpose of this example, we will use a somewhat simplistic definition: a social interaction is an email, incoming or outgoing phone call, or a personal visit, from a friend or a family member. To track this, we will need to pull from three separate data sources: an email log, phone log, as well as the identity of individuals in the home from the LTI context cube. The email log would look something like the excerpt in Table 1.

Table 1. Part of an email log showing likely fields

Subject	Sender	Date	Size
breakfast?	Ivxlcdm@aol.com	9:42 PM	2k
Grow more zinnias	ntwt@buyme.net	8:34 AM	24k
heard from Zoe	Phwerty@blah.com	11:43 PM Saturday	3k
missed you	curio@carlosm.org	8:22 PM Saturday	1k

The phone log would have a similar structure. There are several ways of capturing phone call information. Many local phone companies now will provide an online service; there are standalone devices that interface with caller-id systems as well as computer based systems. Instead of size, the phone log provides the duration of the call (Table 2).

Table 2. Part of a phone log showing likely fields

Number	Caller	Time	Duration
(404)xxx-1234	Elizabeth Bennet	9:42 PM	43 min
(404)xxx-5678	Fitzwilliam Darcy	8:34 AM	6 min
(770)xxx-9123	Charles Bingley	11:43 PM Saturday	22 min
1-800-xxx-7734	Out of Area	8:22 PM Saturday	1 min

The existing LTI context cube can provide information about people located in the home. Assuming that we would possess identity information for every visitor, email, or phone call is unreasonable. Instead, we assume that appropriate identifying information will be available for those individuals that are socially close to the target subject.

The next step is to process or clean the data: email addresses, phone numbers, and caller Ids need to be interpreted as names. For each of the interaction types, we will create a cube with two dimensions (Identity, Initiation Time) and one derived context, the duration of the interaction. For visits and phone calls, the duration is a measure of time, for email the duration is a measure of size. We then perform a roll-up on each of the cubes using an Interaction context dimension hierarchy that classifies they type of interactions based on Identity. This is best done in conjunction with the Person context attribute table shown in Figure 2.

In the next step, we perform a slice on Social on each of the three context cubes and aggregate the results into a new context cube. This cube now contains only the interactions that we are interested in to track social interaction. If we perform a roll-up on the Initiation Time dimension to DAY, we will have a listing of social interactions by day. The data is now ready for analysis. A simple clustering algorithm can be used to classify a set of daily interactions as low, normal, or high. We could also use other systems, dedicated to trending analysis, to construct the qualitative rule, such as the Coordinates system at Microsoft [17]. Once we have a classification to use as a qualitative rule, we can apply that rule to each day and add a new derived context, Social Interaction Level, to the cube.

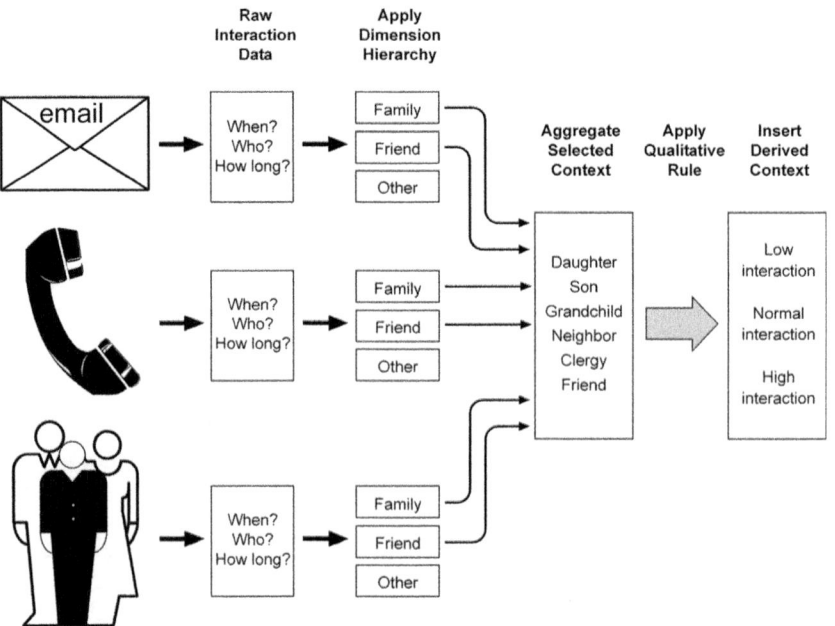

Fig. 12. Creating a social interaction cube from raw interaction data.

4   Related Work

This works lies at an intersection between the research domains of context-aware computing and that of data warehousing and data mining. In the area of context-aware computing, our work relates to the area of context infrastructures. Context cubes are not an independent infrastructure, instead they provide a data warehousing component needed to leverage the historical context. As referenced earlier, context cubes are an adaptation of data cubes from the data warehousing community. In addition to these, the context dimension hierarchies are related to work done with ontologies [1] and inference engines like Cyc [4] as well as lattices used in machine learning [5, 6].

4.1 Context-Aware Computing

In many cases, developers create *ad hoc* solutions for individual context-aware applications. However, as exploration in the field of context-aware computing has broadened, a desire to move away from single application development has emerged. The works of Schilit [7], Dey [8], and Brown [9], all provide a framework for developing multiple applications using shared context resources. The Context Fabric is a more recent infrastructure approach from Hong and Landay [10]. The Context Toolkit [8], Context Fabric, and the Event Heap [11] all provide the ability to store context data. The Context Fabric extends those capabilities by providing a logical context data model, separate from the physical storage model, which they represent using four concepts: entities, attributes, relationships, and aggregates. Hong et. al. created a prototype application, the Communication Information Agent (CIA) [12] that uses stored document access information to predict future document needs. This application is similar to the Remembrance Agent [13] that also recommends documents based on past and current context.

In using an infrastructure framework, middleware transports, manipulates, and manages context data, acquired from the context sources and on behalf of the applications receiving the information. Aggregation, higher-level interpretation, synthesis and fusion, and analysis are all activities performed by middleware applications. Middleware that processes lower-level context are relatively new in context computing. The location management work in QoSDREAM [14] from Cambridge that models and combines location information from multiple heterogeneous sources is an examples of an infrastructure that support immediate context applications. At this time, there are few models or systems that support applications that need context beyond the current state or that require sophisticated data manipulation to produce the necessary information. Of those that exist, the Context Cube is closest in philosophy to the "database" approach of the Context Fabric model. That infrastructure, however, strives for a homogenous data model for capturing, storing and accessing context data. The Context Cube model is more of a "data warehouse" model, expecting context data to have a heterogeneous representation. We designed the Context Cube to be an extension of the previous architectures, agnostic as to the underlying infrastructure used to capture data. Theoretically, a Context Cube system could function simultaneously over all of the current context infrastructures that store context data. At present, the Context Cube is using data provided by the Context Toolkit and from several UPNP devices.

The structured data model of the Context Cube also supports trending and other forms of data analysis. The previously referenced Coordinates system from Microsoft [17, 19] and the work of Hudson et. al. [18] are examples of systems that use analysis in order to interpret or predict human behavior.

4.2 Datawarehousing

Inmon describes a data warehouse as a "subject-oriented, integrated, nonvolatile, time-variant collection of data in support of management's decisions." [15] Traditional databases are designed to be transactional (relational, object-oriented, network, or hierarchical). In contrast, data warehouses are mainly intended for decision-support applications. Traditional databases support *on-line transaction processing* (OLTP), which includes insertion, updates, and deletions. They generally function over a small sub-set of the available data, and that data is usually small, with historic data moved to external archives. Data warehouses, on the other hand, generally contain large amounts of data from multiple sources and they are inherently multidimensional. They are optimized to support on-line analytical process (OLAP), decision support systems (DSS), and data mining operations that need large collections of data. This may include data constructed under different data models and acquired from a collection of external, independent sources [16]. Since context data is also multi-dimensional, with few strong relations among data within context archives, (time being one of those), and since the access to the system will be similar to DSS and OLAP style interactions, storing the archived data as a *context warehouse* is a natural choice.

Unlike traditional data, context data is unidirectional with respect to time. As a result, context storage systems do not require update or delete functions. In time, we will need a "correction" function to replace inaccurate context information, but such a function would still preserve the forward advance of time. This means that context data, once captured, is static and only incoming context is effected by changes in the data environment. This allows us to adapt traditionally static techniques into dynamic ones.

In constructing a context warehouse, the raw data will be stored in *context tables*. Our context tables are constructed directly by the widgets of the Context Toolkit, using the store functions provided by default, and also from UPNP devices in the Aware Home environment. A context table may be a single source of context data or an integration of the related data from the context environment.

5   Conclusions

There is a gap between the context that is being provided by sensors and that needed by more sophisticated, context-aware applications. Adapting data warehousing techniques to create a context warehouse and the Context Cube is one step in bridging that gap. By using the raw data stored in the context tables, referencing that data through information provided by the context dimension hierarchies and with the cube operations, we are able to provide information about individuals (like Grandma) or collections of people (like the family), provide summary information, perform aggregations based on relationships among the context, determine collocation over some period of time, or provide information suitable for higher-level analysis.

Storage space is an important issue when dealing with cube technologies. There are three major implementation alternatives when implementing a cube [3]. Physically materializing the whole cube takes the most space but provides the fastest response to queries. The storage space consumed by a cube has a direct impact on the time necessary to construct a cube. At the other end of the spectrum, a representation may be generated without materializing any of the cubes. In this case, every cell is computed from raw data on request. It requires no extra space, but has extremely slow query response time. The final method materializes a small subset of cells in a cube and then generates the remaining "dependent" cells on request. Dependent cells are either calculated or approximated based on the values of the materialized cells. In our initial prototype, we chose to materialize the entire cube in order to optimize query response time, and to simplify implementation. However, we will need to consider more efficient cube materialization as the amount of data grows. A full LTI cube over all 300,000+ data points takes several minutes to calculate, 1GB of system memory, and produces a materialized representation of over 500,000 lines.

A future goal of our work is to use data mining and knowledge discovery techniques to generate new context through synthesis and analysis. The system developed by Elite Care [20] to determine quality of sleep through an instrumented bed is an example of work that requires context analysis. Though aggregating and interpreting context data can provide a lot of information, more context may be acquired by appropriately analyzing the stored data. For example, though the simple measure of activity used in the current implementation of the Digital Family Portrait is effective, it is not very informative. The system represents a day in which Grandma moves repeatedly from the bedroom to the bathroom and back in the same way as a more normal day, which includes the same number of room changes. If the DFP had more qualitative information available, like the social interaction example, it could in turn provide more information to the family. We can produce this kind of information through analysis of the stored context. In the case of Grandma's movement patterns, a day in which she only moved between two rooms would be identified as different from a day in which her movements covered several rooms. Having the ability to recognize significant changes in physical behavior provides important information for a variety of applications. For our initial work, we are using k-clustering to divide a collection of context data into groups that have strong similarity. Other clustering techniques as well as associative techniques will expand the nature and sophistication of the generated context.

Acknowledgements. This work is sponsored in part by the Aware Home Research Initiative, the Digital Media Lab at Georgia Tech, and the National Science Foundation (grant 0121661). Our thanks to Kris Nagel for her efforts in securing our IRB approval for gathering and storing the context data and to Thomas O'Connell and his team for supporting the Aware Home Living Laboratory. IRB Protocol Title: Aware Home Context-Aware Services Experiment, PI: Abowd, Gregory D., Protocol Number: H02047

References

1. E. D. Mynatt, J. Rowan, A. Jacobs, S. Craighill, *Digital Family Portrait: Supporting peace of mind for extended family members*. In *CHI2001*. 2001, Seattle, WA.
2. J. Gray, S. Chaudhuri, A. Bosworth, A. Layman, D. Reichart, M. Venkatrao, F. Pellow, H. Pirahesh, *Data Cube: A Relational Aggregation Operator Generalizing Group-By, Cross-Tab, and Sub-Totals*, In *Data Mining and Knowledge Discovery*, 1, 29-53, 1997, Kluwer Academic Publishers, The Netherlands
3. V. Harinarayan, A. Rajaraman, D. Ullman, *Implementing data cubes efficiently*, in *ACM SIGMOD International Conference on Management of Data*. 1996
4. D. B. Lenant, *Cyc: A Large-Scale Investment in Knowledge Infrastructure*, In *Communications of the ACM*, 1995, 38(11)
5. R. Poell, *Notion Systems*, http://www.notionsystem.com
6. S. Newcomb, M. Biezunski, *Topic Maps for the Web*, IEEE Multimedia, 2001.

7. B. Schilit, *System architecture for context-aware computing,* Ph.D. Thesis, 1995, Columbia, New York
8. A. K. Dey, G. D. Abowd, D. Salber, *A conceptual framework and a toolkit for supporting the rapid prototyping of context-aware applications,* In *HCI Journal,* 2001, 16(2-4), pp. 97-166
9. P. J. Brown, J. D. Bovey, X. Chen, *Context-aware applications: From the laboratory to the marketplace,* In *IEEE Personal Communications,* 1997, 4(5), pp. 58-64
10. J. I. Hong, J. A. Landay, *An Infrastructure Approach to Context-Aware Computing,* In *Human-Computer Interaction,* 2001, Vol. 16,
11. B. Johanson, A. Fox, *The Event Heap: A Coordination Infrastructure for Interactive Workspaces,* In *Fourth IEEE Workshop o Mobile Computing Systems and Applications (WMCSA 02),* Callicoon, New York, June 2002
12. J. I. Hong, J. A. Landay, *A Context/Communication Information Agent,* In *Personal and Ubiquitous Computing: Special Issue on Situated Interaction and Context-Aware Computing,* 2001, 5(1), pp. 78-81
13. B. Rhodes, T. Starner, *Remembrance Agent: A continuously running automated information retrieval system,* In *Proceedings of Practical Applications of Intelligent Agents and Multi-Agent Tech (PAAM),* 1996, London, UK
14. H. Naguib, G. Coulouris, *Location Information Management,* In *UBICOMP 2001,* Atlanta, GA, 2001
15. W. H. Inmon, *Building the Data Warehouse,* 1992, Wiley
16. R. Elmasari, S. B. Navathe, *Data Warehousing and Data Mining,* In *Fundamentals of Database Systems,* 2000, Addison-Wesley, pp. 841-872
17. E. Horvitz, P. Koch, C. Kadie, A. Jacobs, "Coordinate: Probabilistic Forecasting of Presence and Availability", Proccedings of the 2002 Conference on Uncertainty and Artificial Intelligence, July 2002, AAAI Press, pp. 224-233
18. S. Hudson, J. Fogarty, C. Atkeson, J. Forlizzi, S. Kiesler, J. Lee, J. Yang, "Predicting Human Interruptibility with Sensors: A Wizard of Oz Feasibility Study", *Proceedings of CHI 2003,* ACM Press, 2003
19. N. Oliver, E. Horvitz, A. Garg. "Layered representations for human activity recognition." In Fourth IEEE Int. Conf. on Multimodal Interfaces, pages 3--8, 2002
20. Elite Care, http://www.elite-care.com
21. Y. Ding, S. Foo, "Ontology research and development. Part 1: A review of ontology generation", Journal of Information Sciences, 2002 28 (2)123-136

A Context-Aware Communication Platform for Smart Objects

Frank Siegemund

Institute for Pervasive Computing
Department of Computer Science
ETH Zurich, Switzerland
siegemund@inf.ethz.ch

Abstract. When smart objects participate in context-aware applications, changes in their real-world environment can have a significant impact on underlying networking structures. This paper presents a communication platform for smart objects that takes an object's current real-world context into account and adapts networking structures accordingly. The platform provides (1) mechanisms for specifying and implementing context-aware communication services, (2) a tuplespace-based communication abstraction for inter-object collaboration, and (3) support for linking communication and context-recognition layers. For specifying context-aware communication services, a high-level description language, called SICL, and a compiler that generates corresponding code for smart objects were realized. The tuplespace-based infrastructure layer for inter-object collaboration hides low-level communication issues from higher layers and facilitates collaborative context recognition between cooperating objects. The paper also presents examples of context-aware communication services and evaluates the platform on the basis of a concrete implementation on an embedded device platform.

1 Introduction

As pointed out by Weiser and Brown[12], Pervasive Computing "is fundamentally characterized by the connection of things in the world with computation". Smart everyday objects exemplify this vision of Pervasive Computing because they link everyday items with information technology by augmenting ordinary objects with small sensor-based computing platforms (cf. Fig. 1). A smart object can perceive its environment through sensors and communicates wirelessly with other objects in its vicinity. Given these capabilities, smart objects can collaboratively determine the situational context of nearby users and adapt application behavior accordingly. By providing such context-aware services, smart objects have the potential to revolutionize the way in which people deal with objects in their everyday environment.

But networks of collaborating smart objects are extremely difficult to manage: some objects are mobile, often with distinct resource restrictions, communication is dynamic, takes place in a highly heterogeneous environment, and

A. Ferscha and F. Mattern (Eds.): PERVASIVE 2004, LNCS 3001, pp. 69–86, 2004.

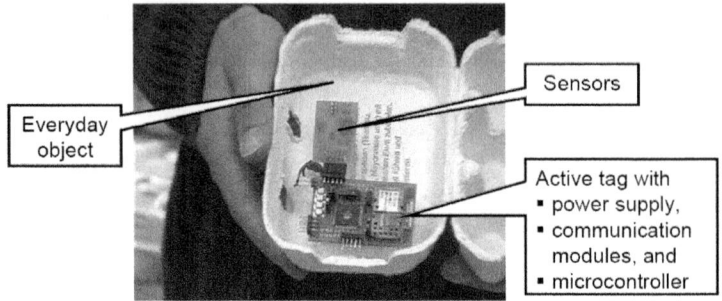

Fig. 1. A smart object: an everyday item augmented with a sensor-based computing platform.

must not rely on a constantly available supporting background infrastructure. Furthermore, as the term Pervasive Computing suggests, there are potentially huge numbers of smart objects in communication range of each other, which even aggravates the problem of keeping communication efficient.

Our approach to addressing this problem is to consider the real-world context of smart objects in the design of communication services. The underlying motivation is that with the integration of computation into everyday items, the real-world situation of smart objects increasingly affects their communications in the virtual world. Thus, everything that happens to an object in the real world (whether and how, when and how often it is used, its current location and whether there are potential users in range) might influence its behavior in the virtual world.

Location-based routing protocols [1] already consider sensory information to improve the effectiveness of networking services. Also, Katz [6] points out that mobile systems must be able to "exploit knowledge about their current situation to improve the quality of communications". Context-aware communication services, however, take such adaptiveness to a new level, because the variety of sensors carried by smart objects makes it possible to determine their current situation with unprecedented accuracy. In context-aware environments, changes in the real-world situation of smart objects also increasingly affect their behavior in the virtual world.

In this paper, we present a platform for building context-aware communication services. It consists of (1) components for the description and generation of context-aware communication services, (2) a tuplespace-based communication abstraction for inter-object collaboration, and (3) mechanisms that enable communication services to access context information. Regarding service generation, we developed a description language, called SICL, for the design of context-aware services, and a compiler that, given an SICL program, automatically generates code for smart objects. The tuplespace-based infrastructure layer for inter-object collaboration is the basis for deriving context information from multiple sensor sources. It also hides low-level communication issues from context-recognition and application layers. Furthermore, we present concrete examples of context-aware communication services that show how to link communication and context layers.

The rest of this paper is structured as follows: In the next section we review related work. Section 3 gives an overview of our platform for building context-aware communication services. Section 4 presents a description language for context-aware services. In section 5, we describe the tuplespace-based infrastructure layer for inter-object collaboration. Section 6 gives concrete examples of context-aware communication services, and section 7 concludes the paper.

2 Related Work

In [8] and [10], we proposed to consider the real-world context of smart objects in the design of specific networking protocols. In this paper, we build on our previous work, generalize our approach, and describe necessary infrastructural components that facilitate context-aware communication services.

Beigl et. al [3] reports on a networking protocol stack for context communication, and suggests to consider situational context in order to improve its efficiency. Thereby, contexts such as the battery level, the current processor load, and the link quality serve as indicators for adapting networking parameters. We do not focus on such virtual-world parameters, but are more interested in adaptiveness based on the real-world situation of objects. Furthermore, we concentrate on providing the distributed infrastructure and high-level services that allow such kind of adaptiveness.

The Stanford Interactive Workspaces [5] project introduces a tuplespace-based infrastructure layer for coordinating devices in a room. This tuplespace is centralized and runs on a stationary server, whereas we distribute the tuplespace among autonomous smart objects and do not assume that there is always a server in wireless transmission range.

Want et al. [11] augments physical objects with passive RFID tags to provide them with a representation in the virtual world. A similar approach is taken in the Cooltown project [14]. Our approach to building smart objects, however, is that of the Smart-Its [15] and MediaCup [2] projects, where active instead of passive tags are used to augment everyday objects. This allows them to implement more sophisticated applications and to autonomously process information. Consequently, smart objects do not depend on always available background infrastructure services that implement applications on their behalf.

3 Architecture Overview

Smart objects usually implement context-aware applications. Thereby, they derive the context of nearby users in collaboration with other smart objects and other sensor sources distributed throughout their environment. A context-aware application then adapts its behavior on the basis of the derived context. A typical software architecture for smart objects therefore consists of a communication layer, a context and perception layer, and an application layer. Fig. 2 shows how the main tasks of these separate layers have been addressed in our platform. In the following, we give a more detailed overview of these layers and show how they influence the design of context-aware communication services.

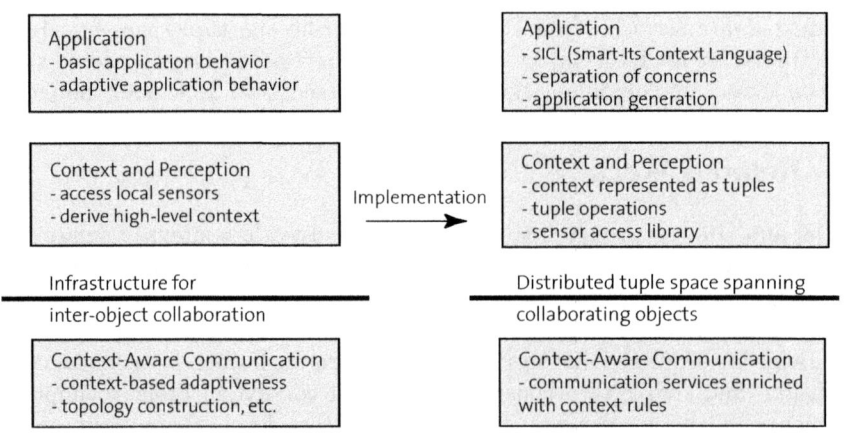

Fig. 2. A software architecture for smart objects and its implementation.

Application Layer. The application layer supports programmers in implementing context-aware services. Generally, context-aware applications have four parts: (1) basic application behavior, i.e., functionality that does not depend on context changes, (2) adaptive application behavior, which describes how an application reacts to changing situational context, (3) a section that specifies how to derive context from local and remote sensor readings, and (4) a part that specifies how to access sensors. We have implemented the Smart-Its Context Language (SICL) that separates these concerns and helps programmers to write more legible and structured code (cf. Fig. 2). Given a description of the four main parts of a context-aware service in SICL, a compiler automatically generates low-level C-code for an embedded platform. Our goal was to implement context-aware communication services using the same description language, and to link communication and application layer by exporting communication modules as sensors (cf. Sect. 6).

Context and Perception Layer. The context and perception layer derives the current situational context of a nearby person or of a smart object itself. Sensor access and context recognition is specified in SICL as part of an application. However, the SICL-compiler generates the actual code for the context and perception subsystem, which operates on the smart objects. The subset of the generated code responsible for querying sensors and fusing data makes up the context and perception layer. It provides information about an object's context to the application layer, which is then able to adapt according to this information (cf. Fig. 2). In our current implementation, context information is represented as tuples and the derivation of higher-level context information as tuple transformations. Thereby, SICL statements are translated into corresponding rules for merging and finding context tuples. Such tuples might originate from other smart objects and the results of a tuple operation might also be of potential interest to other smart objects in vicinity. The challenge is therefore to implement multi-sensor context awareness on sets of cooperating smart objects.

Support for Inter-Object Collaboration. We address this challenge of multi-sensor context awareness with an infrastructure layer providing cooperating entities with a distributed data space (cf. Fig. 2). This data space allows objects to share data as well as resources, and to fuse sensory data originating from different smart objects. This data space has been implemented as a distributed tuplespace. The distributed tuplespace hides low-level communication issues from higher layers, and it is also a means to group sensor nodes. As the accuracy of sensory data sharply decreases with increasing distance – e.g., a sensor that is only two meters away from a user is usually much better suited for deriving his/her context than a sensor that is 20 meters away – smart objects that collaborate to determine a certain context are often in close proximity of each other, for example in the same room or vicinity of a user. Such objects can therefore be grouped into a tuplespace that operates on a dedicated wireless channel, which makes tuplespace operations very efficient (cf. Sect. 5).

Communication Layer. The communication layer is responsible for low-level communication issues, such as device discovery, topology construction, connection establishment, and data transfer. As a Bluetooth-enabled device platform [4] is used in our experiments, the communication layer consists in its basic form of a Bluetooth protocol stack. The context-aware communication platform considerably extends this protocol stack with mechanisms that adapt according to results from the context layer. Thereby, communication services contain rules for the context-layer that describe how to derive relevant context and how to change networking parameters in certain situations. For example, when smart objects find out that they are in the same room (information that is derived by the context layer), the wireless network topology can be automatically changed such that communication between these devices is more efficient (cf. Sect. 6.3). We call communication services that take input from the context layer into account *context-aware*.

4   A Description Language for Context-Aware Services

Developing context-aware communication services in a language that natively supports embedded platforms (usually C) often leads to unnecessarily complex and error-prone code. Consequently, we have designed a high-level description language – the Smart-Its Context Language (SICL) – which facilitates the development of context-aware services and applications.

While developing a range of such context-aware applications [4,9], we found that they all have a similar recurrent structure. This structure consists of four parts that deal with (1) basic application behavior, (2) the context-aware adaptiveness of an application, (3) access to local sensors, and (4) context recognition. Unfortunately, the C programming language does not encourage application programmers to clearly separate these different concerns, which often leads to illegible, unstructured and error-prone code. For these reasons, we decided to automatically generate C code from a more concise higher-level description of the specific application parts. Thereby, SICL aims at simplifying the implementation of a context-aware application in the following ways:

- It separates basic application behavior from adaptive application behavior and therefore leads to better structured code for context-aware applications.
- It allows an application programmer to specify the context-recognition process in a clear and legible way in the form of simple tuple transformations.
- The adaptive part of an application, that is, how the application reacts to results from the context recognition process, can be clearly formulated.
- SICL enables a programmer to specify how to access sensors.

```
1:  smart object name;

    %{
4:  C declarations
    %}

    /* sensor statements */
8:  define [remote|local] sensor name {
9:      tuple type declaration;
10:     sensor access function;
11:     sensor access policy;
12: };

    /* tuple transformations */
```

$$15: \frac{func_1(arg_{11},\ldots,arg_{1m_1}),\ldots,func_n(arg_{n1},\ldots,arg_{nm_n})}{tup_1 < field_{11},\ldots,field_{1p_1} > +\ldots+ tup_q < field_{q1},\ldots,field_{qp_q} >}$$
$$\rightarrow tup_{q+1} < field_{(q+1)1},\ldots,field_{(q+1)p_{q+1}} >;$$

```
    /* adaptation rules */
```
$$20: tup < field_1,\ldots,field_p > \rightarrow func(arg_1,\ldots,argn);$$

```
    %%

24: C code
```

Fig. 3. The basic structure of an SICL program.

As a result, it becomes easier, faster, and less error-prone to implement context-aware applications – and context-aware communication services in particular – on an embedded sensor node platform. Fig. 3 shows the basic structure of an SICL program and Fig. 4 presents a concrete example. An SICL program consists of the following main parts that reflect the previously identified tasks of a typical context-aware application:

Sensor access. Sensor statements (cf. lines 8-12 in Fig. 3) describe how to deal with local and remote sensors. In our program model, sensor readings are embedded into tuples and used in transformations to derive new tuples that finally represent higher-level context information. As sensor tuples are not generated in tuple transformations, their type – i.e., the type of every field in a sensor tuple – must be specified (cf. line 9 in Fig. 3). Also, if a sensor is local, a function must be given that reads out a sensor, creates a tuple with the specified type, and writes it into the distributed tuplespace. It is also possible to determine when to read out a sensor by either providing the amount of time between consecutive sensor readings or by leaving it to the system to access the sensor as often as possible.

```
smart object egg_box;

%{
#include "sensors.h";
#include "gsm.h";
%}

define sensor acceleration {
    <type, subt, leng, accX, accY>;
    void get_and_write_accel();
    /* read out as often as possible */
    best effort;
};

eggs_damaged(x,y)
================================================
acceleration<t, s, l, x, y> => damaged<x, y>;

damaged<x, y> -> send_damaged_message_to_phone(x, y);

%%
```

Fig. 4. A very simple SICL program that monitors the state of a smart product by means of an acceleration sensor and sends a message to a mobile phone when the product has been damaged.

Context recognition. Tuple transformations (cf. line 15 in Fig. 3) describe how tuples are transformed and new tuples are created. This reflects the process of deriving high-level context information by fusing sensor readings from different nodes. Thereby, it is unimportant at which specific smart object a tuple has been created, because an infrastructure for inter-object collaboration makes it available to all collaborating nodes.

The basic structure of a transformation rule is as follows. The left-hand side of a rule lists the tuples that must be available for the rule to be executed, and the right-hand side specifies the tuple that is to be generated. A set of functions can be specified that operate on the fields of left-hand side tuples. The tuple transformation is then only carried out when all these functions return true. When more than one rule can be executed, the rule first specified in the SICL description is executed first.

Adaptive application behavior. Adaptation rules (cf. line 20 in Fig. 3) represent the adaptive part of a context-aware application in that they describe how an application needs to react when a certain context is derived. An adaptation rule simply consists of a tuple on the left hand side, which stands for a certain situational context, and a function on the right hand side that is executed when a corresponding context is derived.

Basic application behavior. The basic application behavior of a smart object is provided in form of ordinary C code. SICL allows to embed corresponding C declarations and definitions into an SICL program. The SICL compiler generates C code from the other parts of an SICL description, which is then compiled and linked with the C code representing the basic application behavior.

The SICL compiler was implemented using lex and yacc and is exceptionally slim, consisting of only around 2'500 lines of code.

5   Tuplespace-Based Inter-object Collaboration

As mentioned before, context-aware services – and context-aware communication services in particular – need a platform to disseminate information to other nodes, to access sensors at various smart objects, and to process these sensory data. This section describes the infrastructure component that facilitates this kind of cooperation among smart objects: a distributed tuplespace on context-based broadcast groups.

5.1   Basic Approach

The distributed tuplespace handles all basic communication-related issues required for inter-object collaboration. It hides low-level communication issues from higher layers and serves as a shared data structure for collaborating smart objects. Thereby, every object provides a certain amount of memory for storing tuples, and the tuplespace implementation handles the distribution of data among nodes. The context and perception layers access local sensors and embed the corresponding data into tuples, which are written into the space and are therefore accessible from all smart objects participating in the distributed tuplespace.

We extend the basic concept of distributed tuple spaces by grouping nodes into a space according to their current context, and by assigning such groups a dedicated broadcast channel (cf. Fig. 5). For example, nodes that are in the same office room can be automatically grouped into one distributed tuplespace. We argue, that such networking structures, which take the real world context of smart objects into account, are better suited for the demands of context-aware applications.

Because Pervasive Computing networks are likely to be dense, nodes in the same distributed tuplespace communicate on a dedicated wireless channel. This minimizes the amount of collisions with other nodes in the same area and considerably improves the effectiveness of tuplespace operations (cf. Sect. 5.4). In case of frequency-hopping spread spectrum techniques a channel is defined by a unique frequency hopping pattern. For multi-frequency communication modules,

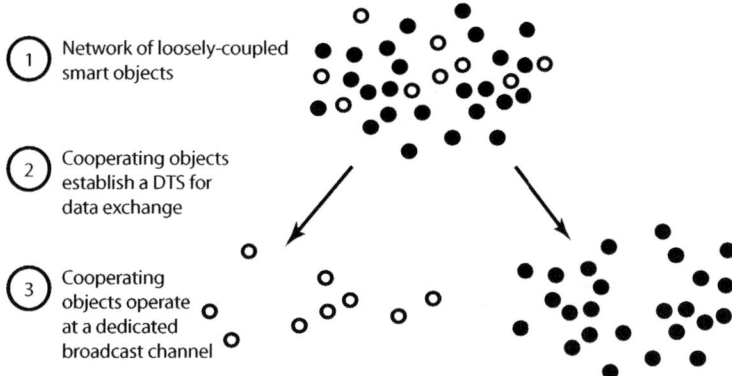

1 Network of loosely-coupled smart objects

2 Cooperating objects establish a DTS for data exchange

3 Cooperating objects operate at a dedicated broadcast channel

Fig. 5. Grouping of collaborating smart objects into distributed tuple spaces that operate on different communication channels.

a channel is usually defined by a single frequency. The communication modules used on the Berkeley Motes [13], for example, support multiple frequencies but do not implement frequency hopping patterns.

5.2 Efficiency Discussion

An infrastructure for inter-object collaboration must consider the energy consumption of nodes and the effectiveness of data exchange (i.e., how fast data can be transferred between objects). We argue that our approach is effective with respect to both aspects in typical Pervasive Computing networks, which share the following characteristics:

1. **Tight collaboration between adjacent nodes.** Sensor readings originating from smart objects that are in close proximity of a user are likely to provide the most accurate information for characterizing the user's situation. Therefore, in order to realize context-aware applications, primarily objects that are in close vicinity need to exchange data. Consequently, cooperating objects are often in range of each other.
2. **Short-range communication.** Because energy-consumption is a major concern in the envisioned settings, smart objects exchange data using short-range wireless communication technologies. A major property of corresponding communication modules is that their energy consumption for sending data is approximately as high as that for receiving data.

Criterion (1) suggests that collaborating smart objects are often in range of each other. Consequently, collaborating nodes that are grouped in a distributed tuplespace can reach each other by broadcasting data. Because nodes in a tuplespace transmit on dedicated channels they are also the only nodes that can receive such data. Tuplespace operations can therefore be implemented by broadcast protocols, which are more efficient than unicast-based solutions (cf. Sect. 5.4).

Because sending data is approximately as energy-consuming as receiving data (cf. criterion (2)), nodes can directly transmit data to other nodes in range without wasting energy. That means, objects would not save energy by reducing their transmission range and routing data over intermediate nodes, because receiving is so expensive. Consequently, when grouping nodes on broadcast channels their actual distance is not important as long as they can reach each other. Grouping of nodes into broadcast channels does therefore not have to consider node distance but can take other parameters – such as context – into account. As a result, our approach to grouping nodes according to their context rather than distance is energy-efficient. Please note, that this is not true for general wireless networks because for long-range communication technologies it is usually more energy-efficient to reduce the transmission range and to transmit data over multiple short distance hops instead of over one long-distance hop [7].

5.3   Implementation

In the following, we describe an implementation of such a distributed tuplespace on context-based broadcast groups for the BTnode device platform [4].

BTnodes communicate with each other using the Bluetooth communication standard. Nearby Bluetooth units can form a piconet, which is a network with a star topology of at most 8 active nodes. Thereby, Bluetooth distinguishes *master* and *slave* roles. The master of a piconet is the node in the centre of the star topology; slaves cannot communicate directly with each other but only over the master. The master also determines the frequency hopping pattern of a piconet and implements a TDD (time division duplex) scheme for communicating with slaves. Consequently, a slave can only transmit data when it was previously addressed by the master, and only the master can broadcast data to all units in a piconet. The frequency hopping pattern of a piconet implements a dedicated channel, which results in very few collisions with other piconets in the same area. Multiple piconets can form so called scatternets. Thereby, bridge nodes participate in more than one piconet and switch between piconet channels in a time division multiplex (TDM) scheme.

Our implementation tries to group collaborating smart objects into one piconet. This corresponds to the aim, according to which tightly cooperating objects shall be grouped such that they operate on a dedicated broadcast channel. Because of its pseudo-random frequency hopping pattern, a piconet constitutes such a dedicated channel on which the master can broadcast data. As Bluetooth nodes in a piconet can be as much as 20m away from each other, in the envisioned settings cooperating objects can be grouped into a piconet most of the time. However, due to the low number of nodes in a piconet, it might become necessary to organize nodes into a scatternet.

Every smart object provides a certain amount of memory for the distributed tuplespace implementation (cf. Fig. 6). As embedded systems often do not provide a reliable dynamic memory allocation mechanism, the memory for the tuplespace is statically allocated. We have implemented our own memory management mechanism for tuples, which poses some restrictions on the storage of

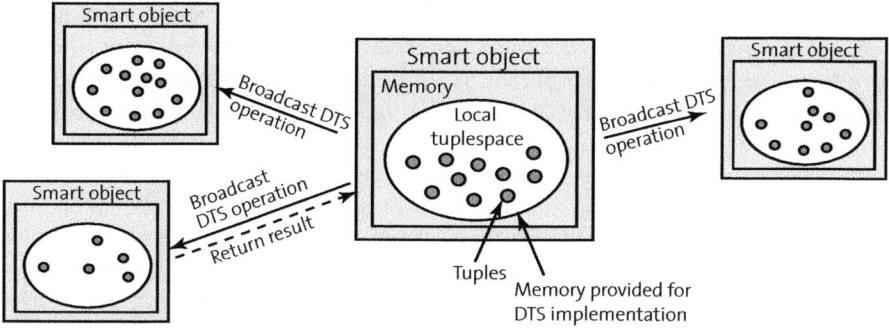

Fig. 6. Overview of the distributed tuplespace implementation: multiple local tuple spaces are connected to form a distributed tuplespace.

tuples. Because of the memory restrictions of embedded systems, for example, tuple fields have always a predefined size. That is, although a tuple may consist of an arbitrary number of fields, the type information and the actual value of a field are restricted in size. The memory restrictions also imply that after a time – especially when sensors are read out often – the tuplespace will be full. In this case, the oldest tuples are deleted in order to allow new tuples to be stored. However, it is also possible to protect tuples that should not be automatically deleted. We call such tuples 'owned'; they have to be manually removed from the tuplespace.

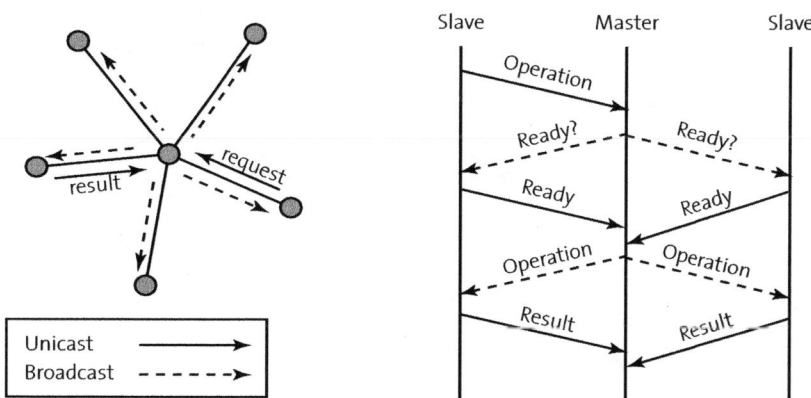

Fig. 7. Basic protocol for distributed tuplespace operations.

The DTS implements typical tuplespace operations such as *write*, *read*, or *take*, but also offers functions such as *scan*, *consumingScan*, and *count*, which originate from other widely-used tuplespace implementations. In our implementation, there are different versions of these functions that operate on different spaces: the tuplespace local to an object, an explicitly specified remote tu-plespace, or the distributed tuplespace as a whole. In the following, however, we focus on functions that operate on all nodes in a distributed tuplespace and describe the underlying protocol based on an example for a *read* operation (cf. Fig. 7). When a slave wants to execute a *read* operation on the distributed tu-plespace, it must send the request first to the master of its piconet. Next, the master broadcasts a *ready?* message to all nodes on the broadcast channel and waits for their response. The explicit response is necessary because broadcast traffic in wireless networks is usually unacknowledged. When all nodes acknowl-edged the request, the master broadcasts the actual operation together with all necessary parameters to its slave and awaits the results (cf. Fig. 7).

We would like to emphasize that the concepts of our implementation are in no means restricted to a specific communication standard. In fact, porting our code to other device platforms, as for example the Berkeley motes [13], would be possible because they do also support multiple communication channels.

5.4   Evaluation

Based on measurements, this section tries to evaluate the performance of the presented implementation.

A central question is whether the broadcast-based implementation for data sharing is superior to a more simpler, unicast-based solution. The *read* opera-tion presented in the previous section, for example, queries all nodes for tuples matching a given template. However, *read* only requires one matching tuple. In Fig. 8 we compared the broadcast-based solution with a unicast-based read operation, which queries all nodes after another and stops as soon as it receives a result from one node. In the experiment, there was always a matching tuple on

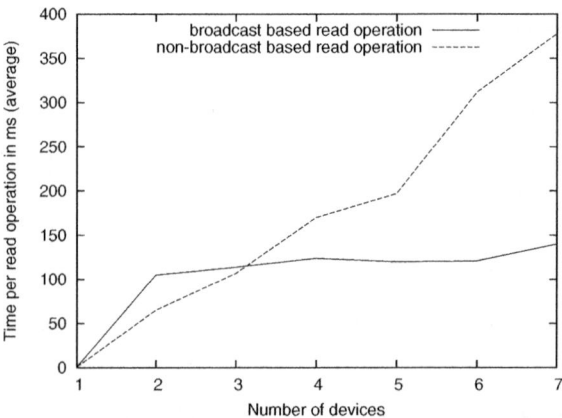

Fig. 8. Performance of a broadcast-based *read* operation compared to a unicast-based implementation.

one of the nodes. This node was picked with equal probability from all available nodes, and the *read* operation was invoked directly at the master.

As can be seen, the broadcast-based implementation clearly outperforms the unicast-based version, especially with increasing number of nodes. A reason for this is that because of the broadcast, fewer packets are transmitted over the medium when there are many nodes on the broadcast channel. However, this fact alone does not fully explain the graphs in Fig. 8. We also noticed that the whole process of sending a unicast packet and receiving a response from one node always takes about 35 ms. On the other hand, sending a broadcast packet and receiving the results takes between 50 and 70 ms for two to seven nodes. We think that this is due to the Bluetooth modules used on the BTnodes, which seem to delay packets before delivering them to the microcontroller when packets are not streamed. Also, the time division duplex scheme implemented in Bluetooth decreases the efficiency of the unicast-based protocol.

We have also made measurements regarding the performance of tuplespace operations executed during the context recognition process. As mentioned before, SICL rules (cf. Fig. 3 in Sec. 4) are translated by the SICL compiler into tuplespace operations for obtaining relevant tuples, merging them, and writing the corresponding results back to the space. Given a typical SICL program, usually all global tuplespace operations except *consumingScan* are executed during this process. The reason for this is that the context recognition layer cannot decide by the time it is scanning for tuples whether to remove them from the space or not. It can only do this when it has evaluated these tuples, and then only deletes selected tuples by calling the *take* operation instead of *consumingScan*. Please note that SICL offers a special operator for deleting tuples after a rule has been successfully evaluated. Fig. 9 shows the number of tuplespace operations executed during the context-recognition process with respect to the number of collaborating nodes. To test the performance of the tuplespace operations un-

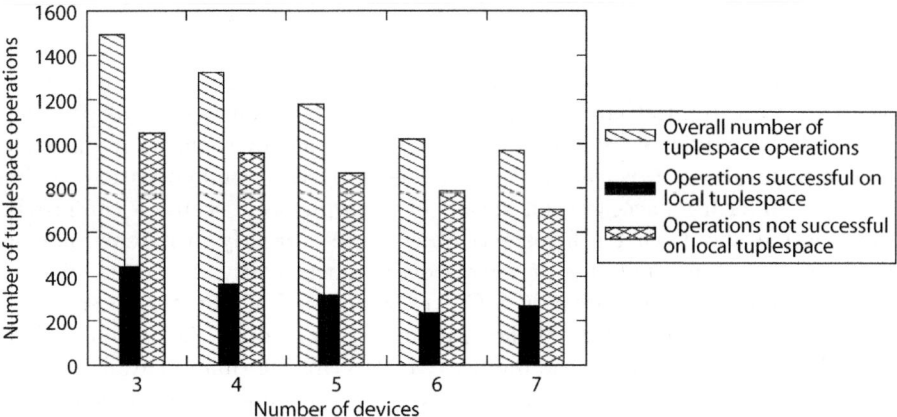

Fig. 9. Number of operations executed on the distributed tuplespace during the context-recognition process under heavy load in t = 3 min.

der heavy load, the underlying context-recognition code was generated from an SICL program such that global tuplespace operations are continuously issued by all participating nodes without periods of low activity. The generated code was uploaded to and executed on all collaborating nodes, not only on the master. The measurements in Fig. 9 show that the number of executed tuplespace operations that require communication with other nodes only decreases from around 1000 to 750 when the number of slaves increases from two to six. Because the presented grouping procedure reduces the number of nodes in a DTS, this is a very promising result. Also, the time needed for executing tuplespace operations on the local space is negligible because of the memory-restrictions of smart objects. The number of local operations only depends on the structure of the SICL program.

6   Context-Aware Communication Services

Context-aware communication services are communication services that take an object's current situational context into account. They can either adapt their behavior according to changes in the real-world environment or parameterize conventional communication services with context information. For example, a service for context-aware device discovery could return not only the devices currently in range but could also determine which of them share the same symbolic location. Context-aware communication services can also completely substitute conventional communication mechanisms (cf. Sect. 6.2). In this section, we describe how we linked the communication and context layers in order to allow communication services to access and process context information. We also present two examples of context-aware communication services.

6.1   Linking Communication and Context Layer

We designed the context-aware communication platform with the aim of being able to realize context-aware communication services in the same way as other context-aware applications. The first step towards this goal is to think of a communication module as a sensor. A sensor monitors its environment and provides information about its surroundings. In this respect, a communication module does not differ from acceleration sensors or microphones, because it can provide information about what other smart objects are currently in its environment. Therefore, we can export the device discovery functionality of a communication module as a sensor statement in SICL (cf. Fig. 10). Thereby, the sensor statement specifies when to read out the communication module – i.e., when to check for other objects in range – and implicitly makes this information available to all other objects by means of the infrastructure layer for inter-object collaboration. Consequently, a tuple that contains the address of a discovered module must also contain the address of the module that actually found it. This can be seen in Fig. 10, where the tuple generated by the sensor contains the lower and upper

```
define sensor communication_module {
    <lapi, uapi, lapd, uapd>;
    void start_inquiry();
    every 60000 ms;
};
```

Fig. 10. Linking communication and context layer by treating communication modules as sensors: sensor statement for a communication module in SICL.

address part of the inquiring unit (lapi and uapi) as well as the corresponding information for the module discovered (lapd and uapd).

By linking communication and context layer in this way, communication services can take advantage of all the services and infrastructural components presented in the previous sections. Thus, they can access derived context information and sensor readings from collaborating objects.

6.2   Context-Aware Device Discovery

In some communication technologies, device discovery can be extremely slow. Especially for frequency-hopping technologies – such as Bluetooth – where there is no dedicated frequency at which devices advertise their presence, a lengthy discovery phase is a major drawback. This problem can be addressed by cooperative approaches to device discovery that are still based on the conventional discovery mechanism [10], or by completely substituting this conventional mechanism with a context-aware communication service. In the following solution, the context-aware discovery mechanism does not use the discovery mechanism of the communication module at all.

```
smart object door;
...
define sensor rfid {
    <lap, uap>;
    void read_device_address_from_rfid();
    best effort;
};

object_already_in_room(lap, uap)
==================================
rfid<lap, uap> => old<lap, uap>;

object_not_in_room(lap, uap)
==================================
rfid<lap, uap> => new<lap, uap>;

old<lap, uap> -> remove(lap, uap);
new<lap, uap> -> join(lap,uap);
...
```

Fig. 11. The code for the smart door in SICL.

In the *smart door* application, a door is augmented with a BTnode and an attached RFID reader, which serves as sensor for the BTnode. Other smart objects have an RFID tag attached that contains their actual device address, such as their Bluetooth address. When they enter a room through the smart door, the device address is read out of the tag by the RFID reader and made available to other objects in the room through the distributed tuplespace shared by all objects inside the room. The smart door also tries to include the arriving object into the distributed tuplespace shared by objects in the room. The corresponding code for the smart door in SICL is depicted in Fig. 11.

The technology break of using another technology for device discovery is of course an overhead because it requires that every smart object is additionally equipped with an RFID tag, but it is faster than conventional Bluetooth device discovery. It is also possible to combine this RFID-based approach with the conventional device discovery mechanism. Thereby, discovery results are enriched with context information by not only identifying the objects in range, but also those that share a certain symbolic location.

6.3   Context-Aware Topology Construction

As already discussed in section 5, it is a goal of the platform for inter-object collaboration to group smart objects according to their current context. Thereby, a network topology is created such that nodes that are likely to cooperate are grouped on one channel (in Bluetooth, this corresponds to grouping nodes in different piconets). Because smart objects provide context-aware services to nearby users, the recognition of the user's context is one of the main reasons for communication when a user is present. Consequently, nodes that are in the same room are more likely to cooperate because their sensory data is better suited for characterizing the situation of a nearby person. In contrast, collaboration across room borders is less likely because sensor readings from another room are often less accurate. The context-aware service for topology construction presented in this section therefore groups nodes in clusters that share the same symbolic location. The goal is to show how the presented platform components support the implementation of such a service.

At first, the communication service assumes that relevant objects participate in one distributed tuplespace. The location of smart objects is then determined by audio input from a small microphone attached to smart objects. The microphone is read out once every minute on every object in the tuplespace. Thereby, a sampling rate of about 5 kHz for the duration of one second was used (cf. Fig. 12). As the access to the microphone has to be synchronized – i.e., it has to take place at the same time – a node triggers the sampling by writing a special tuple in the tuplespace. As a result, the microphone is then simultaneously read out at every node in the tuplespace.

The corresponding results are again made available to all participating nodes via the infrastructure for inter-object collaboration. Because transmitting the whole data stream is impossible, a small number of features are extracted from the audio stream that are embedded into a tuple and written into the space. In our implementation, we simply use the average volume as the main feature.

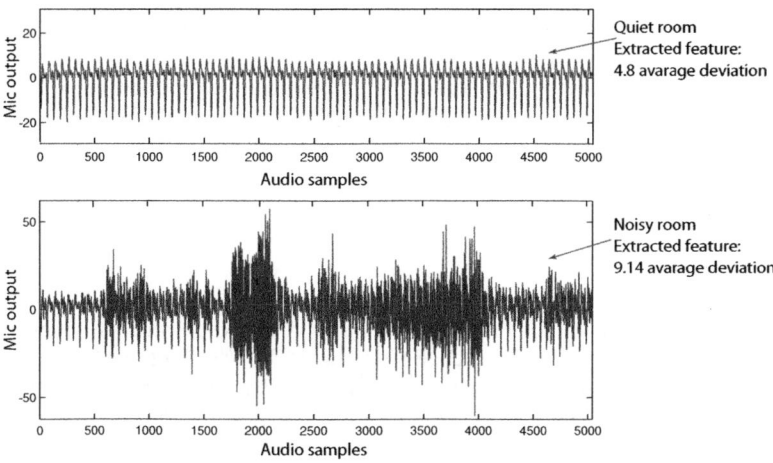

Fig. 12. Audio output of a microphone sensor and the extracted features shared via the infrastructure for inter-object collaboration.

Because this feature provides only limited information and synchronization problems can lead to inaccurate information, one sensor reading is not sufficient to really deduce which objects are in the same room. Instead, only after several consecutive sensor readings is it possible to draw a conclusion about the location of a node. The cluster head (in Bluetooth the master node) accesses this sort of history information by querying the distributed tuplespace and, if necessary, issues a command for organizing nodes into different clusters.

7   Conclusion

In this paper, we presented a context-aware communication platform for resource-restricted, wirelessly-communicating smart objects. The described platform consists of three parts: (1) a high-level description language, called SICL, for the development of context-aware communication services, (2) a tuple-based communication abstraction that hides low-level communication issues from higher layers, and (3) examples of context-aware communication services.

The description language SICL helps application programmers to clearly separate the main aspects of a context-aware application: basic and adaptive application behavior, context-recognition, and sensor access. So far, our experiences with the language are very promising, and the examples given in this paper show that it is practical for the development of context-aware communication services. However, the main advantage of SICL is that it relieves programmers of implementing the low-level communication issues necessary to fuse sensory data. Instead, it provides a simple tuple-based approach for data handling. This approach is realized by a distributed tuplespace that groups collaborating objects based on their current context. The distributed tuplespace is also the main platform for inter-object cooperation and hides low-level communication issues from higher layers. According to the measurements presented in Sect. 5, it supports efficient collaboration among autonomous smart objects.

The concepts described in this paper have been implemented on a resource-restricted device platform, which has been used in several applications to augment everyday objects with computation. Experiments have shown that our implementation is practical and works well in the envisioned settings.

References

1. S. Basagni, I. Chlamtac, V. R. Syrotiuk, and B. A. Woodward. A distance routing effect algorithm for mobility (DREAM). In *ACM/IEEE International Conference on Mobile Computing and Networking (Mobicom)*, pages 76–84, Dallas, USA, 1998.
2. M. Beigl, H.W. Gellersen, and A. Schmidt. MediaCups: Experience with Design and Use of Computer-Augmented Everyday Objects. *Computer Networks, Special Issue on Pervasive Computing*, 25(4):401–409, March 2001.
3. M. Beigl, A. Krohn, T. Zimmer, C. Decker, and P. Robinson. AwareCon: Situation Aware Context Communication. In *Fifth International Conference on Ubiquitous Computing (Ubicomp 2003)*, Seattle, USA, October 2003.
4. J. Beutel, O. Kasten, F. Mattern, K. Roemer, F. Siegemund, and L. Thiele. Prototyping Sensor Network Applications with BTnodes. In *IEEE European Workshop on Wireless Sensor Networks (EWSN)*, Berlin, Germany, January 2004.
5. B. Johanson and A. Fox. Tuplespaces as Coordination Infrastructure for Interactive Workspaces. In *UbiTools '01 Workshop at Ubicomp 2001*, Atlanta, USA, 2001.
6. R. H. Katz. Adaptation and Mobility in Wireless Information Systems. *IEEE Personal Communications*, 1(1):6–17, 1994.
7. V. Rodoplu and T. Meng. Minimum energy mobile wireless networks. In *Proceedings of the 1998 IEEE International Conference on Communications (ICC 98)*, volume 3, pages 1633–1639, Atlanta, USA, June 1998.
8. F. Siegemund. Kontextbasierte Bluetooth-Scatternetz-Formierung in ubiquitaeren Systemen. In *Michael Weber; Frank Kargl (Eds): Proc. First German Workshop on Mobile Ad Hoc Networks*, pages 79–90, Ulm, Germany, March 2002.
9. F. Siegemund and C. Floerkemeier. Interaction in Pervasive Computing Settings using Bluetooth-enabled Active Tags and Passive RFID Technology together with Mobile Phones. In *IEEE Intl. Conference on Pervasive Computing and Communications (PerCom 2003)*, pages 378–387, March 2003.
10. F. Siegemund and M. Rohs. Rendezvous layer protocols for Bluetooth-enabled smart devices. *Personal Ubiquitous Computing*, 2003(7):91–101, 2003.
11. R. Want, K. Fishkin, A. Gujar, and B. Harrison. Bridging Physical and Virtual Worlds with Electronic Tags. In *ACM Conference on Human Factors in Computing Systems (CHI 99)*, Pittsburgh, USA, May 1999.
12. M. Weiser and J. S. Brown. The Coming Age of Calm Technology, October 1996.
13. Berkeley Motes. www.xbow.com/Products/Wireless_Sensor_Networks.htm.
14. The Cooltown project. www.cooltown.com.
15. The Smart-Its Project: Unobtrusive, deeply interconnected smart devices. www.smart-its.org.

Siren: Context-Aware Computing for Firefighting

Xiaodong Jiang[1], Nicholas Y. Chen[1], Jason I. Hong[1], Kevin Wang[1]
Leila Takayama[2], and James A. Landay[3]

[1]Computer Science Division
University of California, Berkeley
Berkeley, CA 94720
{xdjiang,nchen,kwang,jasonh}@cs.berkeley.edu
[2]Department of Communication
Stanford University
Stanford, CA 94305-3230, USA
takayama@stanford.edu
[3]DUB Group, Computer Science and Engineering
University of Washington
Seattle, WA 98195-2350
landay@cs.washington.edu

Abstract. Based on an extensive field study of current firefighting practices, we have developed a system called Siren to support tacit communication between firefighters with multiple levels of redundancy in both communication and user alerts. Siren provides a foundation for gathering, integrating, and distributing contextual data, such as location and temperature. It also simplifies the development of firefighting applications using a peer-to-peer network of embedded devices through a uniform programming interface based on the information space abstraction. As a proof of concept, we have developed a prototype context-aware messaging application in the firefighting domain. We have evaluated this application with firefighters and they have found it to be useful for improving many aspects of their current work practices.

1 Introduction

Each year, fires kill about 4,000 civilians and 100 firefighters in the United States alone [1]. Firefighting is a dangerous profession that calls for quick decisions in high-stress environments, constant reassessment of dynamic situations, and close co-ordination within teams. Furthermore, the smoke, heat, and noise in a structure fire mask the environment and force firefighters to operate with an incomplete picture of the situation. One firefighter we interviewed summarized it best: "Firefighting is making a lot of decisions on little information."

Improvements in information gathering, processing and integration can help fire-fighters work more effectively to prevent injury and loss of life, as well as minimize

A. Ferscha and F. Mattern (Eds.): PERVASIVE 2004, LNCS 3001, pp. 87–105, 2004.

property damage. The pervasive computing community itself can also benefit from research in this area. The nature of emergency response is fundamentally different from office environments, in terms of physical risk, psychological state, and operating conditions. This poses unique challenges for designers and researchers investigating context awareness, new interaction techniques, and information visualization, to name a few. If we can make an impact in this highly stressful domain, where the systems we offer are secondary to the primary task, we might also be able to apply these results in less extreme environments for a wider audience, such as computing while driving.

From an extensive four-month field study of current firefighting practices [23], we found that firefighters often need to exchange information about their situation and their surrounding environment in a spontaneous and opportunistic manner. This type of interaction needs to be *spontaneous* because the time when information exchange will occur depends on the dynamically changing situation and often has to be done without direct human initiation. It also needs to be *opportunistic* because the constant movement of firefighters in a complex urban structure makes it difficult to maintain an always-on communication channel among them. Such interaction is especially useful when firefighters need to be alerted about imminent dangers.

The problem is that spontaneous and opportunistic interactions among firefighters are not well-served by current systems. Today, most firefighters rely on two communication channels on the scene of a fire. The first is a broadcast channel for voice communication. The second is a data broadcast channel for status updates between an incident commander—the person in overall charge of an emergency—and dispatchers at a centralized emergency response center. Both channels use the 800MHz to 900MHz radio band. Often, only the voice channel or data channel can be used at a given time. Moreover, both channels are broadcast driven and manually operated to support explicit communication rather than the tacit communication needs between firefighters.

Advances in pervasive computing technologies are providing us with an opportunity to let firefighters "see through the eyes of fellow firefighters" and provide a greater understanding of the overall situation. Small, cheap, wirelessly networked sensors (such as smart dust [2]) can be deployed on firefighters and in buildings, capturing contextual information—such as temperature, sound, movement, toxicity, and a person's location—at a level never seen before. The wealth of sensor data about firefighters and the environment can be exchanged between firefighters to help improve safety and effectiveness.

Towards this end, we have developed Siren, a peer-to-peer context-aware computing architecture that gathers, integrates, and distributes context data on fire scenes. To make tacit communication between firefighters more robust in the face of an inherently unreliable transport, Siren offers multiple levels of redundancy in communication and feedback. Siren also simplifies development of emergency response applications by providing a uniform programming interface based on the information space abstraction [3]. Using Siren, we have developed a prototype context-aware messaging application and conducted an informal evaluation of this application with firefighters. They have found it useful for improving many aspects of their current work practices.

The rest of this paper is organized as follows. We discuss related work in section 2. An example search and rescue scenario is given in section 3 to illustrate how Siren-enabled applications may assist firefighters. In section 4, we describe key findings from our field study of current firefighting practices, and how they motivate the design of Siren. Section 5 describes key components of the Siren architecture, including a programming model based on the information space abstraction, storage management, communication, and the context rule engine. We describe a prototype context-aware messaging application developed using Siren in section 6, and our evaluation of it with firefighters in section 7. We discuss some lessons we have learned about designing for mission-critical pervasive computing applications in section 8. We conclude in section 9 and discuss future work.

2 Related Work

There has also been a great deal of work at providing programming support for various aspects of pervasive context-aware computing. This includes the PARCTab system [4], Cooltown [5], the Context Toolkit [6], Contextors [7], Limbo [8], Sentient Computing [9], Stick-E notes [10], MUSE [11], SpeakEasy [12], Solar [13], XWeb [14], GAIA [15], one.world [16], and iRoom [17]. Most of these are designed to support office work in traditional work environments. Siren, on the other hand, aims to support field work practices of mobile firefighters.

Context Fabric [18] is a generalized service infrastructure for context-aware computing that implements an information space abstraction [3] and a P2P infrastructure. Siren implements Context Fabric for a peer-to-peer network of PDAs, and provides new capabilities for discovery, multi-hop communication and rule-based adaptation.

Others have also looked at supporting emergency responders using mobile ad-hoc mesh networks (MANETs). For example, Draco [19] aims to develop rapidly deployable networking technologies that can support voice and video communications between emergency responders. MeshNetworks is developing a "self-forming, self-healing technology" that automatically creates a wireless broadband network at an incident [20]. In Siren, we employ a simple mesh network on top of a standard 802.11b wireless network. However, our primary focus is on integrating sensing, communication, and feedback in a single architecture to support the tacit communication needs of firefighters.

The Command Post of the Future [21] is a set of projects investigating command in battlefield situations. These projects focus on developing technologies for better decision-making by battlefield commanders through improvements in multimodal interaction, information visualization, and knowledge-based reasoning. We complement this work by looking instead at tacit interactions between firefighters working inside a building, focusing on how the underlying software infrastructure can be designed to better support spontaneous and opportunistic communication for firefighters on the scene of an incident.

Our work is also related to Camp et al, who looked at communication issues in emergencies and prototyped a radio system that would reduce voice congestion while maintaining situational awareness [22]. In contrast, we concentrate on supporting data communication needs between firefighters.

3   Siren-Enabled Firefighting: A Usage Scenario

In this section, we describe a scenario motivated by our field studies to illustrate what tacit communications needs we intend to support.

Three firefighters conduct a search and rescue task in an office building. Each firefighter carries a PDA. Wireless-enabled sensors have either been placed at strategic locations (e.g. on smoke detectors) throughout the building, or have been deployed on the fly by the initial fire response team whose duty is to size up the situation.

As firefighter F1 walks into the building, his PDA continuously monitors the surrounding temperature measured by nearby sensors. F1 approaches the south exit of the building when his PDA detects an unusually high heat level and alerts him of imminent danger. As a result, F1 avoids the south exit. At the same time, his PDA also detects the presence of firefighter F2, who is conducting a search and rescue task in an adjacent room. F1's PDA automatically sends a message containing information about the south exit to F2's PDA.

A few minutes later, F2 receives an immediate evacuation order from the incident commander. F2 knows that there is a growing fire hazard around the south exit, so he rushes toward north. On his way to the north exit, F2's PDA detects the presence of firefighter F3, who is running toward the south exit, unaware of the potential danger. F2's PDA automatically forwards the message received from F1 to F3's PDA, which immediately alerts him of the danger. As a result, F3 turns back to run toward the north exit, safely leaving the building.

4   Field Study of Firefighting Practices

To inform the design of Siren, we conducted a field study of urban firefighting practices in the United States that spanned four months and included over 30 hours of interviews with 14 firefighters in three fire departments. The firefighters were from many levels of the organizations, including one assistant chief, four battalion chiefs, two captains, two engineers and five firefighters.

We employed several investigation methods for our study, including interviews, training observations, and field observations. We conducted interviews at fire stations while the firefighters were on duty and learned about their organizational structure, tools, routines, regular interactions, and typical environment. One such environment is depicted in Figure 1. By observing one field exercise at a training tower, we received a first hand sense of how firefighters tackle urban structure fires. We also accompanied firefighters on two emergency calls to see how they accomplish their tasks.

A full account of our field study can be found in [23]. In this section, we focus on those findings that helped motivate the design of Siren.

Fig. 1. We conducted a four-month field study with firefighters in their work environment. This figure illustrates one such environment: a mobile incident command located at the back of a command vehicle.

4.1 Key Findings

First, we found that firefighters often have an incomplete picture of the situation. Currently, collecting information on scene is difficult because it involves sending firefighters into unfamiliar areas with many potential hazards. Furthermore, the dynamic nature of fires often quickly reduces the certainty and validity of collected information. Experienced firefighters compensate for this lack of information by making quick visual assessments of the status of a fire. However, fires in walls, attics, roof lines, and other obscured areas are hard to detect. Other hidden hazards, such as a structurally weak floor or a potential backdraft, may cause the floor to collapse or an oxygen-starved fire to suddenly explode. Effective detection and notification of these kinds of hazards are critical to avoiding injuries and fatalities.

Our second finding is that there are several weaknesses in the existing communication systems used by firefighters. Communication is currently done face-to-face or through radio on pre-specified frequencies. However, our interviewees noted that noise intensity is a serious problem. One firefighter we interviewed said, "There is a lot of noise on the fire ground. You're inside; the fire is burning; it makes noise; there's breaking glass; there's chain saws above your head where they're cutting a hole in the roof; there's other rigs coming in with sirens blaring; lots of radio traffic; everybody trying to radio at the same time." Interviewees also noted that, since radios operate as a broadcast channel where everyone can hear everyone else, radio congestion is a problem. One firefighter explained, "I'm usually listening to at least three [radios]. It's tough, and then you've got people calling on the cell phone at the same time." Lastly, there is also the problem of radio dead zones. Radio signals often get blocked in certain parts of structures, such as basements or in high rises. Communications taking place in those areas often cannot get through. This problem caused serious problems for World Trade Center responders, where many firefighters missed the evacuation order [24, 25].

Our third finding is that firefighters operate in extremely harsh environments during a fire. A typical firefighter wears 40 to 60 lbs of equipment. As one firefighter put it, "Crawling down a hallway blackened with smoke, 1500 degrees, wearing fire gear, lens fogged over with steam, one hand up and one hand down – it isn't as easy as you think." Given such a harsh and changing environment, any system needs to support reasonably robust communication from an inherently unreliable underlying transport.

4.2 From the Field to Design

The findings from our field study motivated several key design decisions in Siren, including a peering architecture, a "store-and-forward" communication model, and multiple levels of redundancy in communication and user alerts.

Using a Peer-to-Peer Architecture to Improve Reliability

All emergency response systems have an underlying issue of reliability. They need to withstand extreme physical conditions as well as survive failures like radio dead zones and dead sensors. This suggests that context-aware systems should operate independently without having to rely on any external infrastructure, but also take advantage of additional support whenever possible. For example, a firefighter that is cut off from his companions should still have some base level of context support from whatever sensors and old data he is carrying with him, and get better sensor data and processing if it is available. Architecturally, this suggests a peering model, where each firefighter carries an autonomous peer that can federate with others to increase the level and accuracy of shared information.

Using Store-and-Forward Communication to Support Tacit Interactions

Our field study also revealed the need for supporting spontaneous and opportunistic interactions between firefighters on a fire scene. Such interactions are especially useful when firefighters need to be alerted about the potential danger to themselves, and to their fellow firefighters. This interaction model has the following characteristics:

1. Interactions are more often triggered by external events (e.g. actions by fellow firefighters or changing environmental conditions) than by explicit human requests
2. Such interactions are usually brief (e.g. exchange of small amount of information about firefighters and the environment they operate in) and do not have the strict real time delivery constraint of voice or video communication
3. The times when such interactions take place are hard to predict. Many factors, such as proximity between firefighters and complex environmental impact on radio channel, determine when such an interaction has the opportunity to occur.

To support tacit interactions among firefighters, devices in a Siren network can discover each other when they are within 150 feet of each other and initiate message exchange. Moreover, a "store-and-forward" model is adopted by queuing all intermediate messages until they expire. This prevents losing messages simply due to the absence of network connectivity. All messages are time stamped and assigned an expiration period, which are checked periodically by all receiving devices. This helps ensure the freshness of data that some applications may demand.

Using Redundancy to Support Robustness
Through our field study, we found that redundancy becomes an important requirement for firefighting applications to overcome environmental volatility and an inherently unreliable communication channel. Redundancy should be supported on multiple levels.

First, a robust system needs to support *communication redundancy* on multiple levels. Messages are exchanged between devices over multiple hops in a peer-to-peer network. As such, messages generated by one device may reach another device through multiple paths and a different set of intermediate nodes. This "path redundancy" maximizes the chance that an important message will eventually be received by a device that may need it.

The second redundancy feature is *storage redundancy*. Because of the "store-and-forward" model, any message may reside on multiple devices at a given time. This in turn maximizes the opportunity for other devices to receive a message.

We also need to support *version redundancy*. Multiple versions of information about the same firefighter captured at different time are stored on a receiving device. This way, important information about the environment is not lost.

Firefighting applications also support redundancy in user alerts. Instructions and warnings issued by firefighters or incident command will be sent over a multi-hop data communication channel, in addition to being broadcast on the traditional voice channel. This helps address the problems of noisiness, congestion, and dead zones on traditional broadcast radio channels whose signals originate from outside of the building. Moreover, alert messages generated by Siren can also be displayed in different ways, and in different modalities.

5   Siren Architecture

The Siren architecture consists of three interacting components: a *storage manager* that implements the information space abstraction; a multi-hop *communication manager* that handles message passing between different devices in a peer-to-peer network; and a context rule engine that controls situation-aware user feedback (see Figure 2). All three components and applications running on top of Siren interact via the InfoSpace API, a tuplespace-like [26] programming interface that loosely couples the whole system. Applications running on top of Siren gather information from sensors carried by firefighters or embedded throughout the environment, communicate with each other in a P2P fashion, and provide context-dependent feedback to firefighters. Siren is implemented in 5000 lines of C++ code and currently runs on WiFi-enabled Pocket PCs.

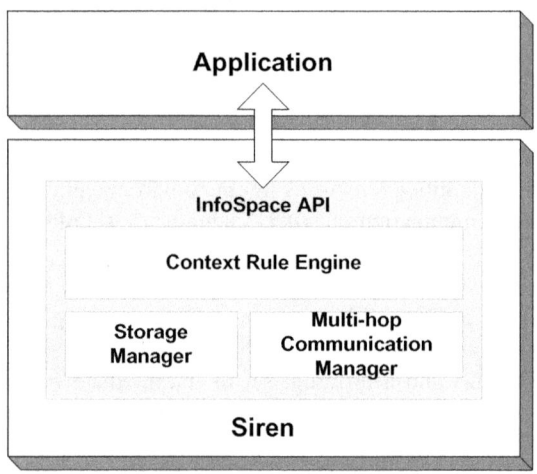

Fig. 2. Siren uses the InfoSpace API to unify storage, communication, and adaptive feedback based on context rules.

5.1   The InfoSpace Programming Model

In Siren, firefighters, places, and devices are assigned network-addressable logical storage units called Information Spaces (infospaces). Sources of context data, such as sensor networks, can populate infospaces to make their data available for use and retrieval (see Figure 3). Applications retrieve and manipulate infospace data to accomplish context-aware tasks. For example, a firefighter's InfoSpace may include temperature data that is collected and stored on multiple devices. InfoSpace operators carry out processing operations on tuples in that InfoSpace, such as the insertion or retrieval of tuples.

Siren provides a unified programming model for developers of emergency response applications by treating sensing, storage, communication, and feedback as *InfoSpace operators*. These operators interact with each other through either local procedure calls or remote message passing if they belong to InfoSpaces residing on different devices. As such, the InfoSpace metaphor provides a decentralized way of combining diverse functionality into a unified, feature-rich system.

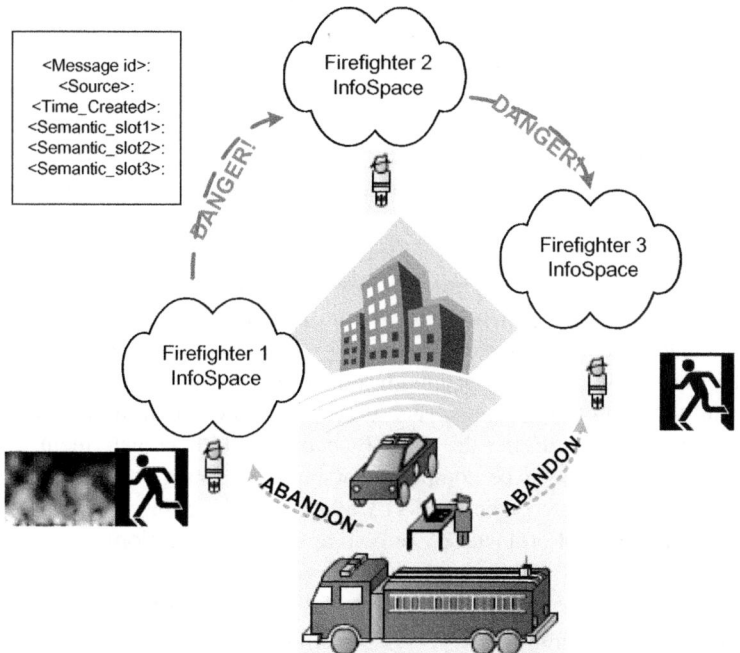

Fig. 3. Each firefighter infospace contains sensor data collected from the environment, and passing of alert messages is done through invoking operators on other infospaces. b

5.2 The Storage Manager

The dynamic nature of firefighting requires loosely-coupled systems that are easily scalable. Siren's storage manager is an adapted implementation of the Context Fabric [18] on a peer-to-peer network of embedded devices and meets the scalability needs using an XML-based communication protocol. This component provides the API used to manipulate InfoSpaces, and manages the storage, querying, and retrieval of tuples.

The InfoSpace API consists of four operations: *inserting* a tuple, *retrieving* a tuple, *subscribing* to an InfoSpace, and *unsubscribing* from an InfoSpace. InfoSpaces are described using an HTTP-like URL with the associated device's name, communication port, and InfoSpace name. Tuples can be retrieved via a unique handle or by an XPath query [27]. For example, an application may retrieve all tuples that are tagged as high in temperature with the query, "`//Temp[@value="high"]`". Any operator, application or component in Siren, can "subscribe" to an InfoSpace to request to be notified when new tuples arrive and can also specify certain types of tuples of interest with a long-standing query.

Siren's InfoSpaces are used to support both storage and communication through two types of tuples: *permanently stored tuples* (such as a firefighter's identity) and *temporary tuples* (such as current temperature) for communication purpose. By default, all tuples are permanently stored upon insert, but their tags can specify that they will only be stored for a limited period of time.

Each firefighter owns an InfoSpace that contains data about his location, surrounding temperature, and the level of remaining oxygen supply. All this data is gathered in real time from sensors embedded in the environment or attached to firefighters. Each device in a Siren network stores both an InfoSpace of its own and a "snapshot" of InfoSpaces owned by other devices.

An *InfoSpace snapshot* is a point-in-time copy of the aggregation of data contained in an InfoSpace during a given time window. It is a view maintained by each device of other peers in a Siren network. Suppose an InfoSpace contain a firefighter's location, surrounding temperature and remaining oxygen supply. Each piece of data is gathered from sensors on a per-second basis. To obtain the snapshot of the InfoSpace for a time window of 10 seconds, each device will check data it has received about that InfoSpace's status during the past 10 seconds, and aggregate them. During aggregation, an average will be computed for numeric sensor readings (such as temperature and remaining oxygen supply) while majority voting will be used to determine the aggregation of ordinate sensor readings (such as location).

Snapshots are done independently by each device in the network at a predefined time interval. It represents a device's "view" of the status of other devices at a given time. Because of the unreliable nature of the radio network in an urban structure, not all messages will be received by all devices during a given time window. Thus snapshots of the same InfoSpace may be different for different receiving device at any given time. We do not attempt to maintain consistency between snapshots taken by different devices because a weakly consistent system will improve overall availability and be more resilient to network partitions.

5.3 The Multi-hop Communication Manager

When Siren receives a request to insert to or retrieve from a remote InfoSpace, the sending device delegates the request to the appropriate remote host by formatting the request into an XML message and sending it to the peer over HTTP. A receiving device parses the incoming XML message and handles the request in a way similar to a local request, returning the result over HTTP. Siren operators can broadcast select tuples (such as "help" messages) to its known hosts, which can then be rebroadcast at the remote device, effectively broadcasting tuples over multiple hops of peers.

Siren supports discovery and time-based message queuing. Each firefighter's device discovers all neighbors currently in range, and broadcasts to all neighbors a message containing the current InfoSpace owned by the device. Each message is structured into the following format:

```
Message id: ….
Source: ….
Time_Created:
Semantic Slot 1:
Semantic Slot 2:
Semantic Slot 3:
```

Each semantic slot represents a different type of sensor reading, such as location or temperature, and contains the following fields:

```
Type: location
Value: 525 Soda Hall
Confidence: 70%
Expiration:
```

The "expiration" field indicates the time after which information contained in this slot becomes invalid. When receiving a new message, a device first checks the `message id` to avoid using a message twice in producing snapshot of an InfoSpace. On the other hand, Siren will continue forwarding a message it has seen before until all semantic slots contained in the message expire.

The device will also check the expiration time of each semantic slot in the current message. If information contained in a semantic slot expires, it will invalidate the slot by rewriting `null` into it before forwarding the message. If all semantic slots in a message expire, the device will stop forwarding the message to others. This way we avoid network flooding by ensuring that no message will travel around the network forever.

Each Siren device maintains its own message queue. Each message queue contains all received messages that originate from the same InfoSpace (device). Messages in each queue are sorted according to the time they are created. A message is marked "sent" when it is forwarded to all neighbors. It will be permanently removed from the queue when a snapshot is taken using this message.

Each message is uniquely identified by its `message_id` and `source` of creation. There is a subtle point here about why we choose not to overwrite older messages as newer versions arrive based on these two attributes. This decision is made based on special requirements of the firefighting domain we want to support. For example, a firefighter may be in room A when a message was generated indicating a high surrounding temperature. A few moments later he may move to another room B with a normal surrounding temperature. Even though sensor readings in the old message no longer accurately reflect the current status of the firefighter, the knowledge that room A has high temperature and therefore may pose danger to other firefighters in the future is nonetheless very useful. This information will be lost if we choose to overwrite all old messages with new ones. This is another example of how communication redundancy is supported in this environment.

Message queuing provides an extended opportunity for interaction between firefighters. Firefighters who are not immediately adjacent to each other can exchange messages at a later time provided that the message being exchanged is still valid.

5.4 The Context Rule Engine

The "brain" of Siren is a context rule engine that takes InfoSpace snapshots as input and interprets them for generating user alerts. The rule engine is a production system that processes application-defined context rules, resolves potential conflicts and outputs reformatted "alert messages" back to the application.

The context rules are structured in the canonical "if...then..." format. The "if" part specifies conditions based on aggregated sensor readings contained in snapshots. The "then" part specifies a "alert message" that is to be output by the rule engine when the condition is true. Alert messages are used by Siren applications to render user feedback in appropriate modalities. Siren currently supports five types of alert messages: (1) "a dangerous place", (2) "danger to oneself", (3) "next to a dangerous place", (4) "others in danger", and (5) "instructions".

"A dangerous place" indicates that a specific location in the urban structure poses potential danger to firefighter safety in the immediate future, such as an unusually high surrounding temperature. "Danger to oneself" signals imminent danger to safety of a firefighter himself, such as a low level of oxygen supply. "Next to a dangerous place" signals that the firefighter is adjacent to a location that is considered a "dangerous place". "Others in danger" signals that there are other firefighters nearby who face "danger to oneself". "Instructions" signal orders or requests from incident command or fellow firefighters such as "abandon" or "help". It should be evident that multiple rules may be triggered by the same snapshot. Consequently, multiple "alert messages" may be produced.

Many context rules in Siren applications depend on access to a location model to decide on spatial notions such as adjacency. In their current practices, firefighters divide different locations within an urban structure along the direction it is facing and name them sides A to D. We have similarly divided each floor into multiple sections, labeled from A to D, and maintain their neighboring information.

Each Siren application defines its own set of context rules. When an application starts, it supplies the rule set to Siren's rule engine and these rules are subsequently applied. Currently, this process is done once at the start of an application.

6 A Prototype Context-Aware Messaging Application for Firefighters

As a proof of concept, we have developed a prototype context-aware messaging application using Siren to support the firefighting scenario we outlined in section 3.

Each firefighter carries a PDA on which the messaging application is running. Each PDA is attached with a Berkeley smart dust mote [28] that is used as both a local sensing device and a communication facility to gather data from other motes embedded in the environment. All PDAs are Wi-Fi enabled and can connect to each other in peer-to-peer mode using the underlying 802.11b protocol.

Our context-aware messaging application consists of two main components: a set of context rules and a user interface for feedback. One example context rule is:

```
IF (firefighter F1 IN room A) AND
   (surrounding temperature > 1500F)
THEN (generate_alert(firefighter F1 in danger)) AND
     (generate_alert(room A is a dangerous place))
```

Through our field study, we have found that two key requirements in designing user feedback in the firefighting domain are minimizing distraction and supporting redundant forms of output. Given the harsh environment firefighters operate in, they cannot be expected to perform sophisticated direct manipulation tasks on a complex interface. Moreover, traditional hands-free interaction techniques such as speech input is hampered by the noisy environment, as well as by the fact that speech input still implies a model in which firefighters are required to attend to a personal device and manually retrieve information.

In our design, we have adopted an alert-driven interaction model in which firefighters need not interact with device except when he issues an instruction to fellow firefighters. All warnings and alerts will be delivered to the firefighter on an as needed basis. In designing the user interface, we avoid all interaction metaphors that would require direct manipulation, such as zooming, selection, scrolling, and text entry.

The visual display of the interface is designed to be as simple as possible to minimize visual distraction to firefighters (see Figure 4). We use simple shapes to represent firefighters (squares represent a firefighter herself, and circles represent other firefighters), and use color to represent their status. Previous research [29] has shown that mobile devices with micro-level form factors can be designed to convey critical information and provide effective notifications. Siren applications are designed for expert users who undergo extensive training in firefighting procedure and equipment. We believe a simple design will be most effective for them.

The PDA screen is divided into three sections (see Figure 4). The upper part of the screen overlays icons representing firefighters on top of a floor plan. The floor plan is divided into different locations according to the location model defined in section 5.3. A location turns red when it is considered to be a "dangerous place". The map is refreshed every 10 seconds to show the current location of firefighters. Each firefighter is represented by a circle labeled by his badge ID.

The bottom-left part of the screen is used to display alert messages generated by the context rule engine. There are three types of alert messages: "danger to oneself" (colored flashing orange), "next to a dangerous place" (colored orange) or "other people in danger" (colored yellow). The screen is divided into multiple panes, where each pane contains a new incoming alert message. Each pane displays the name of the firefighter, his location and the source of potential danger. Every 10 seconds a new alert message is displayed and an old message that has been on the screen for the longest time is discarded.

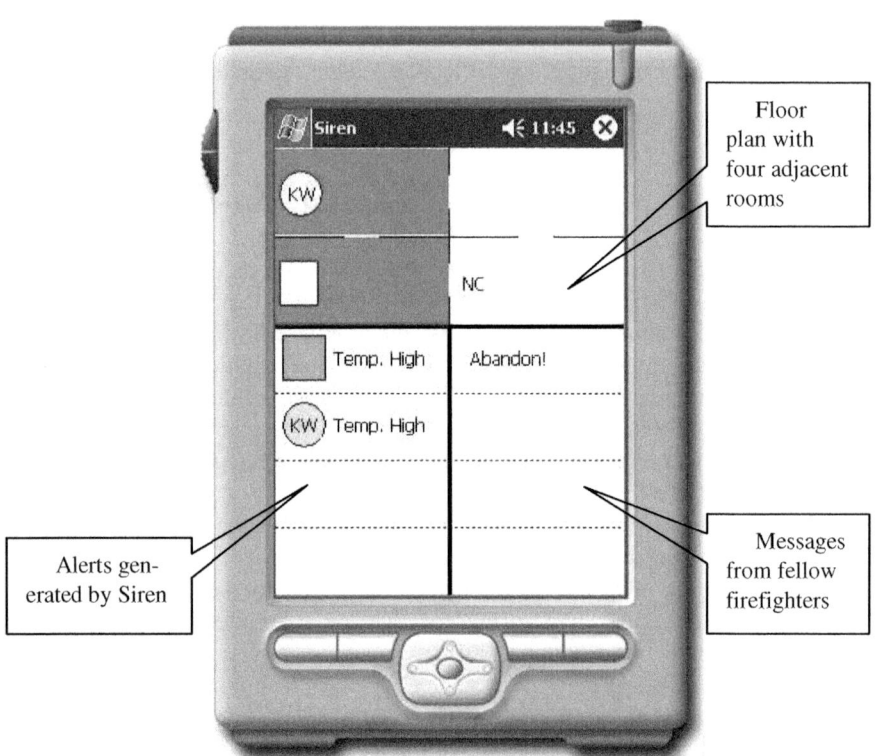

Fig. 4. User Interface for Siren-based Context-aware Messaging Application. The upper part of the interface is a floor plan with four adjacent rooms. The lower left part contains user alerts generated by Siren, and the lower right part contains messages manually generated by fellow firefighters or incident commanders.

A priority queue is used for storing all alert messages generated by the context rule engine. Each incoming message is automatically assigned a priority rating depending on its type: "danger to oneself" is assigned a priority rating of 5, "next to a dangerous place" is assigned 3, while "other people in danger" is assigned 2. Every 10 seconds, the priority ratings of all messages increase by 1. This mechanism ensures that eventually all messages will be displayed, regardless of its original priority assignment.

Information displayed on the bottom-left pane is partly redundant with what is displayed over the floor plan, although in more detail. This design is consistent with our finding that firefighters prefer to see the high-level picture first, and details should be displayed only when needed. In this case, information about individual firefighters is only presented when needed. For example, detailed information about individual personnel is displayed in flashing red if a potentially critical danger is detected, such as high heat level.

The bottom-right pane is used to display manual instructions and warnings issued by firefighters or the incident command, such as "abandon" or "help", as well as identity of the issuer. We have programmed the large physical buttons on Pocket PCs so that each button corresponds to a particular instruction or warning message. Firefighters can issue instructions or warnings by pressing these physical buttons. This frees them from having to perform interaction tasks on a small touch screen with a stylus. While firefighters can also issue these instructions over the traditional radio voice channel, this becomes a redundant form of output to maximize the chance of such instructions being received.

7   Evaluation of Context-Aware Messaging Application with Firefighters

We conducted a formative evaluation of our context-aware messaging application with six firefighters at a large fire department. The firefighters inspected the application, critiqued its functionality and design and brainstormed with us about its potential use. Overall firefighters found the application to be quite useful for enhancing many aspects of their current practices, including improved communication reliability, better firefighter safety, and a better size-up of dynamically changing situations.

Some interviewees commented that our application would be useful for providing adequate redundancy in communication, especially for large and complex urban structure such as hospitals, research labs, high rises and large warehouses. One firefighter commented, "I've been to many hospitals... there are lots of shielded walls and basements for surgeries, my 800MHz radio often doesn't work... The delayed hopping model you have will certainly help here."

Other firefighters would like to leverage the location tracking capabilities of our application to enhance firefighter safety. One firefighter proposed a new way of using our application, "imagine sending one firefighter whose job is to drop your sensors on each floor: in the room, on door wedges and apartment packs [apartment packs are hoses firefighters use to connect to water supplies]... when a firefighter is down, we will be able to know roughly where he is using your system... that is huge for us."

Our interviewees also envisioned our application be used in conjunction with other tools such as a thermal imager to narrow down the area to look in a typical search and rescue task. One firefighter commented, "This way you don't have to comb through the entire area and can avoid potential fire hazard, since your system will tell you which room is likely to be more dangerous."

Firefighters generally liked the simple UI design and the fact that no direct manipulation would be required of them. One battalion chief commented, "This is great.... I am not a techno-wizard....The level of detail here is just about right for me." Some firefighters also suggested that we implement a "multi-dimensional" view of the building and make it easy to switch between views of rooms on different floors.

There are two concerns that our interviewees have expressed regarding the deployment of such application. First is the ruggedness of such technology, including both hardware and software reliability. Second, firefighters cautioned us not to rely solely on sensors embedded in the environment. "The most effective tools for us are always those we can directly access and maintain", said one firefighter. Our interviewees suggested we could also attach smart dust sensors to firefighters' breathing apparatus, because it is carried by all firefighters and is also uniquely assigned for each firefighter.

8   Lessons for Designing Mission-Critical Pervasive Computing Applications

Through designing the prototype context-aware messaging applications for firefighters, we have learned a few key lessons about the design requirements for mission-critical context-aware applications.

First, for users of mission-critical context-aware applications, interacting with a device is often not their primary focus. Their ability to use such devices is also hampered by the harsh environment they operate in. Consequently, mission-critical context-aware applications should be designed to support implicit and opportunistic interactions. In our concept context-aware messaging application for firefighters, we have structured the communication design around the notion of "opportunity for interaction". User alerts are given dynamically based on the current context, and the message queue allows opportunities for interaction to be extended. Also, the interaction paradigm is centered on notification and alert, rather than requiring direct user input.

Second, tasks in mission-critical context-aware applications are often ill-defined. Because of volatile environmental conditions, the actions through which goals can be achieved vary significantly. Through our experience with designing for firefighters, we have found that it is more useful to understand the complex work practice by analyzing the emerging patterns of activities and relationships between actors and artifacts. For example, the task of search and rescue will be completed in a very incident-specific way. Yet the fact that a fire hose is always used to allow firefighters to "feel their way out" in case of evacuation reveals that it's very easy to lose orientation in firefighting.

Third, redundancy is often a less desirable feature for consumer-oriented context-aware applications because it increases complexity and degrades speed of interaction. However, in mission-critical context-aware applications such a feature is almost essential. In our concept context-aware messaging application design, we have employed redundancy in both feedback design and communication design. Redundancy extends the opportunity for interactions and therefore makes the user interaction experience more predictable.

Last but not least, in mission-critical context-aware applications, interaction between users and the context-aware computing system is not through an isolated device. The ability to into the physical environment in which users operate is critical for the success of such an application. In our context-aware messaging application design for firefighters, the point of interaction includes stairways, fire hoses and the rear of a command vehicle. A successful user interface design needs to take these diverse points of interaction into account.

9 Conclusion and Future Work

Based on an extensive field study of current firefighting practices, we have developed a peer-to-peer architecture called Siren that provides the foundation for the gathering, integration, and distribution of context data on fire scenes.

Siren allows the development of emergency response applications on a peer-to-peer network of embedded devices through a uniform programming interface based on the information space abstraction. Siren is designed to support spontaneous and opportunistic interaction between firefighters and provides multiple levels of redundancy to suit the mission-critical nature of firefighting. Using Siren, we have developed a prototype context-aware messaging application to support a search and rescue scenario obtained from our field study. We have evaluated this application with firefighters and they have found it to be useful for improving many aspects of their current work practices.

We are currently investigating supporting alternative sensory modalities such as tactile output in Siren. We plan to improve current user interface design based on feedback from firefighters and test it with firefighters in their training exercises.

Acknowledgements. We thank the Alameda, Berkeley and El Cerrito fire departments for their support and feedback. We also thank Lawrence Leung for helping in the initial design of Siren.

References

1. U.S. Fire Administration, F.E.M.A., Facts on Fire. 2000.
    http://www.usfa.fema.gov/dhtml/public/facts.cfm
2. Estrin, D., Culler, D., Pister, K., and Sukhatme, G.: Connecting the Physical World with Pervasive Networks. IEEE Pervasive Computing 1 (2002) 59-69
3. Jiang, X. and Landay, J. Modeling Privacy Control in Context-aware Systems Using Decentralized Information Spaces. IEEE Pervasive Computing 1(3) (2002)
4. Schilit, B.N., *A Context-Aware System Architecture for Mobile Distributed Computing*, Unpublished PhD, Columbia University, 1995.
    http://seattleweb.intel-research.net/people/schilit/schilit-thesis.pdf
5. Kindberg, T. and J. Barton, A Web-based Nomadic Computing System. *Computer Networks* 2001. **35**: p. 443-456.

6. Dey, A.K., Salber, D. Abowd, G.D. A Conceptual Framework and a Toolkit for Supporting the Rapid Prototyping of Context-Aware Applications, Human-Computer Interaction, Vol. 16(2-4), 2001, pp. 97-166.
7. Crowley, J.L., J. Coutaz, G. Rey, and P. Reignier. Perceptual Components for Context Aware Computing. In Proceedings of *Ubicomp 2002*. Göteborg, Sweden. pp. 117-134 2002.
8. Davies, N., S.P. Wade, A. Friday, and G.S. Blair. Limbo: A tuple space based platform for adaptive mobile applications. In Proceedings of *The International Conference on Open Distributed processing / Distributed Platforms (ICODP/ICDP '97)*. pp. 291-302 1997.
9. Addlesee, M., R. Curwen, S.H. Newman, P. Steggles, A. Ward, and A. Hopper, Implementing a Sentient Computing System. *IEEE Computer* 2001. **34**(8): p. 50-56.
10. Pascoe, J. The Stick-e Note Architecture: Extending the Interface Beyond the User. In Proceedings of *International Conference on Intelligent User Interfaces*. pp. 261-264 1997.
11. Castro, P. and R. Muntz, Managing Context for Smart Spaces. *IEEE Personal Communications* 2000. **5**(5).
12. Edwards, W.K., M.W. Newman, J.Z. Sedivy, T.F. Smith, and S. Izadi. Challenge: Recombinant Computing and the Speakeasy Approach. In Proceedings of *Eighth ACM International Conference on Mobile Computing and Networking (MobiCom 2002)* 2002
13. Chen, G. and D. Kotz. Context Aggregation and Dissemination in Ubiquitous Computing Systems. In Proceedings of *Fourth IEEE Workshop on Mobile Computing Systems and Applications*. pp. 105-114 2002
14. Olsen, D.R., S. Jefferies, T. Nielsen, W. Moyes, and P. Frederickson, Cross-modal Interaction using XWeb. *CHI Letters, The 13th Annual ACM Symposium on User Interface Software and Technology: UIST 2000* 2000. **2**(2)
15. Román, M., C.K. Hess, R. Cerqueira, A. Ranganathan, R.H. Campbell, and K. Nahrstedt, Gaia: A Middleware Infrastructure to Enable Active Spaces. *IEEE Pervasive Computing* 2002. **1**(4): p. 74-83
16. Grimm, R., J. Davis, E. Lemar, A. Macbeth, S. Swanson, T. Anderson, B. Bershad, G. Borriello, S. Gribble, and D. Wetherall, *Programming for pervasive computing environments*. Technical Report UW-CSE-01-06-01, University of Washington Department of Computer Science and Engineering, Seattle, WA 2001
17. Johanson, B., A. Fox, and T. Winograd, The Interactive Workspaces Project: Experiences with Ubiquitous Computing Rooms. *IEEE Pervasive Computing* 2002. **1**(2): p. 67-74
18. Hong, J.I. and J.A. Landay. An Architecture for Privacy-Sensitive Ubiquitous Computing. In Proceedings of the Second International Conference on Mobile Systems, Applications, and Services. Boston, MA. To Appear 2004.
19. Agrawala, A. Draco: Connectivity Beyond Networks (2001)
    http://mindlab.umd.edu/research_draco.html
20. Mesh Netowrks, http://www.meshnetworks.com/
21. DARPA Information Exploitation Office, Command Post of the Future.
    http://dtsn.darpa.mil/ixo/programdetail.asp?progid=18
22. Camp, P.J., *et al.* Supporting Communication and Collaboration Practices in Safety-Critical Situations. In Proceedings of *Human Factors in Computing Systems: CHI 2000*. Fort Lauderdale, FL: ACM Press. pp. 249-250, 2000.
23. Jiang X., Hong J., Takayama, L. Ubiquitous Computing for Firefighting: Field Studies and Large Displays for Incident Command. To appear in CHI 2004. Vienna, Austria
24. McKinsey and Co, Increasing FDNY's Preparedness (2002)
    http://www.nyc.gov/html/fdny/html/mck_report/toc.html
25. Paulison, R.D.: Working for a Fire Safe America: The United States Fire Administration Challenge (2002) http://www.usfa.fema.gov/dhtml/inside-usfa/about.cfm

26. Rowstron, A. Using asynchronous tuple space access primitives (BONITA primitives) for process coordination. In: Garlan, D. and Le Métayer, D. (eds.): Coordination Languages and Models. Lecture Notes in Computer Science. Springer-Verlag , Berlin (1997) 426-429
27. XML Path Language (XPath), W3C, http://www.w3.org/TR/xpath
28. Kahn, J.M., Katz, R.H., and Pister, K.S.J.. Mobile Networking for Smart Dust. Proceedings of MobiCom 1999: The Fifth Annual International Conference on Mobile Computing and Networking. Seattle, WA: ACM Press (1999) 271-278
29. Tarasewich P., Campbell C., Xia T., and Dideles M.. Evaluation of Visual Notification Cues for Ubiquitous Computing. In Proceedings of 5th International Conference on Ubiquitous Computing, Seattle, WA. 2003

Spectacle-Based Design of Wearable See-Through Display for Accommodation-Free Viewing

Marc von Waldkirch, Paul Lukowicz, and Gerhard Tröster

Wearable Computing Lab, ETH Zurich, Gloriastrasse 35, CH-8092 Zurich,
Switzerland,
waldkirch@ife.ee.ethz.ch

Abstract. This paper presents the design and the evaluation of a novel wearable see-through display which provides imaging being nearly independent from the eye's accommodation. This means, that the overlaid display data are perceived to be in-focus independent of the distance the eye is focusing at, making the see-through mode much more convenient in use. The display is based on the direct projection of a miniature LCD onto the user's retina in combination with the use of partially coherent light to further improve the display's defocusing properties.

The paper discusses the display concept and presents various experimental results for verification. Finally, a compact and lightweight design for an unobtrusive integration into normal spectacles is proposed.

1 Introduction

Head-mounted displays (HMD) have gained much popularity in recent years. A special category of HMDs are the so-called see-through displays where the display output is overlaid over the user's real view. These see-through HMDs have two significant advantages: On the one hand, they allow the user to access computer information while (at least) partially paying attention to his environment and performing other tasks. On the other hand, they allow to create a mixture between the real and the virtual world by attaching computer-generated data to real life objects (see Fig. 1). Thus, see-through HMDs are essential for many applications in the area of augmented reality (e.g. in medicine, maintenance and mobile networking) [1].

1.1 Accommodation Problems

This see-through feature leads, however, to some frictions with the optical properties of the human eye: To perceive an object as sharp and of good quality the eye must make sure that all light emitted by a single point of the object ends up in a single point on the retina. This is achieved by adapting the focal length of the eye to the distance from the eye to the corresponding object (referred to as 'eye accommodation'). Thus, strictly speaking, only an object at that precise

A. Ferscha and F. Mattern (Eds.): PERVASIVE 2004, LNCS 3001, pp. 106–123, 2004.

distance in front of the eye is in focus (and thus sharp). As we move away from this distance the perceived image gets blurred. Thanks to this accommodation process, a typical '40-year-old human eye', for instance, is able to focus on an object at any distances between infinity and 20 cm, corresponding to a range of accommodation of about 5 diopters (D)[1].

This property has now direct implications on the see-through feature of the display: The overlaid information is provided by the display at a fixed viewing distance. Thus, as soon as the eye adapts to a real object at an other viewing distance, the overlaid information gets blurred. The rate at which the displayed image gets blurred as the eye changes the viewing distance is a specific property of the given display. The accommodation range within which the overlaid information is not perceived too blurred to be comfortably viewed is referred to as 'depth of focus (DOF)' of the display.

A comfortable use of the see-through feature requires that the overlaid information is perceived sharp and of good quality, independently of the object in the real world the user is accommodating to. Otherwise, severe disturbing effects are caused as demonstrated in experiments by Elgar et al. [2]. This crucial specification, however, is mostly not met by today's see-through HMDs since these provide only a quite restricted depth of focus (DOF) of less than 0.3 D, normally. This means that the overlaid informations gets out of focus when the user is looking to objects being situated closer than about 3.3 m.

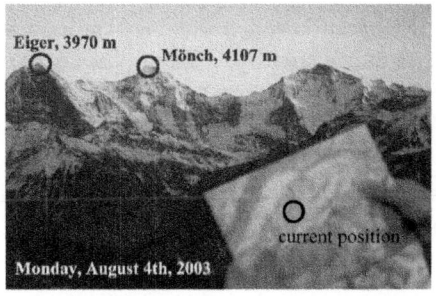
(a) Focusing on the mountains

(b) Focusing on the map

Fig. 1. Simple scenario illustrating the accommodation problems in a see-through display with restricted depth of focus.

Figure 1 illustrates the problem considering a simple scenario: It shows a hiker in the mountain holding a map in his hands and looking at the landscape.

---

[1] Diopters (D) is the measurement unit of the eye accommodation. In our context it signifies the reciprocal of the eye's viewing distance in meters (e.g. accommodation of $2\,D = 50\,cm$ viewing distance)

His see-through HMD provides additional geographical and general information (such as names, current time and position). As long as the hiker is looking at the mountains far-away (corresponding to eye accommodation to infinity) the display information are in-focus and well readable. However, when the hiker is gazing at the map in his hands and thus, accommodates to this closer viewing distance, the overlaid information gets out of focus causing much inconvenience.

1.2   Related Work

The example above illustrates that either the focus of the see-through display should be adapted in real-time to the eye's current accommodation or the display should provide a depth of focus which equals or exceeds the eye's average range of accommodation of about 5 D.

Today's most common HMDs consist of a microdisplay unit illuminated by spatially incoherent backlight. An optical system in front of the user's eye generates a virtual image of the display [3,4]. The most well known method of increasing the depth of focus of such an optical system is the use of a small aperture which has the property of blocking light rays that cause blurring. This is widely used in photo cameras. However it is only of limited use for an HMD since the aperture size required to achieve good depth of focus would reduce the brightness of the display to an unacceptable level.

As a second way to cope with these accommodation problems Sugihara et al. proposed a real-time adaption of the display's focus [5]. Their display provides a movable LCD-screen so that the LCD-position can be varied to match the current eye's accommodation within 0.3 s. However, this requires a fast and accurate tracking of the eye accommodation leading finally to an increased technical complexity.

Finally, Rolland et al. proposed to make the see-through displays multifocal. To this end, they investigated the engineering requirements for adding multi-planes focusing capability to HMDs [6].

1.3   Paper Contribution

We have devised and evaluated a new way of increasing the depth of focus of a system that is suitable for see-through head mounted displays. The paper does also show how our method can be applied to a miniaturized wearable system. In the first part, the paper describes the novel concept and discusses several experimental results to prove it. The second part focuses on the design of a compact and lightweight version which allows an unobtrusive integration into a normal eyeglass for real wearable use.

2   Accommodation-Free Display Concept

2.1   Schematic Setup

Figure 2 depicts schematically the principle of the novel accommodation-free display system. The method is mainly based on the use of partially coherent light to project a mini-LCD directly onto the user's retina. The key to our method is the fact that we have been able to determine the optimal degree of spatial coherence that provides best image quality and depth of focus.

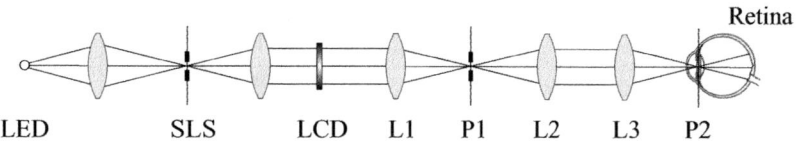

Fig. 2. Principal setup of the retinal projection display. SLS represents the secondary light source for generating partial spatial coherence. L1-L3 indicate the projection lenses. P1 signifies the L1-focal plane with the aperture, P2 the eye iris plane.

The illumination light for the LCD originates from a light emitting diode (LED) whose light is basically spatially incoherent. To get partially coherent light from the incoherent LED-light the rays are focused on a small aperture (referred to as secondary light source (SLS) in Fig. 2), which limits the radiating area and thus, increases the level of spatial coherence [7].

After passing the LCD, the light is focused by a lens L1 to an aperture P1. P1 cuts away higher diffraction orders of the LCD images, which occur due to the use of partially coherent light. These higher diffraction orders would degrade the defocusing properties of the display. The focus plane at P1 is imaged through a relay lens system (consisting of lens L2 and L3) onto the eye's iris plane. Finally, a geometric image of the LCD is formed on the retina.

The large DOF of this setup is mainly caused by two facts: First, due to the retinal projection of the LCD image the light efficiency is much higher what allows to reduce the size of the aperture P1 while preserving high display illuminance. A reduction of the aperture's diameter increases directly the depth of focus as well known from the use of any camera. Secondly, the increase of the spatial coherence level of the illumination light improves the system's defocusing properties as well as enhances the image contrast. This second point will be discussed more thoroughly in the following section.

2.2   Defocusing Properties

Basic physics states that an incoherent system is linear in intensity [8]. This means that for incoherent imaging the illuminance distribution in the image plane is given by the linear superposition of the intensities of any partial waves.

This principle of incoherent imaging defines also its defocusing properties. To make a rough estimate, we simulate the incoherent defocusing behavior of a classical sine pattern whose transmission is modulated sinusoidally with a given spatial frequency.

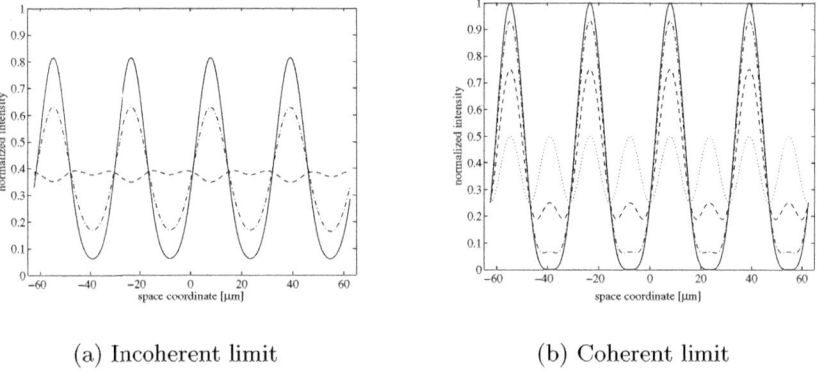

(a) Incoherent limit   (b) Coherent limit

Fig. 3. Defocusing simulation considering a sinusoidal example pattern in order to show the difference in the defocusing properties for a) the incoherent and b) the coherent case. The solid line (—) represents image pattern at best focus ($\Delta D = 0$) while the dash-dotted line ($- \cdot -$) and the dashed line ($--$) indicate the image pattern with $\Delta D = 2.6$ and 5.1 D. The dotted line ($\cdots$) for $\Delta D = 7.6$ D shows the frequency doubling effect for the coherent case, only. For illustration consider also Fig. 6.

Defocusing this sine pattern cause a decline in the sine pattern maxima while the minima rise up, as illustrated in Fig. 3(a). Consequently, the image is more and more blurred and the image contrast lowers more and more. At a specific defocus value the image contrast is finally completely lost.

A fully spatially coherent system, on the other side, is based on the linear superposition of the amplitudes of the partial waves rather than of the intensities. This fundamental difference gives rise to the well-known interference patterns in coherent imaging as well as to a completely altered defocusing behavior. To illustrate this, consider again the same sine pattern as above: In this case defocusing leads to increasing intermediate peaks at the minima of the original sine pattern combined with a decrease of the maxima (see Fig. 3(b)). At a specific defocus value the image sine pattern suffers even a complete frequency doubling. However, of special importance for a display is the observation that the pattern contrast at the same defocus value is much higher in the coherent case than in the incoherent one, as can be seen in the example. For instance at best focus ($\Delta D = 0$) the contrast is equal to 1 in the coherent case while in the incoherent case the contrast starts already from a lowered level.

This rough example shows that the defocusing properties in terms of image contrast can be improved when increasing the spatial coherence of the illumination light. However, the occurring intermediate peaks in the fully coherent case may affect the perceived image quality and thus, may partially compensate the gain in contrast. This should be investigated in the experimental verification.

3   Experimental Verification

Based on an assembled bench model two different experiments were carried out to evaluate the display concept: First, measurements of the image contrast and secondly the evaluation of reading text samples. In both experiments the results were explored by varying defocus, coherence level of the illumination light as well as resolution and text size, respectively.

Measuring the contrast function allows to quantify objectively the gain in contrast due to the increase in the coherence level. However, contrast measurement is a rather technical method which is insensitive to the degrading effect of the coherent intermediate peaks. Thus, this evaluation itself does not allow to infer the perceived image quality.

The evaluation of the text readability, on the other side, are closer to the real use of the display and consider any disturbing coherence effects. However, text quality evaluations are much more subjective. Consequently, the assessment of the readability is carried out by two independent methods: by an image quality metrics on the one hand and by direct visual assessment on the other one.

3.1   Assembled Bench Model

The bench model of the the projection display was assembled on an optical breadboard. The optical layout is depicted in Fig. 4. A detailed discussion of the design criteria is given in an earlier work [9].

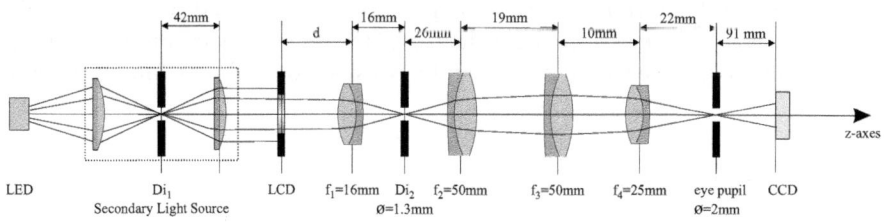

Fig. 4. Optical setup of the assembled bench model. P1 represents the aperture, CCD the CCD camera which replaces the retina. The orientation of the achromat lenses were chosen so that spherical aberration effects are minimized.

As light source a commercial superbright red LED by Agilent Technology was chosen (mean wavelength $\lambda = 626\,\mathrm{nm}$, maximum luminous intensity $I = 18.4\,\mathrm{cd}$). To vary the spatial coherence level of the illumination light a secondary light source (SLS) was created with two plano-convex lenses and a aperture whose diameter can be varied. With this setup the illumination light can be altered between nearly fully spatially incoherent up to nearly coherent. As LCD a transmissive active matrix liquid crystal microdisplay (CyberDisplay 320 Mono by Kopin [10]) was used providing 320 x 240 pixels with a pixel pitch of $15\,\mu m$. The diameter of the aperture P1 in the focal plane of lens L1 is set to 1.3 mm. The last three achromat lenses form the relay optics with magnification $M = 0.63$.

For experimental reasons the retina was replaced by a CCD camera taking images of the projected pattern. A further aperture at the pupil plane acts as an artificial eye pupil (with diameter $\phi = 2\,\mathrm{mm}$). Furthermore, the distance from the eye pupil plane to the CCD was set to 91 mm. With the human eye instead of the CCD camera this distance would be given at about 20.6 mm [11]. Except for the secondary source setup the bench model works with commercial achromat lenses in order to reduce aberration effects.

3.2 Evaluation Method

The degree of spatial coherence will be described by the parameter σ which has an illustrative link to the experimental setup: σ is defined as the numerical aperture of the illumination system divided by the numerical aperture of the objective lens of the system [7,12,13]. Thus, $\sigma \to 0$ indicates the limit of fully spatial coherence while $\sigma \to \infty$ signifies the fully incoherent limit. This σ-parameter is often used in the area of lithography, microscopy and coherent imagery.

Consequently, the coherence level of the illumination light can be varied by changing the diameter $\varnothing_{\mathrm{SLS}}$ of the SLS-aperture:

$$\sigma = \frac{\varnothing_{\mathrm{SLS}}}{\varnothing_1} \frac{f_1}{d_{\mathrm{SLS}}} \qquad (1)$$

where $\varnothing_1$ indicates the diameter of the aperture stop P1 and $d_{\mathrm{SLS}} = 42\,\mathrm{mm}$ stands for the distance from the SLS-aperture to the second SLS-lens. Table 1 shows the $\varnothing_{\mathrm{SLS}}$-values selected for the experiments and their corresponding coherence levels σ. Note that $\sigma > 3.0$ can be considered as almost fully incoherent.

Table 1. Selected spatial coherence levels for the experiments

$\varnothing_{\mathrm{SLS}}$ [mm]	0.5	1.2	1.7	3.5	-
$\sigma(\pm 20\%)$	0.15	0.35	0.5	1.0	>3.0

To control the defocus, the position of the LCD has been shifted by δd instead of an 'artificial eye lens' with variable focal length. If the LCD is displaced by δd then the corresponding change ΔD in the defocus turns out to be [11]:

$$\Delta D = -\frac{\delta d}{f_1^2 M^2} \tag{2}$$

This is valid for any spatial coherence levels σ. M stands for the magnification of the relay lens.

3.3 Contrast Measurements Results

As first evaluation the contrast of the projected image in terms of angular resolution f_α, defocus ΔD and coherence level σ were measured on the basis of the bench model. The contrast function (CF) was defined as the ratio of the image contrast (c') and the object contrast (c):

$$CF(f_\alpha, \Delta D, \sigma) = \frac{c'(f_\alpha, \Delta D, \sigma)}{c} \tag{3}$$

where c and c' stands for the respective Michelson contrast:

$$c = (I_{\max} - I_{\min})/(I_{\max} + I_{\min}) \tag{4}$$
$$c' = (I'_{\max} - I'_{\min})/(I'_{\max} + I'_{\min})$$

Here $I_{\max}$ and $I_{\min}$ indicate the maximum and minimum of the illuminance level in the object. $I'_{\max}$ and $I'_{\min}$ indicate the respective levels in the projected image.

For these contrast measurements the LCD in Fig. 4 was replaced by a transmission mask containing sets of horizontal and vertical bars at different spatial frequencies. The projected image at the CCD was stored as a 8-bit greyscale image. The contrast function was estimated by analyzing the different illuminance levels in the CCD images according to equations (3) and (4).

Figure 5 shows the contrast values as a function of defocus ΔD for different coherence levels σ. These data were taken at vertical bars of an angular resolution $f_\alpha = 5.6\,\mathrm{cyc/deg}$. The data in horizontal direction are quite similar since the the optical system is rotationally symmetric. The curve representing the fully coherent case ($\sigma \to 0$, indicated by open-shaped symbols in Fig. 5) is based on measurements when illuminating the transmission mask by a spatially fully coherent laser diode rather than by a LED.

There are two results to point at: First, at in-focus ($\Delta D = 0$) the CF-values increase and tend to 1 when increasing the spatial coherence, as expected from the rough simulation example in Fig. 3. This holds for any resolution values f_α below the coherent cut-off frequency[2] which is at $\approx 11.5\,\mathrm{cyc/deg}$. Secondly,

---

[2] The cut-off frequency is a specific property of any optical system and signifies the maximum spatial frequency which is transmitted by the system and thus can be displayed.

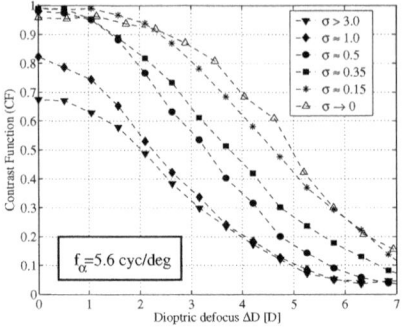

Fig. 5. Contrast function (CF) for an angular resolution of $f_\alpha = 5.6\,\text{cyc/deg}$ in terms of defocus ΔD for various spatial coherence levels σ. The dashed lines should act as eye-guide, only.

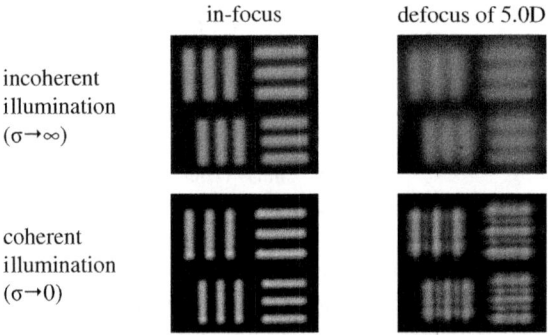

Fig. 6. The projected images of the horizontal and vertical bars. The upper structures in each image have a angular frequency of $f_\alpha = 5.6\,\text{cyc/deg}$, while the lower structures are at $f_\alpha = 6.3\,\text{cyc/deg}$.

the decrease in the CF-values due to defocusing is significantly reduced when increasing the spatial coherence level. This implies, for instance, that the defocusing value where the CF falls below 0.5, raises from $\approx 2\,\text{D}$ up to $\approx 5\,\text{D}$ for the considered angular resolution.

For illustration, Fig. 6 shows some examples of the bar images as taken by the CCD. They represent the projections at best focus as well as at a defocus of 5.0 D for both limits of coherence level σ for two selected angular frequencies. The four images illustrate apparently the gain in contrast when increasing the coherence level from incoherent to coherent illumination. In addition, the occurring intermediate defocusing peaks for coherent illumination are clearly observable.

3.4 Text Readability Evaluation

The CF-measurements indicate a significant gain in image contrast when defocusing by increasing the coherence level of the illumination light. However, they consider only the image contrast, but hardly the perceived quality of the defocused image. The perceived image quality might also be affected by additional effects such as the intermediate peaks which occur for higher σ-values when defocusing. Thus, as second experiment, the readability of text targets was evaluated as a function of defocus ΔD and coherence level σ. This second evaluation is well suited as reading text is considered as one of the most stringent visual tasks in virtual reality [14].

For the text reading experiment the LCD in Fig. 4 was replaced by a binary transmission mask containing text samples of various sizes written in arial font (normal face, capital bright letters on black background). Similarly as for the contrast measurements, the CCD act as 'artificial retina' storing the projected images for different font sizes F_s and coherence levels σ while varying defocus ΔD. Subsequently, the image quality was assessed based on these image series (F_s, σ). However, the judgment of image quality is a rather subjective issue. Thus, to get reliable results, two different assessment methods were applied: First, an objective image quality metrics which assesses the image quality in relation to a reference image and secondly, the direct visual evaluation of the image quality by different test people.

A. Metrics-Based Evaluation

As image quality metrics the algorithm proposed by Wang et al. was used [15, 16]. This algorithm evaluates the quality of a test image Y in comparison to an original reference image X and models any image degradations as a combination of three different factors: loss of correlation, illuminance distortion and contrast distortion. Extended experiments by Wang et al. indicated, that this index performs substantially better than other common image quality metrics and exhibits very consistent correlation with subjective measures for various types of image degradations. For further details see Ref. [15,16].

Fig. 7 shows the image quality index of the 16pt-sized text target as a function of defocus ΔD for various coherence levels σ. The plot indicates that partially coherent illumination (at $\sigma \approx 0.5$) results in a better image quality as almost fully coherent or fully incoherent illumination. This is caused by the discussed trade-off between higher contrast on the one hand and more pronounced intermediate peaks on the other hand when going to more coherent illumination. Similar results are obtained for any other font sizes.

B. Direct Visual Evaluation

For the direct visual evaluation of the image quality, the so called method of limits was used [17]: In this evaluation, five test people were asked to compare

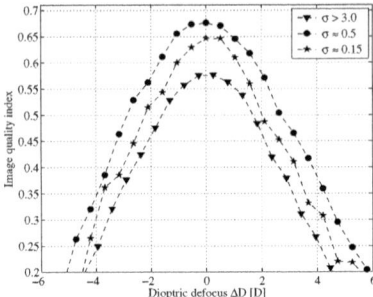

Fig. 7. Wang's image quality index as a function of defocus ΔD for the 16pt-sized text sample. For clarity the plot were limited to three selected coherence levels.

the image quality of the series (F_s, σ) within and between the σ-classes. Based on that the test people determined for each (F_s, σ)-combination the boundaries 1 & 2 on both sides of the focus where the corresponding text sample still seems to be in-focus (see Fig. 8): The difference between these two boundaries corresponds to the searched DOF. This procedure was repeated independently several times for each (F_s, σ)-combination to reduce any subjective errors. Finally, from all these DOF-results the arithmetic mean and the standard deviation were calculated for each (F_s, σ).

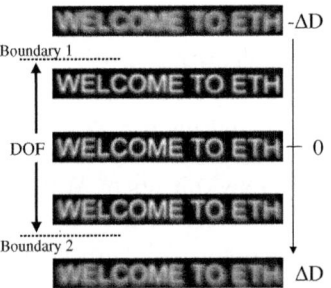

Fig. 8. Part of a (F_s, σ)-series as a function of defocus ΔD. The test people were asked to determine the boundaries 1 & 2 by comparing the image quality as explained in the text.

Figure 9 shows the results of this direct visual assessment for all five evaluated spatial coherence levels. The error bars in y-direction are defined by the standard deviation of all the evaluation runs. The used text samples subtend viewing angles which correspond to the font sizes given on the x-axis when viewed from a distance of 50 cm.

The data in Fig. 9 are in good accordance to the results obtained by the objective quality index method (see Fig. 7 for the case of 16pt-sized text). Both

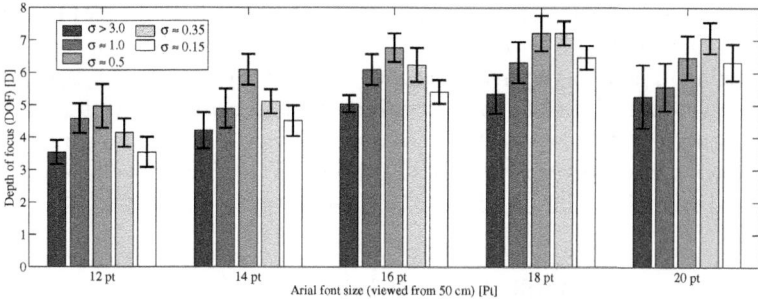

Fig. 9. DOF for various text target sizes when illuminated by light of different coherence levels σ. The data are based on the direct visual assessment. The x-axis shows the arial font size as it would be when viewed from a distance of 50 cm.

methods reveal, that the DOF can be extended by increasing the level of coherence as expected by the contrast analysis. However, both assessment methods show also, that best DOF is not provided by fully coherent, but for partially coherent illumination. For instance, for a font size, corresponding to 14pt when viewed from 50 cm, the maximum DOF of 6 D($\pm$0.5) is reached at a coherence level of $\sigma = 0.5$. For higher coherent illumination the DOF drops to approximately 4.5 D($\pm$0.5) due to the degrading effect of the more pronounced intermediate peaks.

Finally, Fig. 9 shows that the optimum coherence level depends on the font size: For small font sizes the optimum shifts to higher σ-levels while for larger font sizes the coherent illumination gets more appropriate.

4 Compact and Lightweight Design for Wearable Use

The experimental verification has confirmed the promising potential of this display concept for providing quasi accommodation-free imaging. However, for a really wearable use compactness and weight are mandatory. Thus, based on the previous studies further effort has been made to design an lightweight and compact system, but without loosing the excellent defocusing properties. In this section we focus on design specification and conception.

4.1 System Specifications

To prove the concept in the bench experiments a rather limited resolution of 320 x 240 pixels (with a 15 μm-pixel pitch) was adequate. For a convenient use in see-through applications it will be necessary to have at least a VGA-resolution of 640 x 480 pixels. Thus, the miniature display selected is a 9.6-mm backlighting monochrome AMLCD (active-matrix LCD) with VGA-resolution and a 12 μm pixel pitch (provided by 'Kopin's CyberDisplay 640') [18].

Further, as the range of accommodation of a typical '40-year-old human eye' is about 5 D, a DOF of 5 D or more would be required to provide accommodation-free imaging. The experimental part has shown that the display is expected to provide a DOF of 6 D($\pm$0.5) for text targets which correspond to font sizes up to 14-pt when viewed from a distance of 50 cm. Thus, the design should be aimed for displaying text corresponding to 14pt and larger. This specification requires a minimal retinal resolution of ≈ 12.5 cyc/deg.

Table 2. Optical design specifications

Parameter	Specification
1. Retinal resolution @ LCD 12 μm-pixel pitch	12.5 cyc/deg
2. cut-off frequency:	12.5 cyc/deg
3. eye relief d_{eye}	25 mm
4. adaption to eye accommodation:	2.5 D
5. coherence level of the light σ	≈ 0.5

It is the goal to make the wearable display as unobtrusive as possible. This can be best achieved by an integration of the optical system into spectacles. Thus, the eye relief - that is the distance from the last surface lens to the eye's iris pupil - should be set to $d_{eye} = 25$ mm to allow for common types of eyeglasses [19].

Finally, to take best benefit from the large DOF of more than 5 D, the display should be adapted to an eye accommodation which lies in the middle of this DOF-range. That means, that the display should provide best focusing at an eye accommodation of ≈ 2.5 D what corresponds to a viewing distance of about 40 cm. All these system specifications are summarized in Table 2.

4.2   Design Analysis

Within the scope of these specifications the display system can now be designed. Figure 10 illustrates again the setup schematically. To reduce complexity the relay system, which was formed by several lenses in Fig. 2, is considered here to consist of one single lens with focal length f_2, only. A split of the lenses L1 and L2, respectively, into multiple lens system can be made in a later step without lost of generality, since it does not affect the remaining system. Furthermore note, that the secondary light source system (SLS), which provides the illumination light of desired spatial coherence level $\sigma = 0.5$, can be designed completely independently. Therefore, the SLS is depicted in Fig. 10 as a black box.

As shown in Fig. 10, there are principally five parameters which can be varied: the diameter of the aperture P1 (signified by $\varnothing_1$), the focal lengths of the lenses

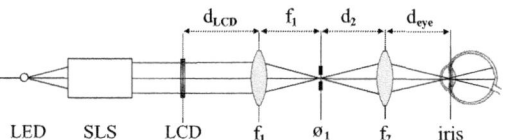

Fig. 10. Five parameters are free to choose in the design process

L1 and L2 (indicated as f_1 and f_2, respectively) as well as the separations d_{LCD} and d_2. The distance from lens L1 to the aperture P1 corresponds to the focal length f_1 as the aperture must lie in the focal plane of lens L1.

However, the specifications defined in Table 2 reduce the degrees of freedom: First, the focal lengths f_1 and f_2 are connected to each other in order to meet the first requirement of a specific retinal resolution for the given pixel pitch of $12\,\mu$m. Secondly, the diameter $\emptyset_1$ is connected to the focal length f_1 by the requirement that the cut-off frequency is given at 12.5 cyc/deg (specification 2). Finally, the separations d_{LCD} and d_2 are linked with the focal lengths f_1, f_2 by the general and well-known lens formula (see equation (5)) in order to fulfill the obligations 3 and 4. Again, the requirement 5 can be treated separately when defining the SLS system.

$$\frac{1}{g} + \frac{1}{b} = \frac{1}{f} \quad \text{with} \quad \begin{cases} \text{g=object distance} \\ \text{b=image distance} \\ \text{f=focal length} \end{cases} \tag{5}$$

That means that considering all specifications the system keeps one degree of freedom, represented - for instance - by the focal length f_2. As soon as f_2 is fixed, all other parameters are given by meeting the specifications above. The goal of this design step is now to find an optimum value for f_2 which minimizes the system length L, defined by the sum of d_{LCD}, f_1 and d_1, but considering practicability limits - especially with regard to the focal lengths f_1 and f_2. The result of this study are illustrated in Fig. 11 showing the values for the parameters f_1, d_{LCD}, d_1 and finally for the overall length L in terms of the focal length f_2. It turns out that the system becomes more compact when reducing the focal lengths f_1 and f_2, accordingly. However, smaller focus lengths cause normally larger aberration effects, if keeping the lens aperture constant. The aperture, on its part, must be kept constant, as it has to be at least the same size as the diagonal dimension of the display device[3]. Consequently, smaller focal lengths require more sophisticated lens systems L1 and L2 to control the growing aberration effects. More complex, multiple lens system, however, reduces the gain in compactness and increases weight and cost.

Consequently, compactness and low aberration effects are two contradictory, but desirable requirements. It is now the challenge to find an optimum balance between these two goals. Concerning the aberration a good guideline is the so

---

[3] for the used Kopin display: 9.6 mm

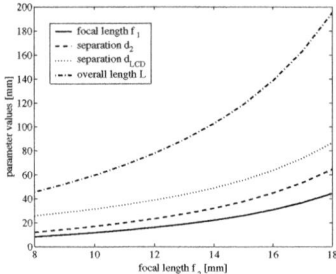

Fig. 11. Parameter values in terms of the focal length f_2 considering the system specifications in Table 2.

called f-number which is defined as the ratio of the lens' focal length to the diameter of the lens' aperture [19]. As a rule of thumb, the f-number should not fall below ≈ 2.0 since otherwise the design complexity jumps up for getting high quality lenses [20].

4.3   Proposed Design

Based on these considerations and the results in Fig. 11 the design as illustrated in Fig. 12 is proposed. It is one possible way for an unobtrusive integration of the display into normal eyeglasses. Other configurations are feasible. In this configuration the focal length f_2 has been set to 13.5 mm. This implies a value of 20 mm for the focal length f_1 and thus a f-number of the lens L1 of about 2.0.

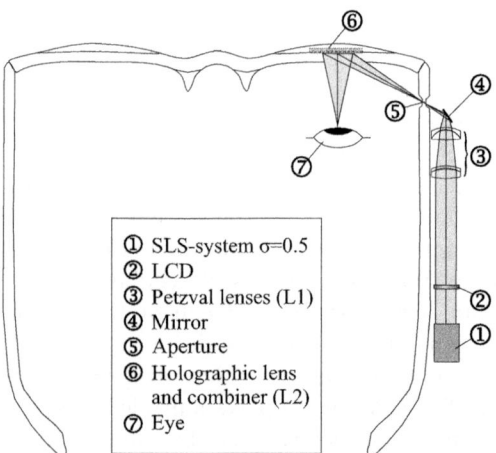

Fig. 12. Possible integration of the novel display system into normal eyeglasses.

Table 3. Parameter values in the proposed configuration

$f_1 = 20\,\text{mm}$	$d_{\text{LCD}} = 44.6\,\text{mm}$
$f_2 = 13.5\,\text{mm}$	$d_1 = 20\,\text{mm}$
$\varnothing_1 = 1.6\,\text{mm}$	$d_2 = 29\,\text{mm}$

To control the aberration effects the lens L1 is proposed to be split into a so-called Petzval lens system consisting of two achromatic doublet lenses. This lens type is known to show significantly less aberrations than any singlet lenses for the same f-number. A more sophisticated, so-called 'double Gauss lens system' would perform even better, leading, however, to more complexity and weight as this system consists of more single lenses than the Petzval design [20].

The lens L2 is formed as a diffractive optical element (DOE or holographic lens). The use of DOEs at this position has some significant advantages over conventional refractive optics: First, they form large-aperture and lightweight optical elements which can be integrated easily into an eyeglass. Secondly, by adopting the first diffraction order the DOE can simultaneously bend and converge the display rays at the eye (as illustrated in Fig.12). In addition, DOEs provide high flexibility on aberration correction, even at high optical power, and so reducing significantly system complexity and cost. Finally, DOE can also act

Table 4. System properties

Parameter	Specification
Display type	Maxwellian-view retinal projection display
display mode	see-through
Field of view (FOV)	31.2 deg diagonally[a] 25.2 deg x 19.0 deg
Resolution	640 x 480 pixels (VGA-resolution)
Color	monochrome red ($\lambda = 626\,\text{nm}$)
Depth of focus (DOF)	> 5 diopters[b] corresponds to viewing distances from ∞ to 20 cm
Number of refractive lenses	2 doublets
Number of diffractive optical elements (DOE)	1 holographic lens

[a] corresponds to a 11"-screen viewed from 50 cm
[b] for 14−pt (or larger) sized fonts viewed from 50 cm

as optical combiner for the see-through display as DOE can accomplished to diffract only specific wavelength rays while be completely transparent for all other visible rays [21].

All other parameters of the design are defined by meeting the requirements 1 − 4, as discussed above. The values are listed in Table 3. Finally, Table 4 summarizes the resulting properties of this new accommodation-free see-through display.

5   Conclusion

We have proposed a novel wearable display system that provides quasi free-accommodation imaging due to the use of partially coherent illumination in combination with a direct retinal projection of the LCD-image. The paper discusses the evaluation results on the basis of contrast as well as text reading measurements and proposes a lightweight, compact design for the unobtrusive integration of the display into normal eyeglasses. The experimental results confirm that this display concept has the potential to provide a depth of focus of more than 5 D for text font sizes corresponding to 14pt (and larger) when viewed from a distance of 50 cm. In fact, taking into account that a typical '40-year-old human eye' has a maximum range of accommodation of 5 D, a practically entirely accommodation-free display is feasible, making real see-through applications much more convenient.

References

1. Azuma, R., Baillot, Y., Behringer, R., Feiner, S., Julier, S., MacIntyre, B.: Recent Advances in Augmented Reality. IEEE Computer Graphics and Applications **21** (2001) 34–47 and references therein.
2. Edgar, G., Pope, J., Craig, I.: Visual accommodation problems with head-up and helmet-mounted displays ? Displays **15** (1994) 68–75
3. Spitzer, M., Rensing, N., McClelland, R., Aquilino, P.: Eyeglass-based Systems for Wearable Computing. In: Proceeding of IEEE ISWC'97, Cambridge, MA, USA (1997) 48–51
4. Kasai, I., Tanijiri, Y., Endo, T., Ueda, H.: A Forgettable near Eye Display. In: Proceeding of IEEE ISWC'00, Atlanta, GA, USA (2000) 115–118
5. Sugihara, T., Miyasato, T.: A Lightweight 3-D HMD with Accommodative Compensation. In: Proc. 29th Soc. Information Display (SID98). Volume XXIX., San Jose, CA (1998) 927–930
6. Rolland, J., Krueger, M., Goon, A.: Multifocal planes head-mounted displays. Appl. Optics **39** (2000) 3209–3215
7. Born, M., Wolf, E.: Principles of Optics. Pergamon Press, Oxford (1975)
8. Goodman, J.W.: Introduction to Fourier Optics. McGraw-Hill, New York (1996)
9. von Waldkirch, M., Lukowicz, P., Tröster, G.: LCD-based Coherent Wearable Projection Display for quasi Accommodation-Free Imaging. Opt. Commun. **217** (2003) 133–140
10. Ong, H., Gale, R.: Small Displays have a big Future. Information-Display **14** (1998) 18–22

11. Bass, M., ed. In: Handbook of Optics. Volume 1. McGraw-Hill, Inc., New York (1995)
12. Hopkins, H.: On the diffraction theory of optical images. Proc. of the Royal Society of London. Series A, Mathematical and Physical Sciences **217** (1953) 408–432
13. Beaudry, N., Milster, T.: Effects of object roughness on partially coherent image formation. Opt. Lett. **25** (2000) 454–456
14. de Wit, G.: A Retinal Scanning Display for Virtual Reality. PhD thesis, TU Delft (1997)
15. Wang, Z., Bovik, A.: A universal image quality index. IEEE Signal Processing Letters **9** (2002) 81–84
16. Wang, Z., Bovik, A.: Why is image quality assessment so difficult ? In: Proc. of IEEE International Conference on Acoustics, Speech, & Signal Processing (ICASSP '02). Volume 4. (2002) 3313–3316
17. Gescheider, G.: Psychophysics, the Fundamentals. Erlbaum Inc, NJ (1997)
18. Homepage of Kopin Corp., Taunton (MA); www.kopin.com.
19. Fischer, R.: Optical System Design. McGraw-Hill (SPIE press monograph; PM87) (2000)
20. Smith, W.: Modern Lens Design: A resource manual. McGraw-Hill (1992)
21. Ando, T., Matsumoto, T., H.Takahasihi, Shimizu, E.: Head mounted display for mixed reality using holographic optical elements. In: Memoirs of the Faculty of Engineering Osaka City University. Volume 40. (1999) 1–7

A Compact Battery-Less Information Terminal for Real World Interaction

Takuichi Nishimura[1,2], Hideo Itoh[1,2,3], Yoshiyuki Nakamura[1],
Yoshinobu Yamamoto[1,2], and Hideyuki Nakashima[1,2,3]

[1] Cyber Assist Research Center, National Institute of Advanced Industrial Science
and Technology, 2-41-6 Aomi, Koto-ku, Tokyo, 135-0064 Japan
taku@ni.aist.go.jp
http://staff.aist.go.jp/takuichi.nishimura/CoBITsystem.htm
[2] Core Research for Evolutional Science and Technology, Japan Science and
Technology Corporation
[3] School of Information Science, Japan Advanced Institute of Science and Technology

Abstract. A ubiquitous computing environment is intended to support
users in their search for necessary information and services in a situation-
dependent form. This paper proposes a location-based information support system using a Compact Battery-less Information Terminal (Co-
BIT) to support users interactively. A CoBIT can communicate with
the environmental system and with the user using only energy supplied
from the environmental system and the user. The environmental system
has functions to detect the terminal position and direction to realize
situational support. This paper newly shows detailed characteristics of
information download and upload using the CoBIT system. And it also
describes various types of CoBITs and their usage in museums and event
shows.

1 Introduction

More and more people will come to enjoy information services while moving
around in the real world. In the fields of "pervasive" [4], "ubiquitous" [3], and
"context-aware" [5] computing, an important goal is realization of a context-
aware information service system which supplies proper information "here, now,
and for me." [6] This paper specifically addresses the use of location as a salient
clue to the user's context. Examples of location-independent information include
electronic books, e-mail, and movies that are intended to be delivered "anytime,
anywhere, and to any person." On the other hand, users sometimes need information that is strongly dependent on user location and orientation. This paper
presents discussion of the latter information service system with walking and
browsing users as the primary target, rather than quickly-moving users riding
vehicles. Moreover, we aim to create a mobile terminal that could be used easily
by anyone.

Two widely used media that connect mobile terminals and interface systems
are radio waves and light beams. Radio waves spread omni-directionally and can

A. Ferscha and F. Mattern (Eds.): PERVASIVE 2004, LNCS 3001, pp. 124–139, 2004.

broadcast over a wide area by use of strong radio wave sources. In contrast, light beams offer directivity and can be controlled. For that reason, we can set a service area more easily even though light is occluded by objects. Next, we compare those two types of media, uni- and omni-directional, from the perspective of use as a location-aware information service for pedestrians.

Cellular phones, personal handy-phone systems (PHS), and wireless LANs use wide range radio waves which extend beyond 10 m. When they are used for locating a device, their accuracy is usually within a cell range unless other methods are used in combination with it. A positioning device, such as a global positioning system (GPS) or terminal orientation sensor such as terrestrial magnetism sensor, must be installed in the mobile terminal to make such mobile terminals location-aware for pedestrians [8]. Such users usually require positional accuracy of 1 m or less for use in guided mobility.

Alternatively, using weak radio waves, positional accuracy is achieved by use of many interface terminals. For example, weak FM radio wave systems can support a user who has an FM radio. If a user approaches a base station, she receives location-dependent information. Very promising ubiquitous items, RF-tags and IC-tags, use very weak radio waves which travel less than 1 m. They will be implanted to various objects holding each characteristic or history [9].

Generally, beam light travels shorter distances than radio waves and is easily occluded. Notwithstanding, we believe that a salient characteristic of light, its straight path, is important because we can control the service area and a mobile terminal's orientation conveniently. The most important merit is that the mobile terminal requires no sensors in a ubiquitous environment with many base stations to achieve location- and orientation-based information support.

When information is projected to a certain area, only information terminals with the correct location and orientation can receive information. One information system that utilizes this feature is the Talking Signs [7] system. The Talking Signs system allows a user to receive sound information through infrared rays from an electronic label (a source of light) installed in an environment such as a public space or mode of transportation. One typical application is to announce the traffic signal color to visually handicapped people. Direction is critical for this use. A Talking Signs terminal regenerates sound information with a built-in speaker or an earphone from frequency modulated (FM) signals. The Active Badge system[9] uses an infra-red LED badge and sensors in each room detect the user position.

However, Talking Signs is not intended to realize interaction between the user and the sign. Interactive information support is preferable because the system cannot estimate the user condition easily. Furthermore, signals from the user are an efficient method to estimate the user condition. Many systems, such as Real World Interaction[2], C-MAP [10], Location-Aware Shopping Assistance [11], and Cyberguide [12], have achieved interactive location-aware information services using high function terminals like personal data assistants (PDAs) that have small displays and rich communication devices.

It is important for a mobile information terminal to be both intuitive and easy to use in consideration of those people who have never used a computer or a PDA. Lack of latency is also important because people can walk out of the usage area while the terminal wakes up in a few seconds. A mobile user will not use a terminal if it is not compact. A design with no plug-in battery is also desirable to ensure low maintenance. Maintenance-free operation for at least several years, as with a watch, would facilitate widespread use of this system. A mobile terminal with all the characteristics mentioned above (intuitive and interactive interface, no-latency, compact, and no-battery) would be a feasible creation for a future ubiquitous world that has many base stations, display devices, sensor devices, and computers connected to the internet.

Based on the above considerations, we developed the CoBIT [15], - a compact, battery-less information terminal - which can provide situated information based on a user's position and orientation. The CoBIT is a small, low cost communication terminal that functions using only energy from the information carrier and the user. We realized one example of CoBIT, which downloads sound information and uploads the user's position and orientation, and potentially, signs from the user.

However, previous work realized only earphone-type CoBITs; no detailed characteristics of communication between CoBIT and the interface system were discussed. Therefore, we investigated detailed characteristics of information download and upload using the CoBIT system. We also developed two types of CoBIT and used them in demonstration services that have been used by more than 200,000 people. The next section illustrates information support system using CoBITs. Subsequently, sound download and position upload characteristics are investigated in Sections 3 and 4 respectively. Section 5 presents two demonstrations which lasted more than one month. Section 6 concludes this paper.

2   CoBIT System

2.1   Basic CoBIT

Figure 1 shows a CoBIT system architecture diagram. The CoBIT comprises the three following parts.

1. A solar cell: This photovoltaic device converts available light into electrical energy. In our usage, it serves an antenna function because it receives radiant beams from an environmental light source device and reproduces sound information from the beam. It also serves as an electrical energy device – the original function of the solar cell. In other words, this device uses a single-beam channel as a simultaneous source of energy and information.

2. A speaker: This acoustic device converts energy and information, which are received by the solar cell, into sound. It is connected directly to a solar cell and produces audio information for the user.

3. A reflective sheet: This is an optical material that transmits the user position and sign as a form of gesture signal to the interface system. It is reasonable

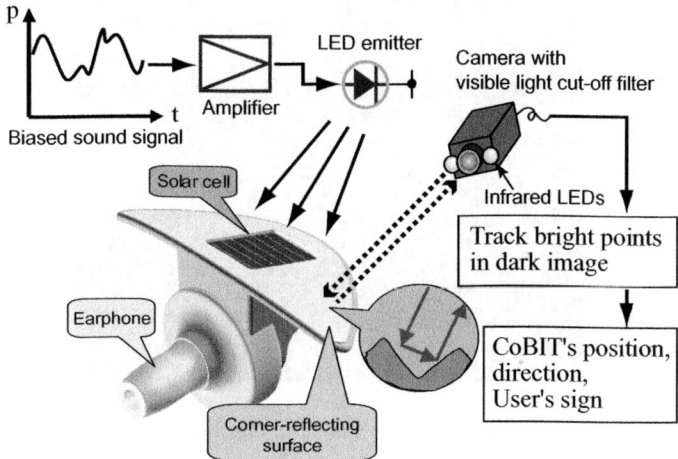

Fig. 1. Basic CoBIT system architecture.

to use materials such as corner-cube reflectors, which reflect much light toward the incident direction of light, to process images easily. The infrared projection camera, which is the first element of the interface system we will mention, can detect only reflective sheets of the CoBIT system as a bright spot by installing a visible-light cutting optical filter, as shown in Fig. 2.

Segmentation of objects in the image is simplified accordingly. The interface system observes movement of the reflective sheet through an infrared projection camera. This observation allows the system to recognize user signals and thereby estimate a user's position and direction.

1. An infrared projection camera: This sensor captures movement of a CoBIT reflective sheet: it defines the user's position and orientation. A gesture recognition engine analyzes this information and identifies user signals.

2. A light source: This optical device sends energy and sound information to the user as a form of light beam. This device enables location-based and orientation-based communication using light direction.

As explained previously, the user can send her position and signals to the interface system by moving the CoBIT or changing its reflection rate. Advantages of this visible-light cutting technique include the following:

privacy problems do not occur because the system records no personal facial image;

the system can recognize a user's position, direction, and signal more robustly than recognizing a person from a normal camera image;

pointer operation is possible by linking movement of the reflective sheet to a mouse;

it can easily interact with plural users and count their number; and

a user can "hear" information as a form of a sound just by looking at interesting objects.

Fig. 2. A sample image of the camera. The bright point is a CoBIT

Taken together, these features constitute a location-and-orientation-based information service.

2.2   Related Work

Alexander Graham Bell invented the Photophone[16] in 1880. It modulates a solar beam and a receiver amplifies the signal for a loud speaker. Recently, high school teachers have executed sound communication experiments using a small light bulb and a solar cell with a speaker. The output signal of the solar cell is directly fed into the speaker. Therefore, the download technology of CoBIT is not original or even recent. Nevertheless, other users have not used the technology for real world interaction. In fact, their method for information support is not useful in the real world because sound quality is very low and the service distance is about 10 – 50 cm at most.

For that reason, we developed a CoBIT driver, which gives bias to the audio wave, and amplifies it sufficiently to light 24 LEDs. We also employed a ceramic phone, which consumes little energy compared to a headphone. Using them, the CoBIT achieved good sound quality and 2 – 3 m service distance.

Moreover, we examined the frequency response of LEDs and solar cells. The LED character is far better than that of a light bulb. This is the second reason for the sound quality increase. Because solar cells have not been used conventionally as sensors, we examine their frequency response. In addition, we show that audible sound (20 Hz – 20 kHz) is detectable with little sound degradation. Detailed examination is explained in Section 3.

The uploading method of CoBIT is identical to the motion capture technology which uses a reflective ball on a subject. However, the system is not designed for information support. Section 4 will examine those characteristics.

The most original contribution of our proposal is the use of uploading and downloading technology to develop a compact battery-less information terminal while realizing orientation and position dependent real world interaction.

Fig. 3. Sound information is provided by looking in a certain direction at a proper position with a CoBIT.

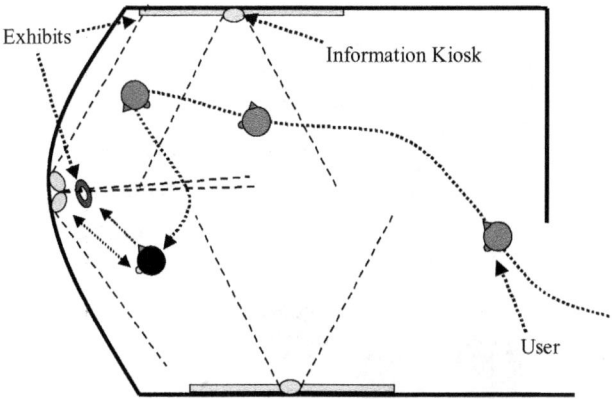

Fig. 4. Information Support Environment using CoBITs

2.3 Information Support Using CoBIT

As shown in Fig. 3, a CoBIT user can hear audio content from an interesting exhibit or objects if the user wears a CoBIT with the solar cell facing forward. Several signs can be sent from the user to the system by moving one's head or occluding the CoBIT for a certain period of time. The light source is usually attached to a wall about 2 m high and facing 30deg downward. The CoBIT solar cell is usually set 30deg upward to allow exposure to the signal light.

Figure 4 shows the information support environment. Event sites or museums have many content areas, such as exhibits or panels. Users walk and browse those content areas and sometimes pick up a CoBIT, thereby getting sound information. The site may have one or no conventional sounds, but a user can hear much more sound content, and personalized content, using the CoBIT.

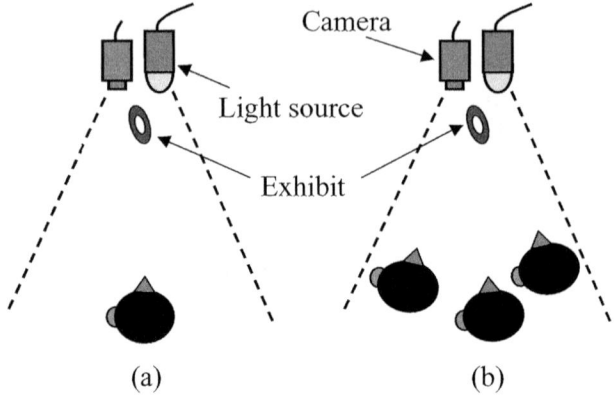

Fig. 5. Information kiosk with one user (a) and multiple users (b).

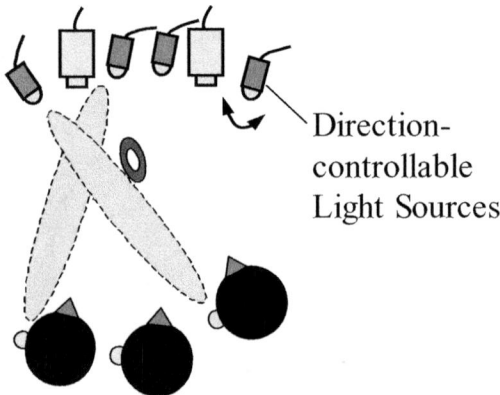

Fig. 6. An information kiosk with multiple direction-controllable light sources and cameras.

Figure 5(a) shows the basic configuration of an information kiosk. The information kiosk recognizes the user with the camera and begins information support. The area for support can be arranged easily by controlling the camera and CoBIT light. If there more than two users, as shown in Fig. 5(b), the system can recognize their signs simultaneously and determine its reaction to serve the majority of users.

Figure 6 shows that the system with more than two cameras and estimates the CoBIT 3D position. In that case, CoBIT light can be directed only to the targeted user, thereby sending private sound content.

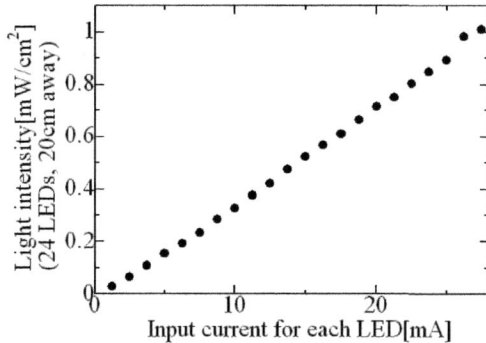

Fig. 7. Characteristics of the CoBIT light source.

3   Sound Download Characteristics

3.1   Light Emitting Module for the CoBIT

A light emitting module for sound download of CoBITs comprises three parts: an audio signal, an amplifier, and a light emitter that can vary its intensity quickly, as LEDs can. We implemented the light module as follows: the audio signal was taken from a general audio device such as a CD, PC, or radio. We amplified that signal and added a 20 mA bias current. We chose an infrared (870 nm, 20 mW at 50 mA) LED for the light emitter, DNK318U, produced by Stanley Electric Co., Ltd.

We produced an LED array unit containing 24 LEDs and placed it 20 cm away from the light intensity sensor for measurement. Experimental results of light intensity vs. input current are shown in Fig. 7. Light intensity of the LED unit is roughly proportional to the input current; intensity is 1 mW/cm^2 at 25 mA current input. We verified the experimental results. The power of one LED is estimated as 10 mW at 25 mA current input dividing the value in Table 1 by half. Assuming that the light intensity is uniform inside the half intensity angle, light intensity [cm] away from the LED unit is about [mW/cm^2]. Light intensity is inferred to be 1.44 mW/cm^2 theoretically because the sensor was 20 cm away from the LED unit, which has 24 LEDs in the experiment. However, Fig. 7 shows that intensity decreased to 1 mW/cm^2, which is about 70% of the theoretical value. One reason may be that the direction of LEDs was not in parallel. Still, as explained in the following sections, linearity and output power of the light emitting module are sufficient for CoBIT use.

3.2   Solar Cell and Earphone

Solar cells and earphones are used in CoBIT to download and present sound information. Earphones can be obtained commercially. Ceramic phones use a

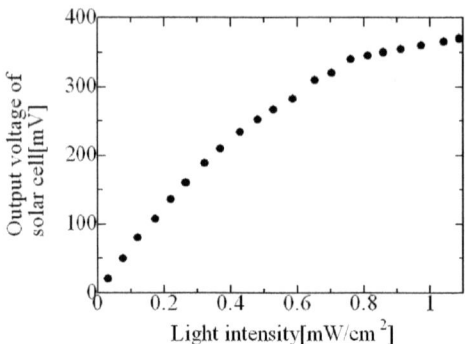

Fig. 8. Open-circuit voltage of the solar cell.

piezoelectric sheet; dynamic phones use a coil and magnet. Sound quality of a dynamic phone is better than that of the ceramic phone, but the dynamic phone requires much more energy because the ceramic phone produces acoustic waves according to the input voltage with almost no current. The dynamic phone makes a sound directly proportional to the current input. Therefore, we evaluated two types of solar cell characteristics: open circuit voltage, and current with a resistor versus light intensity. There are various types of solar cells, but amorphous and crystal silicon types are sensitive to infrared light and are easily acquired. This study investigated characteristics of the latter type of solar cell, with components that were 2 cm wide and 4 cm high.

Figure 8 shows open circuit voltage with light intensity on the horizontal axis, which is useful to estimate acoustic characteristics when the solar cell is connected to a ceramic phone. Non-linearity increases with light intensity of more than 0.4 mW/cm². However, to achieve the same light intensity, one should approach as near as 30 cm from the LED unit. There is no serious decrease of sound quality because users usually walk more than 30 cm distant from the LED units.

Figure 9 shows the output current of the solar cell vs. light intensity regarding the dynamic phone. Resistance was set to 30 Ω. These results also show that linearity is high with respect to actual CoBIT use. Loudness is about 93 dB if a dynamic phone with 106 dB/mW sensitivity is used.

Directional sensitivity is vital for CoBIT users. It depends on solar cell directivity. Figure 10 shows those characteristics. The horizontal axis indicates the incident angle of the input light, whereas the vertical axis indicates output voltage of the solar cell. The half-intensity angle was found to be about 60deg. The reason must be that the effective size of the solar cell decreases to half when the incident angle is 60deg. Physical occlusion or an optical lens would be useful if the user should desire sharper directivity. In contrast, two or more solar cells can be attached to CoBIT if the user desires information from a wider direction.

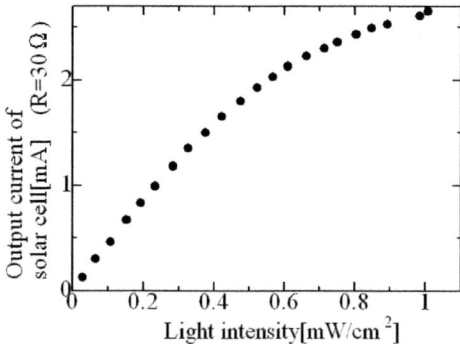

Fig. 9. Output current of the solar cell.

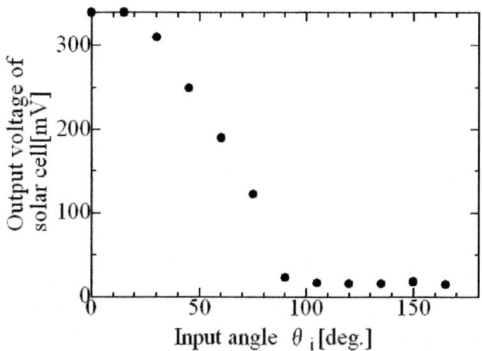

Fig. 10. Directivity of the solar cell.

3.3   Environmental Noise

Light from CoBIT's light source is a "signal", whereas light from other light sources becomes "noise" for a CoBIT. Therefore, we measured light intensity in an office. Offices typically have stronger lighting than that installed in public spaces such as museums or underground shopping malls. Here, we presume that CoBIT will not receive sunlight directly because the power is more than 103 times stronger than the CoBIT's light source. Light intensity was measured at heights of 180 cm and 90 cm turning the detector from an upward position to a downward position in 15deg steps. Figure 11 shows those results. The horizontal axis indicates the solar cell inclination, while the vertical axis shows the detected noise intensity. At both heights, intensity decreased as the detector turned downward. This is true because the ceiling light was the main light in the office. The amount of change was smaller for the lower height because light

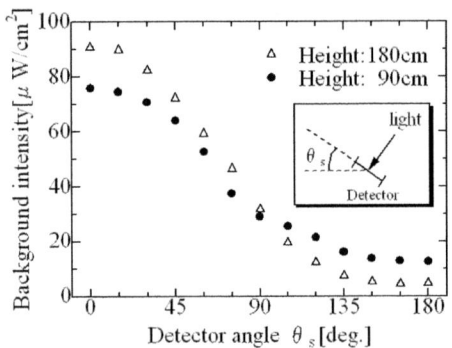

Fig. 11. Environmental noise.

intensity decreases. That from the floor increases because of diffusion on the floor.

To prevent occlusion by people standing between users and a display area, Co-BIT's light source is usually installed at a height of ca. 2 m; it emits light at about 30deg downward. Also, we orient the CoBIT solar cell at about 60degupward. Thus, environmental noise would be about 60 W/cm^2. Flickering noise differs according to the types of fluorescent light, but maximum noise is less than the above value.

High frequency inverted fluorescent lighting does not have sound frequency noise, but a noise decreasing method is generally necessary. Therefore, we tested a visible light cutting filter because the signal light is infrared with 870 nm wavelength. We used a filter, IR84 (Fuji Photo Film Co., Ltd.) on the solar cell and performed similar measurement. Fig. 11 shows those results. Noise was cut to 2 W/cm^2, which is about 1/30 that of the filterless measurement.

On the other hand, the signal decreased to only 77%, improving the signal-to-noise ratio (SNR) by about 25 times. Therefore, a proper optical filter improves the S/N ratio.

3.4   Sound Download Evaluation

This section describes evaluation of synthetic characteristics, from light module input to solar cell output. Figure 12 shows results with frequency on the horizontal axis and with output voltage of the solar cell on the vertical axis. Gain peaked when a 1 kHz signal was input; the gain decrease was less than 1/2 in the range from 200 Hz to 20 kHz. The condenser effect of the solar cell decreased the gain in the high frequency range. In addition, the amplifier decreased the gain at low frequencies because it was designed wrongly in this prototype system. Nevertheless, the sound quality was sufficient for voice information support.

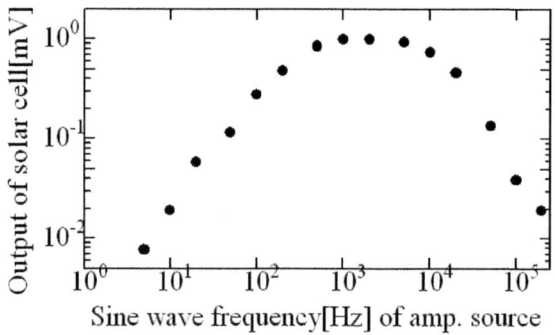

Fig. 12. Gain of different frequency.

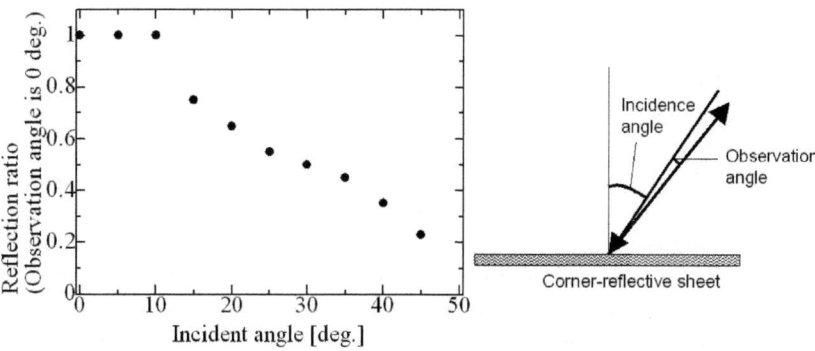

Fig. 13. Reflective ratio of the reflective sheet when the incident angle is changed.

4   Detection of the Reflective Sheet

Position and signal upload is achieved by detecting the reflective sheet on CoBIT using a light-emitting infrared camera. We implemented CoBIT using a reflective sheet, 3970G (3M Co.), which has small corner cubes; it reflects back in the incident direction. Here, we denote the observation angle as that between the incident and reflected beam as shown in the right portion of Fig. 13. The light figure shows the reflective sheet according to the incident angle, for which the observation angle is fixed to 0deg. A mirror has a reflection only when the incident angle is 0deg. This reflective sheet has about a 30deg half-intensity angle. Figure 14 shows the reflection ratio when the observation angle is changed fixing the incident angle to 0deg. The dotted point is the data value. A curve was fitted for further use. The half value angle is about 0.3deg and the fitted curve is where the observation angle is $\exp(-(2.38x)^2)$.

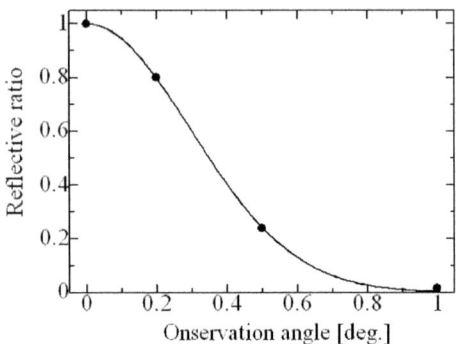

Fig. 14. Reflective ratio of the reflective sheet when the observation angle is changed.(Incident angle is set to 0.)

In this experiment, we attached a circle light emitter (18 mm radius) with 12 infrared LEDs onto the camera. We calculated the reflective intensity from the above information. First, l[mm] denotes the distance between the camera and the reflective sheet; r[mm]denotes the light emitter radius. Here, the observation angle x deg is approximately . As mentioned above, reflective intensity is $x = r/l \cdot 180/\pi (r << l)$. On the other hand, illumination from the circle light emitter decreases in inverse proportion to square of $\exp(-(2.38x)^2)$. Therefore, the reflective intensity is l . Figure 15 shows the reflective intensity; the horizontal axis is . The reflective sheet can be detected easily at distances between 1.5 m and 5 m because the sheet is observed as brighter by the camera. If users walk further from a camera, then the larger radius of the circle light emitter should be used. If the user approaches nearer to the camera, then the radius should be almost 0 using a half mirror.

The next important factor is the size of the reflective sheet on the image taken by the camera. Let us denote θ[rad]angle of view, R[mm] as the real size of the reflective sheet and $I \times I$[pixel]image size. Then, the size of reflective sheet on the image is $IR/2l \tan \theta$[pixel]. This would be 3×3 [pixel] when $R = 20$[mm], $I = 640$[pixel], and $R = 20$ [mm]. Therefore, the CoBIT is detected easily by the camera if the user remains 3 m distant from the camera.

5   Demonstrations

5.1   Ceramic-Phone Type CoBIT

Figure 16 shows the ceramic-phone type CoBIT and its sample usage. The ceramic phone is connected directly to the solar cell. The solar cell is crystal silicon; it is 12 mm wide and 17.5 mm high. It has a visible-light cutting filter (IR84) on its surface. This type of CoBIT was adopted by a designer, Tatsuya Matsui.

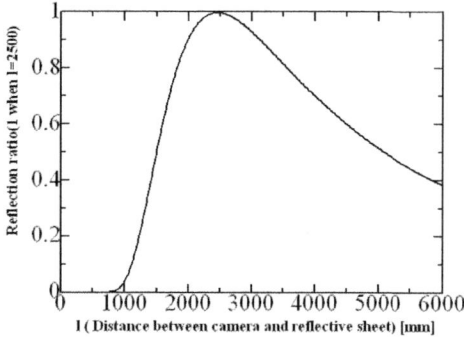

Fig. 15. Reflective ratio of the reflective sheet when the distance from camera is changed.

Fig. 16. Ceramic phone typo CoBIT

It was used by about 200,000 people in The Doraemon, 13 July – 23 September, 2002 in Osaka, Japan. Two CoBIT light sources were installed in a translucent window and in a closet. People with one CoBIT in each ear heard sounds of cars and peoples' conversation from the window and a sound of vacuuming the floor and washing in a bathroom from the closet.

More than 10 CoBITs were damaged during that period, mainly because people dropped them. The ceramic phone is designed to be inserted into the ear canal and is supported only at the aural opening. Therefore, it is difficult for those with a smaller aural opening to securely place a CoBIT. Moreover, more than 20 CoBITs were destroyed by those who were interested in the inside of the CoBIT. A museum employee had to wipe CoBITs after each use and wash each CoBIT every day by dismantling the ear plug part. About 10 CoBITs were damaged during those operations. This ceramic-phone type device offers low

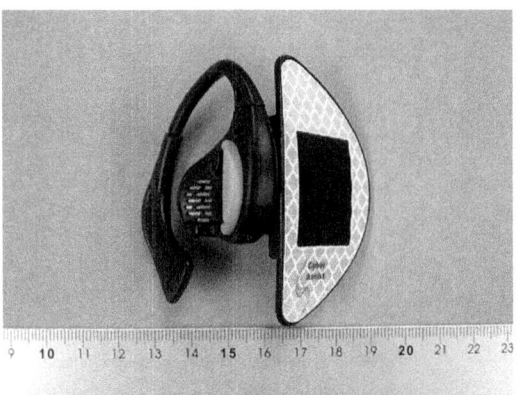

Fig. 17. Headphone type CoBIT

cost, but the time required for maintenance work by museum employees was great and management of rentals would be very difficult: giving it to event or museum attendees as a base terminal to obtain information service would be an ideal use for the CoBIT.

5.2   Dynamic-Phone Type CoBIT

We produced a dynamic phone type CoBIT offering better sound quality and secure ear placement to allow rental of CoBITs with less maintenance and damage. Figure 17 shows the dynamic-phone type CoBIT. It can be hung on the earlobe and is robust because the thickness of the solar cell component was increased, thereby realizing lower maintenance and damage. The headphone is a SE-E03II (Pioneer Co.); the solar cell is 30 mm wide and 32 mm high. This type of CoBIT was used at an exhibition called "After 5 years'", October 4 – October 30 2002, in Tokyo, Japan. There were 25 CoBIT light sources installed at the exhibition. Most were above video monitors emitting audio information to CoBIT users. No CoBIT was damaged at that exhibition. Some users complained about the CoBIT loudness because of a nearby loud speaker. But in quiet rooms, the sound quality was high without mixing with the nearby CobIT light source. This system was also used at The Doraemon, March 15 – May 15 2003, in Yokohama, Japan.

6   Conclusion

The CoBIT system was explained and detailed characteristics of the download and upload were described. Two types of CoBIT were produced and used for demonstrations, underscoring the effectiveness of CoBIT information support. Future interface systems will comprise more sensors and display devices. They will be connected to various AI modules achieving intelligent information support via extremely simple mobile terminals.

References

1. T. Masui and I. Siio, "Real-World Graphical User Interfaces," *Proceedings of the First International Symposium on Handheld and Ubiquitous Computing (2000)*, no. 1927, pp. 72-84.
2. I. Siio, T. Masui, and K. Fukuchi, "Real-world Interaction using the FieldMouse," *Proceedings of the ACM Symposium on User Interface Software and Technology (UIST'99)*, November 1999, pp. 113-119.
3. M. Weiser, "Some Computer Science Issues in Ubiquitous Computing," *CACM*, 1993, vol. 36, no. 7, pp. 75-84.
4. M. Satyanarayanan, "Pervasive Computing: Vision and Challenges," *IEEE Personal Communications*, 2001, pp. 10-17.
5. B. Schilit, N. Adams, and R. Want, "Context-Aware Computing Applications," *IEEE Workshop on Mobile Computing Systems and Applications*, 1994, pp.85–90.
6. H. Nakashima and K. Hasida, "Location-based communication infrastructure for situated human support," In *Proc. SCI 2001*, 2001, pp. 47-51.
7. J. Brabyn and L. Brabyn, "Speech Intelligibility of the Talking Signs," *Journal of Visual Impairment & Blindness, JVIB*, 1982 vol. 76, pp. 77-78.
8. R.G. Golledge, et al., "Personal Guidance System for the Visually Impaired," *Proc. First Annual International ACM/SIGCAPH Conf. on Assistive Technologies*, 1994.
9. R. Want, A. Hopper, V. Falcao, and J. Gibbons, "The Active Badge Location System," *ACM Transactions on Information Systems*, 1992, vol. 10, no. 1, pp. 91-102.
10. Y. Sumi, T. Etani, S. Fels, N. Simonet, K. Kobayashi, and K. Mase, "C-MAP: Building a context-aware mobile assistant for exhibition tours," *The First Kyoto Meeting on Social Interaction and Community ware*, June 1998.
11. T. Bohnenberger, A. Jameson, A. Kruger, and A. Butz, "Location-Aware Shopping Assistance: Evaluation of a Decision-Theoretic Approach," In *Proceedings of the Fourth International Symposium on Human Computer Interaction Mobile-HCI-02*, ACM Press, 2002.
12. G.D. Anowd, C.G. Atkeson, J. Hong, S. Long, R. Kooper, and M. Pinkerton, "Cyberguide: A mobile context-aware tour guide," *Wireless Networks*, vol. 3, no. 5, pp. 421-433, 1977.
13. J. Rekimoto and K. Nagao, "The World through the Computer: Computer Augmented Interaction with Real World Environments," *User Interface Software and Technology (UIST '95)*, 1995.
14. S. Feiner, B. MacIntyre, T. Hollerer, and T. Webster, "A touring machine: Prototyping 3D mobile augmented reality systems for exploring the urban environment," In *Proc. ISWC'97* , 1997, pp. 13-14.
15. T. Nishimura, H. Itoh, Y. Yamamoto, and H. Nakashima, "A Compact Battery-less Information Terminal (CoBIT) for Location-based Support System," In *Proc. SPIE*, 2002, vol. 4863, pp. 80-86.
16. The Photophone: The First Wireless Telephone, http://www.alecbell.org/Invent-Photophone.html , 1880.

INCA: A Software Infrastructure to Facilitate the Construction and Evolution of Ubiquitous Capture and Access Applications

Khai N. Truong and Gregory D. Abowd

Georgia Institute of Technology
College of Computing & GVU Center
Atlanta, GA 30332-0280, USA
{khai,abowd}@cc.gatech.edu

Abstract. People's daily lives provide them with memories and records that they often want to review later. They must expend time and effort to record these experiences manually for future retrieval. To address this issue, applications that automatically capture details of a live experience and provide future access to that experience have become an increasingly common theme of research in ubiquitous computing. In this paper, we present our experience building a number of capture and access applications, sharing insights on the successes and difficulties we encountered. These lessons inform the design of the INCA toolkit (Infrastructure for Capture and Access), which supports the construction of applications in this class. We will demonstrate how INCA facilitates the rapid prototyping and simplified evolution of increasingly complex capture and access applications.

1 Introduction

There are many everyday examples of people capturing information for later use. People often take pictures to capture a moment or write notes to record the important information from an experience. Reliance on manual methods of capturing information is not foolproof; people often fail to capture necessary details in a timely fashion. Many are not good at creating accurate records of an experience; as a result, these records are often biased, incomplete, and in some cases may even contain errors. The act of manual capture can distract people from fully engaging in the experience.

Increasingly, researchers are applying ubiquitous computing technology to capture details of a live experience automatically and to provide future access to those records. Automated capture and access applications leverage what computers do best — record information. In return, humans are free to fully engage in the activity and to synthesize the experience, without having to worry about tediously exerting effort to preserve specific details for later perusal.

There are many examples of automated capture and access applications, but they have explored only a few domains, such as the classroom, meetings, and other generalized experiences. Though there are significant social and cultural barriers that

A. Ferscha and F. Mattern (Eds.): PERVASIVE 2004, LNCS 3001, pp. 140–157, 2004.
© Springer-Verlag Berlin Heidelberg 2004

dictate against a world of continuous capture, experience shows many limited situations in which the value of capture can outweigh its cost. Therefore, continued rich exploration is appropriate, especially in a research context. Unfortunately, many new capture systems simply revisit ideas already explored in earlier work, and existing applications typically are not leveraged as platforms for further investigation because of the challenges of managing and evolving them. These issues also have prevented most researchers from evaluating their prototypes under authentic use and then modifying them to include interesting functionalities according to continual feedback from the user population.

To facilitate research in ubiquitous computing, advances are necessary to improve the tools we provide ourselves and other creative designers who wish to improve upon the vision of Mark Weiser. Many researchers have begun to provide such support for the development of physical [12], tangible [15], and smart devices/applications [11] and the collaboration between heterogeneous devices [28]. Previously, we created tools to support context-aware computing [9] and human-assisted error correction resulting from recognition-based interfaces [19]. In these previous cases, the common method has been to present the relevant design abstractions for a well-defined class of applications, develop an architectural solution to support the design and construction of these applications, implement an infrastructure/toolkit that embodies these abstractions and then validate the abstractions, architecture and toolkit by developing interesting and complex applications within the design space and exploring critical issues for deployed applications.

Using this research method, we introduce the Infrastructure for Capture and Access (INCA) toolkit, which encourages a simplified model for designing, implementing and evolving capture and access applications. We validate this infrastructure by demonstrating how it addresses design challenges, and more importantly how it supports the evolution of increasingly complex capture and access applications.

2 Related Work: The Capture and Access Design Space

We surveyed many of the existing and past projects that support the capture and access of various experiences (for a full review, consult [30]). This body of work can be organized based on the main domains/areas that have been explored: the classroom, meetings, and other generalized experiences.

2.1 Capture and Access in the Classroom

Classroom capture systems have experienced much success because they automatically record the activities of the instructor so a large number of students can directly benefit from the work. The eClass/Classroom 2000 system [1,5], the Cornell Lecture Browser [20], MANIC [24], AutoAuditorium [4], STREAMS [7], Authoring on the Fly [2], and work from Microsoft Research [13,17] all capture with varying degrees of automation significant streams of information presented during the lecture. Commonly captured streams include the instructor's presentations, audio, video, ink written on a physical or electronic whiteboard, visited Web pages, and arbitrary program executions. Access to these notes is typically provided through a Web interface that integrates the various captured streams and allows users to index into specific portions of the audio or video of the live experience.

A small number of projects support the capture of personal notes during lectures. The Audio Notebook [25] is an augmented paper notebook that records and integrates audio with ink written in the notebook. StuPad [29] and DEBBIE [3] are systems that use video display tablets to examine the integration of public lecture notes with private annotations. NotePals [8] is an example of a collaborative access system in which different users take separate notes during the experience and those notes are then merged during the access phase with the separately captured public presentation; this sharing can enable users to easily recognize the important points presented during class (as observed by multiple users at a time).

2.2 Capture and Access in Meetings

Meeting capture also has been a frequent subject of research in capture and access. Like the classroom domain, many meeting systems provide similar capabilities for reviewing presentations given during a meeting. Most support the public capture of meetings. Some of these systems support the collaborative capture of information using a shared whiteboard that a group of users may place and interact with artifacts. Examples of this class of system are DOLPHIN [26], TeamSpace [23] and Tivoli [22]. Dynomite [33] and FiloChat [32] are systems that support the recording and integration of personal notes with audio or video streams of the meeting. The Note-Look system [6] provides users control of an array of cameras to grab the images of the meeting they wish to store in their personal notebook and perhaps annotate.

2.3 Capture and Access in Other Environments

The potential benefits of capture and access have been in a few other settings such as offices, conferences, and museums. The Forget-Me-Not application [16] was perhaps the first to demonstrate the continuous capture of information for a user as she moves about an instrumented capture environment, the office, exhibiting the use of capture and access as a general memory prosthesis. This concept of personal mobile capture in a sensor-rich environment has been revisited in a number of recent applications. The Conference Assistant [10] allows the user to capture personal notes using a mobile device at a conference. Her notes are later integrated with content publicly captured based on her automatically sensed location information. The Comic Diary [27] automatically generates a comic strip recounting the conference attendee's experience based on sensed and manually inputted content. Similarly, the HP Remember [14] system allows a museum visitor to author an automatically generated Web page recounting her experience through both sensed and manually added content. A museum visitor is provided the ability to control cameras for capturing images of her during a museum experience is much like how a NoteLook user can specify what image from a meeting to include in her notes.

2.4 Summary

This review of research systems shows that although there are many existing and past projects in capture and access, only a small number of domains have been explored. Software products distributed with the Mimio (http://www.mimio.com) and Silicon Chalk (http://www.silicon-chalk.com/) reveal the same domains are investigated commercially as well. Despite the number of research and commercial efforts, there has been relatively little innovation in the past 5 years. Many new applications simply revisit ideas that have been previously explored. Furthermore, there has been relatively little research contribution in the way of understanding/evaluating these systems under authentic use, with the notable exception of Tivoli and eClass.

3   Lessons Learned from Classroom Capture and Access Systems

In 1995, we began our investigation of the automated capture of live university lectures so that students and teachers may later access them. In this section, we present the lessons learned from the successes and difficulties we experienced building these applications and evolving them to include necessary (and potentially complex) behaviors over the course of a longitudinal study of use.

3.1 eClass: A Successful Motivation for Classroom Capture and Access

The eClass project (formerly known as Classroom 2000) was an experiment in which we created a classroom environment to capture details from the university lecture experience on behalf of the students, automatically generating a set of Web accessible notes immediately available after class for student review [1, 5]. The task of capturing the various streams of information was divided among several specialized machines. An electronic whiteboard (such as the LiveBoard or SmartBoard) was used in place of a traditional whiteboard, recording slides presented in class as well as the instructor's handwriting. A different machine running a proxy server to log HTTP requests recorded the Web pages visited by the instructor during lecture. Finally, a separate machine recorded the audio inside the classrooms.

A central server connected to each of these capture services to provide coordination for each lecture, or capture session. This coordination included the collection of prepared materials prior to a lecture, initiating and terminating the recording for all services for a given lecture, and the integration and post-production of all captured materials to create the Web-accessible notes. The application required some initial set up and maintenance, which included the specification of all machines involved before runtime. However, the system succeeded largely because this coordination was transparent to the users, requiring very little extra instructor or student effort.

Over time, requests from users (both teachers and students) resulted in several changes to the system, including an extended whiteboard application (*i.e.*, additional display surfaces showing the history of a lecture), video capture, and a database of captured lectures to support server-side, dynamically generated Web notes and a search function. This evolution in eClass was possible largely because initially we adopted a structure to the capture problem that separated concerns into four phases:

- pre-production to prepare materials for a captured lecture;
- live recording to capture and timestamp all relevant streams;
- post-production to gather and temporally integrate all captured streams; and
- access to allow end-users to view the captured information.

Clear boundaries between the phases allowed us to evolve the prototype to include the improved capabilities described above as small isolated changes to the software.

3.2 StuPad: Challenges in Extending eClass with Personal Capture and Access

One goal of eClass was to relieve students from the tedious task of copying notes during class. However, we observed that some students continued to take small amounts of notes on paper. To support the integration of each student's notes with the eClass notes, we added the Student Notepad (StuPad) system to the existing eClass system [29]. StuPad provided students with an interface integrating the prepared presentation, digital ink annotations and Web pages browsed from the public classroom notes into each student's private notebook for added personal annotations.

Unfortunately, certain aspects of the existing eClass system lead to challenges implementing StuPad. Although the structuring of eClass into four phases facilitated much evolution to the system, these extensions to the system resulted in inconsistencies between how the eClass server communicated with the different clients —further weakening a communications scheme that already was not multithreaded. This communications structure prevented us from implementing StuPad in the most obvious way, where each student notepad directly obtains the various information streams in the classroom. Ultimately, we were able to create a working StuPad system, but it required redistributing the different data streams in a non-uniform manner that was considerably more difficult than anticipated.

Despite the student motivation to integrate their in-class personal notes with the public capture of eClass, StuPad turned out to be a less useful application than expected. When study occurred outside of class, additional note taking remained difficult for students to integrate with the captured notes. At that time, the IBM CrossPad presented an affordable solution that allowed one to work with pen and paper while also capturing an electronic record. Such a platform would have enabled students to capture inside or outside of the classroom. However, the eClass system stored captured information in a rigid hierarchical structure that consisted of course numbers, terms, and dates of the lectures. Records captured using the CrossPad required a more flexible organizational scheme, so forcing them to fit inside the same directory structure was not a logical solution. Furthermore, the storage scheme employed by eClass suggested specific ways information is accessed; students could identify the course and then the particular lecture date that they wish to review. Personal notes created during a lecture easily can be synchronized with the notes captured by eClass, but notes created outside of a lecture pertaining to topics addressed in lecture needed to be integrated as well. Providing flexible methods for reviewing the captured data proved to be a second difficult challenge, requiring the integration of the private notes and the classroom notes to happen through other contextual relationships beyond temporal synchronization.

3.3 Lessons Learned

The StuPad project eventually ended because the hurdles described above over-whelmed the development efforts. Through many makeshift solutions, we were able to avoid risky architectural changes involved in directly addressing the problems presented by the underlying communication structure of eClass. However, this issue could have been avoided if the essential application features were decoupled from this concern, making the system easier to build and extend. The rigid hierarchical data structure employed by eClass resulted in storage and access challenges that were too difficult to overcome. We learned the importance of information storage being flexi-ble in order to support a growing set of captured information as the system continues to evolve. Information integration also occurs due to different contextual relationships between captured streams beyond just time. Both lessons may seem obvious, but in the case of eClass, they were overlooked as key design issues.

A primary reason why our classroom research has relative success was because we could evolve the system over a long period of evaluation. This was aided considerably by early architectural decisions we made to structure the system into four phases: pre-production, live capture, post-production/integration, and access. However, the phases of eClass imply a sequential ordering of activities that does not always happens. In-stead, it is generally better to consider the functional components of the overall ar-chitectural solution. We also observed that post-production/integration activity can be further separated into: storage of information until it is later accessed; transduction (or transformation) between different data types; and integration, in which relationships between separately captured streams cause the multiple streams to be delivered col-lectively during access. Additionally, access happens on varying time-scales, de-pending on when information needs to be reviewed relative to when it was captured; therefore, different forms of access methods and interfaces are desired.

4 INCA: Infrastructure for Capture and Access

To help designers focus the development effort on the essential features of the capture and access application, we developed the INCA toolkit with a small set of key archi-tectural abstractions in mind.

- Part of the system is responsible for the **capture** of information as streams of data that are tagged with relevant metadata attributes.
- Part of the system is responsible for the **storage** of information along with meta-data.
- When information needs to be converted into different formats and types, part of the system must **transduce** the information.
- Part of the system is responsible for the **access** to multiple, related, or integrated, streams of information that are gathered as response to context-based queries; *i.e.*, support for the integration of information can be wrapped directly into sup-port for the access of the information, such that when information is requested, related streams of information are jointly provided.

For any given application, there may be more than one instance of each of the above functions. From the implementation perspective, INCA provides a direct way

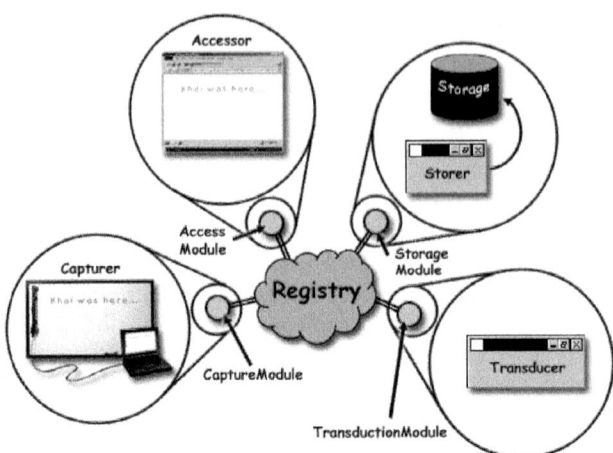

Fig. 1. General architecture for systems built using INCA. A *Registry* runs at some well-known location and any number of applications acting as *Capturers, Accessors, Storers,* or *Transducers* can connect to it and share captured information through instances of specialized networked modules (such as a *CaptureModule, AccessModule, etc.*).

to translate applications designed the following way into executable form. For each functional component above, INCA provides an encapsulated module that a programmer extends or uses as part of the application code. We next present the different components within the infrastructure. There are additional features of INCA that simplify other aspects of application development that stem from the inherent distributed nature of these applications, and common data types and features that allow programmers and end-users to inspect and control the run-time system.

4.1   Capturing and Tagging Information

INCA defines a *CaptureModule* object to support the capture of information; where capture is defined as the act of collecting data from the physical environment. Data is captured and digitized as raw bytes with tagged attributes that describe some properties about the data (such as its data type or format) and the context of the capture activity (such as when and where it was captured). These tags are used in later stages to make the captured data automatically available to those parts of the system that are responsible for storing, transducing, or otherwise accessing it.

The `capture` function defined in the *CaptureModule* is invoked when the application attached to a device (such as a camera, microphone, or electronic whiteboard) has data that is available to be tagged and stored, transduced or provided to some access service. A *CaptureModule* can register various *Tagger* objects to add metadata information automatically to the output objects from the `capture` function. INCA provides a number of reusable *Tagger* objects in its toolkit library for adding attributes specifying people present in a location, the current time, the data type being captured and the location of the captured activity.

4.2 Storing Information

A *StorageModule* provides persistence for captured data. A *StorageModule* can specify a list of attributes for the kind of captured information it is interested in automatically receiving via a `subscribe` function. When the `capture` function in a *CaptureModule* is invoked, the `store` callback function of any *StorageModule* that has registered a satisfied set of attributes is provided with the captured data. Similarly, a `publish` function announces to other components an attribute list describing the kind of information it stores. When access to stored information is needed, the `retrieve` function is called. How this information is actually stored and retrieved is left up to the part of the application that actually extends the *StorageModule*.

A *Repository* service defines all the basic *StorageModule* functionality. It provides a relational database and supports the storage and retrieval of any kind of data tagged with attributes. A *Repository* can be launched and left running, so application developers can have storage performed as an existing service without additional development effort or modification. The *Repository* class can also be extended to meet a specific application need, such as storing only personal information or optimized for a specific captured data type. The *Repository* provided by INCA uses MYSQL as the back-end database. A *FileRepository* service is also available and provides the same functionality as the *Repository* object without using MYSQL.

4.3 Accessing Information

An *AccessModule* supports `handle`, `subscribe` and `request` functions. When information is desired as it is being captured, an access interface can `subscribe` for information it wants (*e.g.*, subscribe for all data created by "John" originating from "Building 4: Room 106"). As information is captured, an *AccessModule*'s `handle` callback function is invoked providing that object with the captured data. A `request` creates a context-based query to the INCA runtime system consisting of attributes to be matched against stored metadata. Upon receiving this query, INCA checks with all existing *StorageModule* and *Repository* instances for data matching the specified query (by invoking the `retrieve` function in the storage components) to obtain all matching data, resolves any cases of redundancy and then returns a list of data found back to the object. Information is integrated based on how it is requested through context-based queries. In its simplest form, the query match is based on attribute name-value pairs and can grow to include more general data retrieval operations that more effectively filter and mine large distributed repositories.

One example of INCA simplifying the programming task is seen in the relationship between an *AccessModule* and the remaining run-time system. An *AccessModule* makes context-based queries for information, but the application programmer does not need to know the location of any of this captured data. The INCA run-time system resolves the query and delivers the information to the requesting *AccessModule*.

4.4 Transducing Information

A *TransductionModule* supports the transformation of information between different data types (such as from a video file to a series of image frames) and formats (such as from a WAV file to an MP3 file). A *TransductionModule* instance `subscribes` with a list of attributes specifying the metadata for information that it can convert. When matching captured data is available, the `transduce` function of each *TransductionModule* is automatically invoked by the INCA runtime system. The transduced information is then passed on to those *StorageModules, AccessModules* or *TransductionModules* that have matching subscriptions for the newly generated data.   Additional tagging of metadata to newly transduced data happens in a way similar to that described for the *CaptureModule*. INCA provides a number of *Transducer* services, such as transforming a video file into a series of image frames and vice versa, transcribing handwritten ink, or converting text to speech.

4.5 Additional Abstractions and Features

The previous features of INCA provided abstractions meant to guide a designer's thinking about how to create a specific capture and access application. In addition to those essential application features, there are a number of other concerns that INCA supports to simplify the development and evolution process:

- Attribute-triggered automated garbage collection allows the system to discard unwanted data.
- An *ObserveModule* provides a detailed description of the run-time state of the system. A *ControlModule* allows for the modification of this run-time state. Together, these features could allow for implementation of privacy and security features, instrumentation for extended evaluation purposes and dynamic adaptation of application features.
- An extensible library of reusable components supports the capture and access common data types, currently including audio, video, ink and Web visits.

As seen with eClass, the communications structure of a system cuts across all the architectural concerns of a capture and access application and is an important factor in evolving these applications. We implemented a network abstraction layer to separate network concerns from the application code. This layer supports a general client-server architecture, where the server binds to two ports. The server uses one of the ports to support connections for synchronous communication between the clients and the server. The second port is used to support asynchronous communication. We build the four functional modules described previously (the *CaptureModule, AccessModule,* etc.) as clients in this architecture. A *Registry* object is built as a server maintaining a list of the available modules that handle the capture, storage, transduction and access of information (see Figure 2). The *Registry* and all the specialized network modules are implemented with a watchdog thread which monitors its network connectivity, increasing reliability and self-maintenance.

To support a variety of applications designed for different domains, the server and it clients exchange serialized message objects. By viewing captured data as only raw bytes with tagged attributes, the infrastructure is able to handle many different kinds of data in the same fashion.

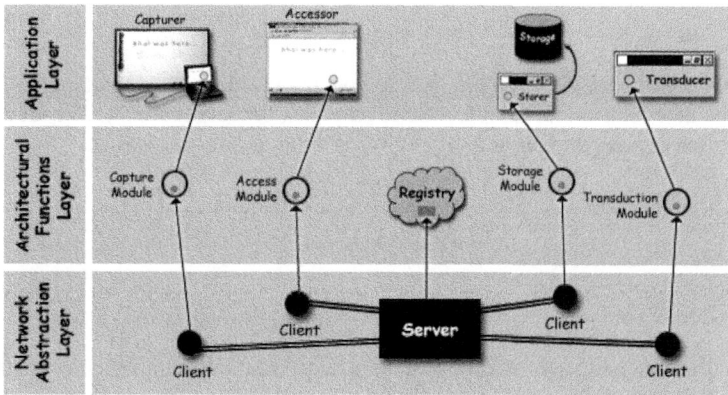

Fig. 2. INCA includes a network abstraction layer on which the specialized networked modules (such as the *CaptureModule, AccessModule, etc.*) and the *Registry* are built. Developers create applications without worrying about details of the underlying network code.

5 Building a Simple Capture and Access Application with INCA

Now that we have defined the key architectural abstractions and other useful services provided by INCA, we will demonstrate how this infrastructure can be used to develop a simple audio capture application that supports the access of near-term recorded conversations (shown in Figure 3). This Personal Audio Loop application (PAL) is intended to run on a single, mobile device that travels with its user. Unlike a tape recorder, this service continues to capture audio even when playback of previously recorded information is accessed. Furthermore, the application automatically discards portions of the captured audio that are older than fifteen minutes.

To begin, we instantiate a *Registry* component in our main program:

```
Registry registry = new Registry();
```

Capture behavior

We use a predefined *WaveCapturer* component, an extended *CaptureModule* from INCA, to capture audio and register *Tagger* objects to add attributes facilitating the future retrieval of the captured audio (e.g. time stamps). Once initialized, we start the *WaveCapturer*.

```
WaveCapturer wave_capturer = new WaveCapturer();
CaptureModule capture_module = new CaptureModule(wave_capturer);
capture_module.addTagger(new TimeStampTagger());
wave_capturer.startCapture();
```

Storage behavior

To store the audio, we simply instantiate a *Repository*.

```
Repository repository = new Repository();
```

Access behavior

We use a predefined *AudioPlayer* component, an extended AccessModule from INCA, to play back requested audio.

```
AudioPlayer audio_player = new AudioPlayer();
```

We developed a GUI for the user to specify the time in seconds, t, back from the present, at which audio should begin playback (see Figure 3). We developed our application to request 1 minute of audio from that point for playback.

```
Query q_start_time = new Query();
q_start_time.greaterThan(new Attribute("TimeStamp",new Long(t).toString()));
Query q_stop_time = new Query();
q_stop_time.lessThan(new Attribute("TimeStamp", new Long(t+(1*60*1000)).toString()));
Query q_main = new Query();
q_main.and(q_start_time);
q_main.and(q_stop_time);
audio_player.playback(q_main);
```

To discard audio, we create a special *GarbageCollector* object that periodically discards information.

```
class AudioGarbageCollector extends GarbageCollectionModule implements Runnable {
    protected Thread thread;
    public AudioGarbageCollector() {
        thread = new Thread(this);
        thread.start();
    }
    public void run()    {
        while(true) {
            try {
                Thread.sleep(1000 * 60);
                Query q_time = new Query();
                q_time.lessThan(new Attribute("TimeStamp",
                    new Long(System.currentTimeMillis() - (15 * 60* 1000)).toString()));
                gc(q_time);  // remove data older than a certain time from storage
            } catch(Exception e) {}
        }
    }
}
```

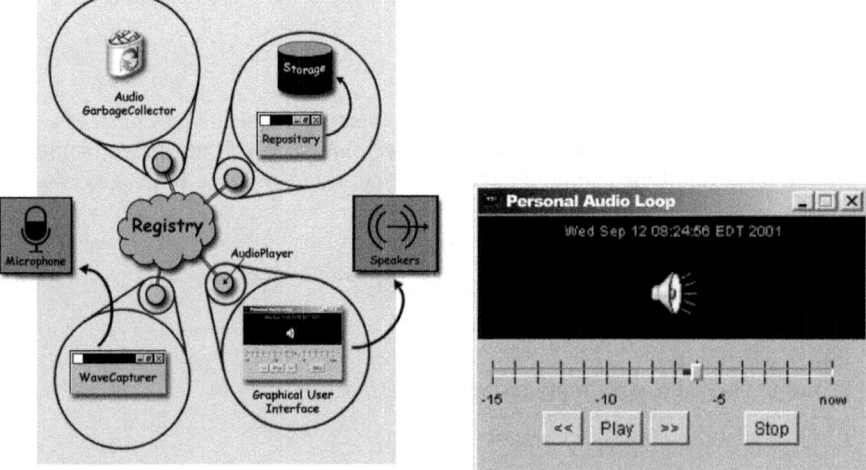

Fig. 3. The Personal Audio Loop application's high-level architecture (left) and user interface (right).

6   Uses of INCA

We now present three more complex systems built with INCA and evolved to investigate interesting issues and new features. Finally, we discuss applications built by others using INCA as evidence to support the general applicability of this toolkit and its value to the ubiquitous computing community.

6.1   Building and Evolving a Classroom System

The foundation of the classroom capture and access experience is that of the public information available before, during, and after the actual lecture. A minimally useful system must capture the instructor's writing and prepared notes, the instructor's speech while communicating with the class, and outside information brought into the lecture in the form of visited web pages. Furthermore, all of this raw data must be stored in a logical manner for easy access at a later time. When users choose to access the information from any particular lecture, they must be able to do so from any other location and using a variety of techniques.

To support the capture of the instructor's writing, we developed a custom application, known as e-Board (see Figure 4a) that is installed on an electronic whiteboard in the classroom. For instructors who teach with prepared presentations, the e-Board application allows users to load a presentation for display on the electronic whiteboard. e-Board is built with a *BoardSurface* component, a specialized *CaptureModule*. *BoardSurface* provides a blank writing surface that listens to pen events and also displays a slide image. As the instructor creates a new blank slide or chooses an existing slide to present, *BoardSurface* captures that slide and automatically tags it with a unique ID (a string containing the name of the machine and the time the slide was loaded or created in the classroom capture and access system). Similarly when an instructor visits a slide, *BoardSurface* captures a visit event tagged with that slide's unique ID. *BoardSurface* also tags ink strokes captured during the slide's presentation with the slide's unique ID. Tagger objects, registered by e-Board, automatically associate time, physical location (*e.g.*, classroom number), relevant course information (*e.g.*, class name and instructor name), and relevant system information (*e.g.*, application name and IP address of the machine running the application) to the captured ink strokes, slides, and slide visits. *BoardSurface* is distributed with INCA, so other developers can use and extend it as desired.

INCA includes *WaveCapturer*, a reusable *CaptureModule* specialized for capture of low-bandwidth audio in the .WAV format. An instance of *WaveCapturer* is installed on a suitable machine attached to a recording device in a particular space (a classroom, in this case). The INCA *Registry* is informed automatically of this instance. The e-Board application uses an *ObserveModule* to determine the list of available nearby capture services and a *ControlModule* to start those services. Starting and stopping the audio recording can be coupled with the starting and stopping of the e-Board application. It also can be tailored to start and stop recording at different times during the use of e-Board, either automatically or through human intervention.

WebMemex is a specialized *CaptureModule* provided by INCA to record Web pages requested by a client browser. Acting as a Web proxy, it handles the HTTP requests and logs pages visited. INCA provides this Web proxy application as a general capture service. In eClass, *WebMemex* helps capture pages viewed in class.

All captured classroom activity is stored in the property-based *Repository*. The access application runs as a standard Web interface (see Figure 4e) allowing users the freedom to access classroom data from any web-enabled machine. The application, composed of custom built Java Server Pages (JSP), instantiates an *AccessModule* to request captured information tagged with the course name and date specified by the student reviewing the lecture. It then presents the lecture as a sequence of discrete slides in the same order used by the instructor during the lecture, regardless of what ordering might have existed in the prepared slides. Users can examine a timeline that indicates the important events taking place during the lecture, such as a slide being created or visited or a Web page being viewed. For further details, a user may click any portion of the timeline or the handwritten ink to begin playback of the audio corresponding to that portion of the lecture. In this case, the *AudioPlayer*, a toolkit component for playing back audio (using an *AccessModule*) requests audio chunks and begins playback until the user clicks the stop button.

Playback of the ink stream at variable speed is also a desirable application feature [21]. INCA supplies this functionality with a reusable *NotesPlayer* component that provides an interface for requesting captured slides and ink annotations and methods for invoking and stopping playback. Using an *AccessModule*, the *NotesPlayer* object

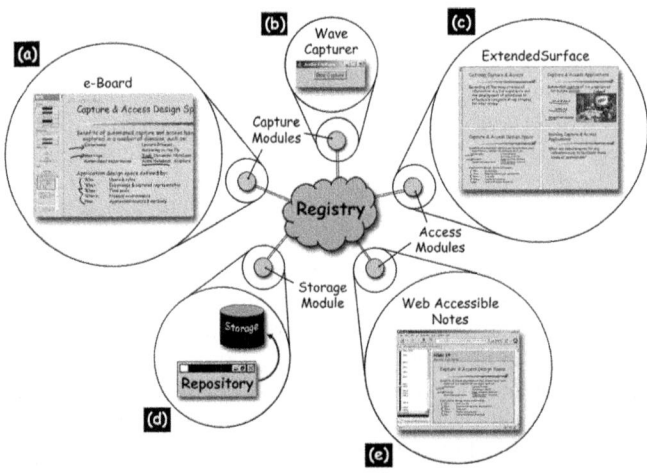

Fig. 4. The simplified architecture of the basic eClass system. (a) The e-Board application built using the *BoardSurface* capture module provides a writable presentation surface that allows the instructor to present prepared slides and/or write on a blank surface. (b) A separate component built using the *WaveCapturer* capture module records audio during the lecture. (c) The *ExtendedSurface* application shows a history of slides captured during the lecture. (d) A Web access interface that includes an audio player, a timeline of the captured lecture, and the slides and their annotations automatically generate using JSP allows students to review the captured lecture. (e) A generic *Repository* object stores all the captured data.

requests the information specified by the user. INCA delivers information back to the *NotesPlayer* in time sequential order allowing custom rendering over time. A *Clock* object is used to control the playback of the captured notes in the NotesPlayer. This *Clock* object can be paused or even programmed to run at different rates. The *Audio-Player* and the *NotesPlayer* can share a *Clock* object to synchronize playback of the captured notes augmented with audio.

A useful near-term access application in the classroom to both students and teachers is an extended whiteboard to show the history of the lecture. INCA provides a reusable *ExtendedSurface* component. This specialized *AccessModule* provides an interface supporting the subscription queries for captured slides and ink annotations. When slides are created or visited, they are added to the surface. The surface can be defined to show the current slide, the previous slide, the current and previous three slides or the previous four slides (see Figure 4c). The surface displays not only the slides but any ink annotations of the slides as well.

A potential evolution of the system is to change the technique to capture presentation slides and ink annotations. While most of the related public capture systems support the explicit capturing of information through electronic whiteboards, the Cornell Lecture Browser demonstrates the ability to do this capture with cameras and vision techniques [20]. INCA supports this behavior through a component that frequently captures frames from a camera or display signal to a projector. This component uses an algorithm that performs image differencing on specified regions of the frames representing the actual presentation slides. By scaling the captured frames to $1/8^{th}$ of their original resolution and computing the percent of pixels that are different between the two frames, this component is able to determine a new slide or a slide visit event. The separation of concerns supported by INCA minimizes the impact of the change in the underlying capture technique to the rest of system.

Not all instructors have electronically prepared presentation slides. As a result, the hurdle for using eClass can be lowered with support for scanning materials and making it available on a blank slide that an instructor could then annotate. This feature is also desirable at times during class when illustrations or examples are written on paper and needs to be integrated with the rest of the captured electronic lecture notes. We developed a custom application, *eScanner* to run on the computer connected to a scanner. This application continuously scans material until a blank image is detected. As images are scanned in, they are captured and made available to any application interested in it. This application publishes that it is a scanner application and it is located in a particular room. The e-Board application is modified to use an *Access-Module* and subscribes for slides created in the classroom. We added to the e-Board application a GUI button that activates the loading of slides from the scanner. The e-Board application uses an *ObserveModule* and *ControlModule* pair to request the scanning of any material the instructor has placed on the scanner.

Finally, most instructors occasionally want the ability to suspend audio recording. We modified the e-Board application to include in its interface a button to stop and resume audio recording when it observes than an audio capture service is available. The button invokes the *ControlModule* to control audio capture in this way.

6.2 Capturing and Sharing Web Experiences

To facilitate a content and/or context based history search mechanism, we used the *WebMemex* component to capture an annotated Web history. We registered a number of existing and custom *Tagger* objects to help associate the user's ID, time and location to each captured Web visit (in addition to the keywords). The access interface consisted of custom Java Server Pages that used an *AccessModule* to handle search queries for previously captured Web visits.

This annotated Web history also enables a number of other access features, such an automatic recommendation capability. As a user browses new Web pages, a different access application suggests related URLs that the user has visited from the past. An *AccessModule* requests the last Web visit handled by the proxy server (which is currently viewed by the client browser). By taking the keywords of this Web page, the access application can query for previous URLs that she visited with matching keywords. This information is displayed in a pop-up window.

Although we originally developed this service to support individual users as they surf the Web, it has since been extended to investigate the sharing of Web histories within a social network [18]. We developed a component that communicates with Yahoo's Messenger service to authenticate the user's login and obtain her list of friends, which we considered the user's social network. This component simply replaced the less sophisticated user authentication component we previously used in the system; *i.e.*, this modification happened as an isolated change from the rest of the capture and access application. We then modified the access behavior to allow information sharing between users if they exist in each other's buddy list.

6.3 Recording and Analyzing Developmental Behavioral Patterns

To better support the collection and analysis of developmental behavioral data of children with autism or other disorders, we developed a mobile capture application that runs on a Tablet PC and integrates the notes taken during an observation session with the corresponding video clip automatically captured [31]. Using an *InkSurface* component, which the *WhiteBoard* component used in the classroom system extends, we recreated the paper forms used by the teachers. We added the ability to tag the markings the teachers create on this form with their semantic meaning (such as if a behavior is observed, not observed, *etc.*) by recognizing the gestures. This electronic form includes a *VideoCapturer* component that automatically records video clips during a session. After each session, the captured information is stored and then made available again when the teacher reviews the data with the parents. We used an *AccessModule* to query for all behaviors captured over time based on their tagged values (if a behavior is observed, not observed, *etc.*). This data is color-coded and then plotted on two timelines to provide a macro view of all the behaviors observed by the teacher and a micro view that shows details of a particular session. Clicking on marks in the micro timeline causes a *VideoPlayer* component embedded in the access interface to access a video clip of that behavior.

By iteratively designing this application with members of the Emory Autism Centers, we uncovered major usability problems in the capture application. Writing on a Tablet PC was too different from writing on paper in two important ways: calibration and resolution. Furthermore, imperfect gesture recognition resulted in too much time and effort spent correcting the data. As a result, we needed to modify the capture interface. We replaced the *InkSurface* components found in the capture interface with buttons that users can push to specify an observed behavior. This event is captured and tagged with the same semantic meaning as the strokes were before.

6.4 Other Uses of INCA

INCA was developed as part of a joint research effort between the Georgia Institute of Technology and Universidade de São Paulo (USP). We provided INCA to developers at our own institution and at the partner university. In addition to the applications described above, developers at Georgia Tech used INCA for the following projects:

- An e-Board application for another department on campus.
- A very large-scale input surface, covering two entire walls of a meeting room using six chained mimio recording devices.
- A video recording application that automatically captures and tags home videos based on room-level entry and exit of individuals in the home.

At the partner university, Universidade de São Paulo, the following uses of INCA were reported:

- The iClass system (http://iclass.icmc.usp.br) to support the capture of lectures and seminars in order to generate a varied of web-based multimedia documents.
- An application to facilitate exchange of notes between a Palm-based PDA and the normal electronic whiteboard.
- A component for Web capture that could do specialized processing on the content of the URL. The same student built an ink capture module linked with a hand-writing recognition engine he implemented. This represents an interesting use of the transduction capabilities of INCA.
- A distributed XML service that handles transparent storage and retrieval of XML documents reporting session-level information of captured data.

7 Conclusions

Previous research demonstrated the value of automated capture and access as a sig-nificant class of ubiquitous computing systems. Despite this, we observe that features of capture and access have not been sufficiently explored in many domains. We re-searched and identified key architectural insights into the creation of this class of applications. Designing in terms of these architectural features—capture of attribute-tagged data streams, storage, transduction and access of related and integrated streams—allows a designer to focus on key distinguishing features of any capture and access applications. We introduce the INCA toolkit as a software infrastructure for transforming high-level designs into implementations while hiding from the pro-grammer details of certain development tasks incidental to the software. INCA sepa-rates various application concerns into individual functional building blocks and de-couples features that cut across all aspects of the application from system code. We validated that INCA simplifies the development and evolution of complex capture and access capabilities through use in a number of applications. The successful uses of

INCA by us and others demonstrated that we identified the proper software structuring for this class of applications and gives us confidence that we created an important tool for others to build upon.

Acknowledgements. This material is based upon work supported by the National Science Foundation under Grants No. 0070345 and 0121661. The authors thank Maria da Graça Pimentel, Tom Barnwell and Lonnie Harvel for their continued collaboration on this work. We also deeply appreciate the endless help and support provided by Anind Dey, Gillian Hayes, Elaine Huang and members of the Georgia Tech's Ubiquitous Computing group.

References

1. Abowd, G.D., Classroom 2000: An Experiment with the Instrumentation of a Living Educational Environment. IBM Systems Journal. **38**(4) (1999) pp.508-530.
2. Bacher, C., Muller, R., Ottmann, T., and Will, M. Authoring on the Fly. A New Way of Integrating Telepresentation and Courseware Production. In the *Proceedings of International Conference on Computers in Education (ICCE'97)*. Kuching, Sarawak, Malaysia (1997)
3. Berque, D. Using a Variation of the WYSIWIS Shared Drawing Surface Paradigm to Support Electronic Classrooms. In the *Proceedings of HCI International 1999*. Munich, Germany (1999)
4. Bianchi, M. AutoAuditorium: A Fully Automatic, Multi-Camera System to Televise Auditorium Presentations. In the *Proceedings of DARPA/NIST Smart Spaces Technology Workshop*. (1998)
5. Brotherton, J.A. Enriching Everyday Activities through the Automated Capture and Access of Live Experiences - eClass: Building, Observing and Understanding the Impact of Capture and Access in an Educational Domain.Ph.D. Thesis, College of Computing, Georgia Institute of Technology (2001)
6. Chiu, P., Kapuskar, A., Reitmeier, S., and Wilcox, L. NoteLook: Taking Notes in Meetings with Digital Video and Ink. In the *Proceedings of ACM Multimedia'99*. Orlando, FL (1999) pp.149-158.
7. Cruz, G. and Hill, R. Capturing and Playing Multimedia Events with STREAMS. In the *Proceedings of ACM Multimedia '94*. San Francisco, CA (1994) pp.193-200.
8. Davis, R.C., Landay, J.A., Chen, V., Huang, J., Lee, R.B., Li, F.C., Lin, J., III, C.B.M., Schleimer, B., Price, M.N., and Schilit, B.N. NotePals: Lightweight Note Sharing by the Group, for the Group. In the *Proceedings of CHI'99*. Pittsburgh, PA (1999) pp.338-345.
9. Dey, A.K., Salber, D. and Abowd, G.D. A Conceptual Framework and a Toolkit for Supporting the Rapid Prototyping of Context-Aware Applications. Human-Computer Interaction (HCI) Journal. **16**(2-4) (2001) pp.97-166.
10. Dey, A.K., Futakawa, M., Salber, D., Abowd, G.D. The Conference Assistant: Combining Context-Awareness with Wearable Computing, In the *Proceedings of the 3rd International Symposium on Wearable Computers (ISWC '99)*. San Francisco, CA (1999) pp.21-28.
11. Gellersen, H.W., Schmidt, A. and Beigl, M. Multi-Sensor Context-Awareness in Mobile Devices and Smart Artefacts.. Mobile Networks and Applications (MONET). **7**(5) (2002) pp.341-351.
12. Greenberg, S. and Fitchett, C. Phidgets: Easy Development of Physical Interfaces through Physical Widgets. In the *Proceedings of UIST 2001*. Orlando, FL (2001) pp.209-218.
13. He, L., Sanocki, E., Gupta, A., and Grudin, J. Auto-Summarization of Audio-Video Presentations. In the *Proceedings of ACM Multimedia 1999*. Orlando, FL (1999) pp.489-498.

14. Fleck, M., Frid, M., Kindberg, T., O'Brien-Strain, E., Rajani, R. and Spasojevic M. Rememberer: A Tool for Capturing Museum Visits. In the *Proceedings of UBICOMP 2002*. Goteberg, Sweden (2002) pp.48-55.

15. Klemmer, S.R., Li, J., Lin, J. and Landay, J.A. Papier-Mâché: Toolkit Support for Tangible Input. In the *Proceedings of CHI 2004*. Vienna, Austria (2004)

16. Lamming, M., and Flynn, M. "Forget-me-not" Intimate Computing in Support of Human Memory. In the *Proceedings of FRIEND21: Symposium on Next Generation Human Interfaces*. Tokyo, Japan (1994)

17. Liu, Q., Rui, Y., Gupta, A., and Cadiz, J.J. Automating Camera Management for Lecture Room Environments. In *Proceedings of CHI 2001*. Seattle, WA (2001)

18. Macedo, A.A., Truong, K.N., Pimentel, M.G.C., and Camacho, J.A. Automatically Sharing Web Experiences through a Hyperdocument Recommender System. In the *Proceedings of ACM HyperText 2003*. Nottingham, UK (2003)

19. Mankoff, J.C., Hudson, S.E. and Abowd, G.D. Interaction Techniques for Ambiguity Resolution in Recognition-Based Interfaces. In the *Proceedings of UIST 2000*. San Diego, CA (2000) pp.11-20.

20. Mukhopadhyay, S. and Smith, B. Passive Capture and Structuring of Lectures. In the *Proceedings of ACM Multimedia'99*. Orlando, FL (1999) pp.477-487.

21. Omoigui, N., He, L., Gupta, A., Grudin, J., and Sanocki, E. Time-Compression: Systems Concerns, Usage, and Benefits. In the *Proceedings of CHI'99*. Pittsburgh, PA (1999) pp.136-143.

22. Pedersen, E. McCall, K., Moran, T.P. and Halasz F. Tivoli: An Electronic Whiteboard for Informal Workgroup Meetings. In the *Proceedings of INTERCHI'93*. Amsterdam, The Netherlands. (1993) pp.391-398.

23. Richter, H., Abowd, G.D., Geyer, W., Fuchs, L., Daijavad, S. and Poltrock, S. Integrating Meeting Capture within a Collaborative Team Environment. In the *Proceedings of UBICOMP 2001*. Atlanta, GA (2001) pp.123-138.

24. Stern, M., Steinberg, J., Lee, H., Padhye, J., and Kurose, J. MANIC: Multimedia Asynchronous Networked Individualized CourseWare. In the *Proceedings of Educational multimedia and Hypermedia*. (1997).

25. Stifelman, L.J. The Audio Notebook.Ph.D. Thesis, Media Laboratory, MIT (1997)

26. Streitz, N.A., Geibler, J., Haarke, J., and Hol. J. DOLPHIN: Integrated meeting Support across Local and Remote Desktop Enviroments and LiveBoards. In the *Proceedings of CSCW'94*. Chapel Hill, NC. (1994) pp. 345-357.

27. Sumi,Y., Sakamoto,R., Nakao,K., and Mase,K. ComicDiary: Representing Individual Experiences in a Comics Style. In the *Proceedings of UBICOMP 2002*. Goteberg, Sweden (2002) pp.16-32.

28. Tandler, P. Software Infrastructure for Ubiquitous Computing Environments: Supporting Synchronous Collaboration with Heterogeneous Devices. In the *Proceedings of UBICOMP 2001*. Atlanta, GA (2001) pp.96-115.

29. Truong, K.N., Abowd, G.D., and Brotherton, J.A. Personalizing the Capture of Public Experiences. In the *Proceedings of UIST'99*. Asheville, NC (1999) pp.121-130.

30. Truong, K.N., Abowd, G.D., and Brotherton, J.A. Who, What, When, Where, How: Design Issues of Capture & Access Applications. In the *Proceedings of UBICOMP 2001*. Atlanta, GA (2001) pp.209-224.

31. White, D.R., Camacho-Guerrero, J.A., Truong, K.N., Abowd, G.D, Morrier, M.J., Vekaria, P.C. and Gromala, D. Mobile Capture and Access for Assessing Language and Social Development in Children with Autism. In the *Extended Abstracts of UBICOMP 2003*. Seattle, WA (2003) pp.137-140.

32. Whittaker, S., Hyland, P. and Wiley, M., FiloChat: Handwritten Notes Provide Access to Recorded Conversations. In the *Proceedings of CHI'94*. Boston, MA (1994) pp.271-276.

33. Wilcox, L., Schilit, B.N., and Sawhney, N. Dynomite: A Dynamically Organized Ink and Audio Notebook. In the *Proceedings of CHI'97*. Atlanta, GA (1997) pp.186-193.

Activity Recognition in the Home Using Simple and Ubiquitous Sensors

Emmanuel Munguia Tapia, Stephen S. Intille, and Kent Larson

Massachusetts Institute of Technology
1 Cambridge Center, 4FL
Cambridge, MA 02142 USA
{emunguia,intille,kll}@mit.edu

Abstract. In this work, a system for recognizing activities in the home setting using a set of small and simple state-change sensors is introduced. The sensors are designed to be "tape on and forget" devices that can be quickly and ubiquitously installed in home environments. The proposed sensing system presents an alternative to sensors that are sometimes perceived as invasive, such as cameras and microphones. Unlike prior work, the system has been deployed in multiple residential environments with non-researcher occupants. Preliminary results on a small dataset show that it is possible to recognize activities of interest to medical professionals such as toileting, bathing, and grooming with detection accuracies ranging from 25% to 89% depending on the evaluation criteria used [1].

1 Introduction

In this paper, a system for recognizing activities in the home setting using a set of small, easy-to-install, and low-cost state-change sensors is introduced. We show early results that suggest that our sensing technology, which users may perceive as less invasive than cameras and microphones, can be used to detect activities in real homes. The results we present are preliminary but show promise. They are unusual because the ubiquitous computing system and results we describe have been tested in multiple real homes with subjects who are not affiliated with the investigators' research group or university.

Our vision is one where a large number of simple, low-cost "tape on and forget" sensors are easily taped on objects throughout an environment and used by a computing system to detect specific activities of the occupant. Computers that can automatically detect the user's behavior could provide new context-aware services in the home. One such service that has motivated this work is proactive care for the aging. Medical professionals believe that one of the best ways to detect emerging medical conditions before they become critical is to look for changes in the activities of daily living (ADLs), instrumental ADLs (IADLs) [17], and enhanced ADLs (EADLs) [24]. These activities include eating, getting

---

[1] This work was supported, in part, by National Science Foundation ITR grant #0112900 and the Changing Places/House_n Consortium.

A. Ferscha and F. Mattern (Eds.): PERVASIVE 2004, LNCS 3001, pp. 158–175, 2004.

in and out of bed, using the toilet, bathing or showering, dressing, using the telephone, shopping, preparing meals, housekeeping, doing laundry, and managing medications. If it is possible to develop computational systems that recognize such activities, researchers may be able to automatically detect changes in patterns of behavior of people at home that indicate declines in health. The system described in this work could potentially be retrofit into existing homes to detect and monitor ADLs.

2   Background

Everyday activities in the home roughly break down into two categories. Some activities require repetitive motion of the human body and are constrained, to a large extent, by the structure of the body. Examples are walking, running, scrubbing, and exercising. These activities may be most easily recognized using sensors that are placed on the body (e.g. [19,11,18]). A second class of activities, however, may be more easily recognized not by watching for patterns in how people move but instead by watching for patterns in how people move things. For instance, the objects that someone touches or manipulates when performing activities such as grooming, cooking, and socializing may exhibit more consistency than the way the person moves the limbs.

In this work we focus on the latter problem and ask the question, "can activities be recognized in complex home settings using simple sensors that detect changes in state of objects and devices?" Although progress is being made on algorithms that monitor a scene and interpret the sensor signals from complex sensors such as cameras or microphones, the recognition inference problem is often seriously underconstrained. Computer vision sensing, for example, often works in the laboratory but fails in real home settings due to clutter, variable lighting, and highly varied activities that take place in natural environments. Little of the work with video and audio processing in the lab has been extensively tested in the field. Perhaps just as importantly, however, because sensors such as microphones and cameras are so general and most commonly used as recording devices, they can be perceived as invasive and threatening by some people.

For these reasons, we are exploring the recognition potential of deploying very large numbers of extremely simple sensors. Simple sensors can often provide powerful clues about activity. For instance, a switch sensor in the bed can strongly suggest sleeping [1], and pressure mat sensors can be used for tracking the movement and position of people [22,2]. Although others have written on the potential of sensor networks (e.g. [7,13,14]), we are unaware of work where large numbers have been deployed in multiple, non-laboratory home environments and used for ADL pattern recognition.

Previous work where sensors have been placed on objects in the environment have typically been used in laboratories or homes of the researchers themselves and their affiliates. Further, all of these systems have required careful (and usually painstaking) installation and maintenance by research staff and students (e.g. [20,1,21]). With few exceptions (e.g. [20]) only a small portion of the homes are sensor-enabled. Prior work, however, has shown the potential of multiple, simple switches for activity detection. In the MARC home, simple sensors in a kitchen (temperature on stove, mat sensors, and cabinet door sensors) have been used to detect meal preparation activities [2]. In that work, mixture models and hierarchical clustering were used to cluster the low-level sensor readings into cooking events using temporal information [2]. However, choosing the number of clusters to use and correlating the clusters of sensor readings to activities may grow more difficult as larger numbers of sensors are added to environments to recognize a more diverse set of activities. RFID tags placed at objects in the environment and combined with unsupervised mining of activity models from the web have also shown promise for activity recognition [23]. Although this approach does not need the subject to label his activities, it could prove difficult to adapt to individual patterns of activities. In this work we explore a supervised learning approach.

Hierarchical hidden semi-Markov models (HHSMMs), a type of dynamic belief network (DBN), have been used to track the daily activities of residents in an assisted living community [15]. The algorithm can distinguish different activities such as "asleep" and "having meals" solely based on noisy information about the location of the residents and when they move. Even though DBNs show some promise, they may not scale to environments that contain hundreds of sensors, particularly if real-time recognition of activity is a goal.

Sequence matching approaches have been applied to predict inhabitant's actions. The SHIP algorithm matches the most recent sequence of events with collected histories of actions to predict inhabitant future actions [4]. This approach, however, does not model ambiguous and noisy information from multiple sensors.

3   Activity Detection Approach

The following design goals motivated the activity recognition algorithms developed in this work.

Supervised learning. Homes and their furnishings have highly variable layouts, and individuals perform activities in many different ways. The same activity (e.g. brushing teeth) may result in a significantly different sensor activation profile based upon the habits, or routines of the home occupant and the layout and organization of the particular home. One approach to handling such variability is to use supervised learning with an explicit training phase.

Probabilistic classification. Probabilistic reasoning offers a way to deal with ambiguous and noisy information from multiple sensors.

Model-based vs instance-based learning. Model-based algorithms use the training examples to construct a mathematical model of the target classification function, which avoids the need to save all examples as raw data. This could help alleviate end-user privacy concerns.

Sensor location and type independent. Ideally, the system would operate effectively even when the algorithm is never explicitly told the location (e.g. kitchen) and type (e.g. drawer) of a particular sensor. This would dramatically reduce installation time.

Real-time performance. A system that recognizes activities in the home setting is most useful if it performs in real-time. Training or model construction time is less of a concern.

Online learning. Ideally the system would be capable of adjusting the internal model in real-time as new examples of activities become available. This will allow the algorithm to adapt to changes in the user's routines over time.

In this work, we chose naive Bayesian classifiers[16] to detect activities using the tape-on sensor system . Naive Bayesian classifiers make strong (and often clearly incorrect) assumptions that each class attribute is independent given the class. They also assume that all attributes that influence a classification decision are observable and represented. For these reasons, they are sometimes assumed to perform poorly in real domains. On the contrary, however, experimental testing has demonstrated that naive Bayes networks are surprisingly good classifiers on some problem domains, despite their strict independence assumptions between attributes and the class. In fact, simple naive networks have proven comparable to much more complex algorithms, such as the C4 decision tree algorithm [16,3, 12,5]. They also meet the design goals listed above.

One theory on why naive Bayes classifiers work so well is that the low variance of the classifier can offset the effect of the high bias that results from the strong independence assumptions [6]. Although in this preliminary work we are limited to small datasets, over time a tape-on sensor system and the experience sampling data collection method described shortly could be used to collect a large sample of activity from a user's home to train a classification system. Even in the results presented in this work, more data appear to lead to better recognition results. To apply naive Bayes classifiers to the activity recognition problem, however, temporal dependencies may need to be considered. Therefore, one approach would be to encode large numbers of low-order temporal relationships in the networks [8]. In this work, the naive Bayes classifier is extended to incorporate temporal relationships among sensor firings and recognize activities in the home setting. These classifiers are easy to train, fast, and seem to improve in performance with larger training sets.

Two versions of the activity recognition classifier were implemented. The first is a multi-class naive classifier in which the class node represents all the activities to recognize and its child nodes consist of one of two types: exist and before attributes. In this configuration, all the activities are considered to be mutually-exclusive, which means that the probabilities for all activities sum up to one at any given time. The second version of the activity recognition classifier implemented is multiple binary naive Bayes classifiers, each of them representing an activity to recognize. The main advantage of this binary decomposition is that the representation does not enforce mutual exclusivity. In this way, detection of listening to music does not preclude detection of preparing breakfast. In

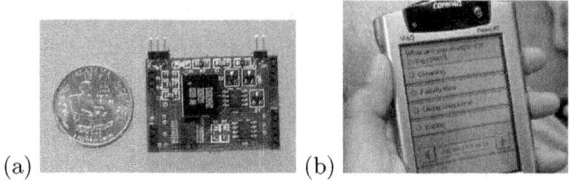

(a)   (b)

Fig. 1. (a) The state-change sensors that can be installed ubiquitously throughout an environment. Each device consists of a data collection board (shown) and a small sensor. (b) One screenshot from the ESM tool used in this work to collect training data on activities in the home setting.

this work, the prior probabilities for all the activities are assumed to be equal. Moreover, maximum likelihood was used to learn the parameters of the networks.

4   Activity Recognition System Architecture

The proposed system consists of three major components: (1) The environmental state-change sensors used to collect information about use of objects in the environment, (2) the context-aware experience sampling tool (ESM) used by the end user to label his or her own activities, and (3) the pattern recognition and classification algorithms for recognizing activities after constructing a model based on a training set.

4.1   Environmental State-Change Sensors

Although other low-cost wireless sensing systems have been developed, notably Berkeley Motes [7] and Smart-ITS [14], their power and cost points still pose a challenge for researchers interested in distributing hundreds of units in a single home to collect synchronized data for several weeks or longer. The cost of these devices is relatively high because they are designed as multi-purpose sensors. Therefore, we have designed a new set of tape-on sensors optimized to perform a single task at low cost: measuring change in the state of an object in home [9]. To achieve well-synchronized measurements, the most precise real-time clock hardware was used in each board. Further, the signals from each board were linearly interpolated to match the reference clock better after the end of the study. These highly-specialized boards are 3-5 times less expensive than Smart-ITS and Motes, which dramatically increases the number that can be installed in homes working within a tight research budget. The estimated battery life of the data collection board is one year if the external sensor is activated an average of 10 times per day for 30 seconds.

Figure 1a shows a sensor device, which actually consists of the sensor itself connected by a thin wire to a 27mm x 38mm x 12mm data collection board. The board fits snugly in a small plastic case of dimensions 37mm x 44mm x 14mm. The boards can use either reed switches, which are activated when brought into

Table 1. An example of the type of data that was acquired by the state-change sensors and ESM. The activity attributes are acquired using experience sampling during a training period. The sensor activations are collected by the state-change sensors distributed all around the environment. In the table, opt stands for optional attribute.

Activity	sensor ID	day	activation time	deactivation time	duration (sec)	room (opt)	object type (opt)
Preparing breakfast	PDA	12/1/02	08:23:01		10 min		
	23	12/1/02	08:23:03	08:23:07	4	kitchen	drawer
	18	12/1/02	08:23:09	08:23:17	8	kitchen	cabinet
	89	12/1/02	08:24:49	08:24:59	10	kitchen	fridge door
	⋮		(many readings)				

contact with a small magnet, or piezoelectric switches, which detect movement of a small plastic strip.

4.2   Context-Aware Experience Sampling

Supervised learning algorithms require training data. In the laboratory, obtaining annotated data is a straightforward process. Researchers can directly observe and label activity in real-time, or later through observation of video sequences. In the home environment, however, direct observation is prohibitively time-consuming and invasive.

One alternative is to use the Experience Sampling Method (ESM) [10,9]. When using ESM, subjects carry a personal digital assistant (PDA) that is used as timing device to trigger self-reported diary entries. The PDA samples (via a beep) for information. Multiple choice questions can then be answered by the user. Figure 1b shows a screen shot from the ESM tool used in this work. The protocol used to collect subject self report labels of activity in this work using ESM is described in section 5.2.

4.3   Activity Recognition Algorithms

The purpose of the state-change sensors and ESM was to provide the necessary data to create machine learning algorithms that can identify routines in activities from sensor activations alone. In order to accomplish this goal, new algorithms that correlate the sensor firings and activity labels and predict activities from new sensor firings are required. Table 1 shows an example of the type of data acquired with the sensors and using the ESM tool.

5   Study and Data Collection

Two studies were run in two homes of people not affiliated with our research group to collect data in order to develop and test the activity recognition algorithms. Both subjects granted informed consent and were compensated with

$15.00 dollars per day of participation in the study. The first subject was a professional 30-year-old woman who spent free time at home, and the second was an 80-year-old woman who spent most of her time at home. Both subjects lived alone in one-bedroom apartments. 77 state-change sensors were installed in the first subject's apartment and 84 in the second subject's apartment. The sensors were left unattended, collecting data for 14 days in each apartment. During the study, the subjects used the context-aware ESM to create a detailed record of their activities.

5.1   State-Change Sensors Installation

The state-changes sensors described in section 4.1 were installed on doors, windows, cabinets, drawers, microwave ovens, refrigerators, stoves, sinks, toilets, showers, light switches, lamps, some containers (e.g water, sugar, and cereal), and electric/electronic appliances (e.g DVDs, stereos, washing machines, dish washers, coffee machines) among other locations. The plastic cases of the data collection boards were simply placed on surfaces or adhered to walls using non-damaging adhesive selected according to the material of the application surface. The sensor components (e.g. reed and magnet) and wire were then taped to the surface so that contact was measured. Figure 2 shows how some of the 77 sensors were installed in the home of the first subject. The devices were quickly installed by a small team of researchers: an average of about 3 hours is required for the sensors installation in a small one-bedroom apartment of typical complexity. When sensors were installed, each data collection board (which has a unique ID) was marked on a plan-view of the environment so that when the sensor data was collected, the location (e.g kitchen) and type (e.g cabinet) of each sensor was known.

5.2   Labelling Subject's Activities

Experience sampling. The subjects were given a PDA running the ESM software at the start of the study. As the state-change sensors recorded data about the movement of objects, the subjects used experience sampling to record information about their activities. A high sampling rate was used, where the subject was beeped once every 15 minutes for 14 days (study duration) while at home. At the beep, the subject received the following series of questions. First the user was asked "what are you doing at the beep (now)?". The subject could select the activity that best matched the one that he/she was doing at the time of the beep from a menu showing up to 35 activities. Next, the following question was "For how long have you been doing this activity?" The subject could select from a list of four choices: less than 2 min., less than 5 min, less than 10 min., and more than 10 min. Then, the user was asked, "Were you doing another activity before the beep?". If the user responded positively, the user was presented with a menu of 35 activities once again. For the studies, an adaptation of the activity categories used by Szalai in the multi-national time-use study [25] were used.

Fig. 2. Examples of some of the 77 sensors that were installed in the home of the first subject. The sensors and data collection boards were literally taped to objects and surfaces for the duration of the data collection period.

Several problems were experienced with the ESM annotation method, some of which were learned about via interviews with subjects. Errors were observed where the user selected the wrong activity from the list by mistake. Short duration activities such as toileting were difficult to capture. There were delays between the sensor firings and the labels of the activities specified in the ESM. Fewer labels were collected than anticipated because subjects sometimes did not answer the ESM questions at the beep. Finally, sometimes subjects specified one activity and carried out a different activity.

Indirect observation of sensors activations. Unfortunately, the number of labels acquired using the ESM method was not sufficient for training the machine learning algorithms. Therefore, we were forced to resort to indirect observation by studying the sensor activations. In this method, the author, with the help of each subject, used self-inference to label the sensor data by visualizing the sensor activations clustered by location, time of activation, and type (description) of each sensor. Photographs of the sensors were also used to help the subject remember her activities during the sensor firings. A few decisions made during the manual annotation step impact the results that follow. First, activities were assumed to occur sequentially. The only activities allowed to occur in parallel with other activities were Listening to Music and Watching TV. Only the primary activity was labeled if a person was multi-tasking. Finally, only activities for which there exist sensor activations were labeled.

Figure 2 shows the number of labels generated by ESM and by indirect observation of sensor activations. For both subjects, the combined number of labeled activities is far less than desirable for a supervised learning algorithm. In current work, we are improving the subject self-annotation methodology to generate

Table 2. Average number of labels collected by the ESM and indirect Observation (I.O) per day during the study.

Measure	Subject 1	Subject 2
Average activities captured per day using ESM	9.5	13
Average activities per day generated by I.O	17.8	15.5
Different activities captured using ESM	22	24
Different activities generated by I.O	21	27
Average ESM Prompts answered to per day	18.7	20.1

better datasets. However, here we described our work with this admittedly small but still useful pilot dataset.

6   Feature Extraction, Training, and Prediction

We assumed that temporal information, in addition to which sensors fired, would be necessary to achieve good recognition using the naive Bayesian network approach. Therefore, one idea we explored in this work was to encode large numbers of low-order binary temporal relationships in the naive Bayesian network classifier. Two temporal features have been used. The first is whether activation of a particular sensor exists during some time period. The second is whether a particular sensor fires before another particular sensor. Table 3 shows the binary features calculated over the sensor data. These features output the evidence entered into the nodes of the naive Bayesian network.

The last two features in the table incorporate high level contextual information about the type of object in which the sensor was installed (e.g cabinet) and location of the sensor (e.g bathroom). The number of exist features that will become nodes in the naive Bayes networks is equal to the number of sensors present in the system (77 and 84 for subject one and two respectively). The number features that become nodes for the before sensorID, before type and before location features is equal to the number of all pairs of sensors, object types, and locations existent in the home environment (77x77=5929, 27x27=729, and 6x6=36 for subject one respectively).

Incorporating activity duration. Different activities have different mean lengths of time. Therefore, in order to incorporate the activity duration, one feature window per activity to recognize was used, and the length of each window corresponded to the activity duration as carried out by the subject. Thus, if M is the number of activities to recognize, there were M different feature windows with lengths $L_i \cdots L_m$. The duration or length L_i for each feature window was the average duration for each activity calculated from all the activity labels generated by ESM and indirect observation. For example, the feature window for toileting for the first subject was estimated to be 7 min, 27 sec. Preparing lunch was estimated to be 37 min, 54 sec.

Table 3. Features calculated and evaluated

Feature description	Example
exist(*sensorA, start, end*)	Sensor A fires within time interval
before(*sensorA, sensorB, start, end*)	Sensor A fires before sensor B within time interval
before(*sensorTypeA, sensorTypeB, start, end*)	Sensor in a drawer fires before a sensor in the fridge within time interval
before(*sensorLocationA, sensorLocationB, start, end*)	Sensor in kitchen fires before sensor in bathroom within time interval

Generation of Training Examples. In the training stage, training examples are generated by calculating the features from the start to the end time of each activity label. Figure 3a shows an example of how training examples are generated. Examples for washing hands, toileting, and grooming are generated whenever a label for washing hands, toileting, and grooming is found in the dataset respectively.

Originally, there was no unknown activity, but examples of this class were created by generating an example of it whenever no activity labels for other activities were found in the dataset. Figure 3a also shows an example of how two examples for the unknown class were generated.

Predicting the activity labels. In the prediction step, each feature window (of length L_i) is positioned at the current time to analyze, t. The features are then calculated from time $t - L_i$ to time t. Once the features are calculated, the probability for the current activity is calculated using the naive Bayes classifier.

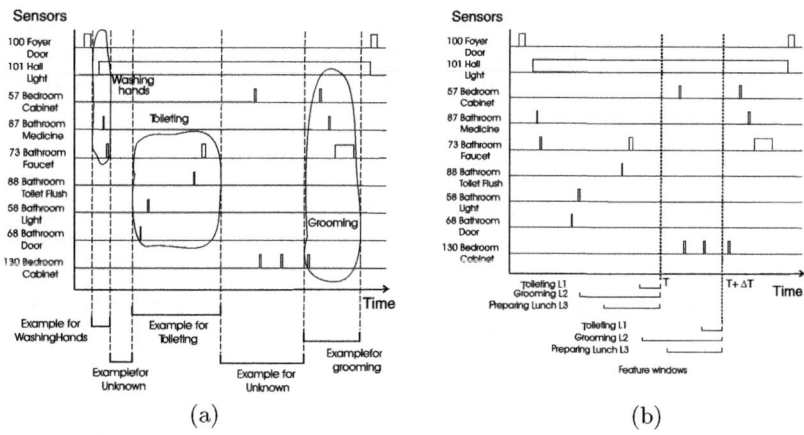

Fig. 3. (a) Example of how training examples are created for "washing hands", "toileting", "grooming" and two "unknown" activities. (b) Example of how features are extracted from sensor firings using different feature window lengths for each activity for time t and the next time to analyze $t+\triangle t$ in the prediction step.

Figure 4 conceptually shows an example of how the probability for each activity is generated in the prediction step by shifting the feature window for each activity over the sensor activations. Note that the probability is maximum when the feature window aligns with the duration of the activity represented by sensor activations (activity label). This indicates that the classifier may (depending upon noise) report the best match at the moment the activity is ending.

Figure 3b shows an example of how the feature windows for each activity are positioned in the current time to analyze t and in the next time to analyze $t + \triangle t$ over simulated sensor data. The $\triangle t$ increment in time used in the experiments was three minutes, which was half of the duration of the quickest activity. In a real-time application, however, the $\triangle t$ can be chosen to be as small as required, for example 5 seconds. While predicting an activity label for new observed sensor firings, the activity with the maximum likelihood at any given time is considered to be the classification result.

7   Algorithm Recognition Performance

Once the ESM and indirect observation labels were available, they were used to train and test the machine learning algorithms. All activities containing less than six examples were eliminated before training.[2]

Unlike other machine learning and pattern recognition problems, there is no "right" answer when recognizing activities. The boundaries when activities begin and end are fuzzy since they can occur sequentially, in parallel, alternating, and even overlapping. Finally, there is significant variation in the way observers would label the same activities.

Three methods were used to evaluate and measure the accuracy of the activity recognition algorithms. Which method is most informative depends upon the type of application that would be using the activity recognition data. The methods consider different features of the system that could be important for different applications, for example: (1) is the activity detected at all? and (2) for how long is the activity detected? Figure 5 shows examples of each of the three evaluation measures.

Percentage of time that activity is detected. This measures the percentage of time that the activity is correctly detected for the duration of the labelled activity.

Activity detected in best interval. This measures whether the activity was detected "around" the end of the real activity or with some delay ϕ. As discussed in section 6, the end of the activity is "the best detection interval". Thus, the right most edge of each activity (E) is analyzed within an interval of $\pm\phi$. It is important to remember that a detection delay is introduced

---

[2] The threshold of six was chosen arbitrarily. Given the complexity of the activities and the large amount of variation possible due to day of the week, time, and other factors, to expect an algorithm to learn patterns with less than six examples did not seem reasonable.

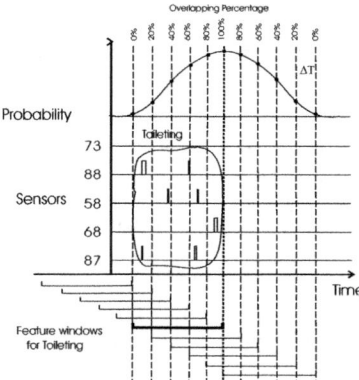

Fig. 4. Example of how the probability for the "toileting" activity is generated in the prediction step by shifting the feature window for "toileting" over the sensor activations with increments of $\triangle t$ (3 minutes for this study). Note that the probability is maximum when the feature window aligns with the duration of the activity represented by the sensor activations.

by the use of the feature windows that capture features back in time in our algorithm. In this work, the interval ϕ was chosen to be 7.5 minutes. Different applications would require different detection delays, thus different values of ϕ could be used.

Activity detected at least once. This measures if an activity was detected at least once for the duration of the activity label (no delay allowed).

Leave-one-out cross-validation was used in each evaluation method in order to calculate the confusion matrix and measure the classification accuracy. Cross-validation permits some classification testing even on small datasets. The activity with the maximum likelihood at a given time was used when determining classification accuracy using each of the three evaluation metrics.

Experiments to determine the discrimination power of the attributes were performed by running the multi-class and multiple binary naive Bayes classifiers with some of the possible combinations of attributes shown in Table 3. Tables 4 and 5 show the accuracies per class for the combination of attributes that performed the best for the multiclass naive Bayes classifier for subject one and two respectively.

8   Discussion

Accuracies vs number of examples. As expected, the activities with higher accuracies were generally those with more examples. For subject one, they were "toileting", "grooming", "bathing", and "doing laundry". For subject two, they were "preparing lunch", "listening to music", "toileting" and "preparing breakfast".

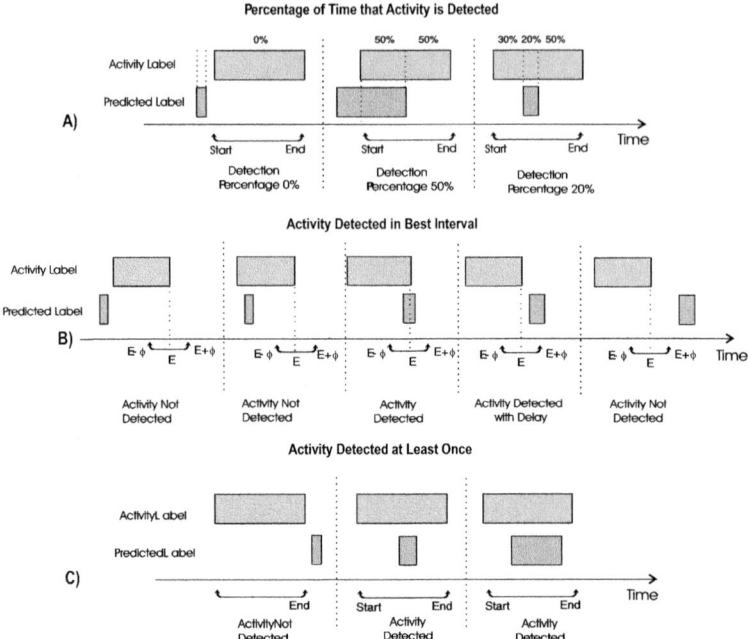

Fig. 5. Example cases of the (a) "percentage of time that activity is detected", (b) "activity detected in best interval" and (c) "activity detected at least once" evaluation methods.

Discriminant attributes. Overall, the exist attribute showed the best discriminant power. In this case, the naive Bayesian network is actually acting as a weighted voting mechanism. Adding temporal features such as before did not provide the discrimination power expected. We attribute this to the relatively small size of our datasets. When "ground truth" video is available for labelling the subject activities (planned for future installations) and more examples and multi-tasking examples are collected, this feature may show a higher discrimination power.

Optional attributes _Type and Location._ Preliminary results show that adding the attributes using the type of object in which the sensor was installed and location information such as the before type and before location features to the exist attribute did not represent a significant improvement in accuracy. This suggests that the development of activity recognition algorithms that do not use the "type" and "location" information may be possible.

Accuracy vs evaluation method used. The activities show lower accuracies for the "percentage of time" evaluation method. Since activities are being detected from sensor firings, some activities such as watching TV, listening to music and dressing are represented by sensors firing only at the beginning or at the end of the activity. This means that there is no information other

Table 4. Leave-one-day-out crossvalidation accuracies per class for the multiclass naive Bayes classifier using the best two combination of features for subject one. E stands for the *exist* feature, and BT stands for the *before type* feature.

Multiclass Naive Bayes Classifier for Subject One					
Activity	No. Examples	E	E+BT	Random Guess	Evaluation
Preparing lunch	17	0.25	0.29	0.07	
Toileting	85	0.27	0.31	0.07	Percentage of Time
Preparing breakfast	14	0.08	0.06	0.07	Activity is Detected
Bathing	18	0.25	0.29	0.07	
Dressing	24	0.07	0.03	0.07	
Grooming	37	0.26	0.26	0.07	
Preparing a beverage	15	0.07	0.13	0.07	
Doing laundry	19	0.09	0.07	0.07	
Preparing lunch	17	0.59	0.78	0.30	
Toileting	85	0.71	0.71	0.30	Activity Detected
Preparing breakfast	14	0.45	0.45	0.30	in Best Interval
Bathing	18	0.87	0.79	0.30	
Dressing	24	0.64	0.41	0.30	
Grooming	37	0.89	0.86	0.30	
Preparing a beverage	15	0.36	0.36	0.30	
Doing laundry	19	0.86	0.78	0.30	
Preparing lunch	17	0.50	0.68	0.17	
Toileting	85	0.42	0.43	0.03	Activity Detected
Preparing breakfast	14	0.20	0.12	0.07	at Least Once
Preparing a snack	14	0.08	0.05	0.03	
Bathing	18	0.70	0.75	0.11	
Going out to work	12	0.12	0.00	0.02	
Dressing	24	0.21	0.07	0.02	
Grooming	37	0.68	0.71	0.05	
Preparing a beverage	15	0.22	0.31	0.04	
Doing laundry	19	0.27	0.23	0.05	
Activities with Less than Six Examples					
Work at home(0), Eating(0), Washing hands(1), Sleeping(0), Taking medication(0), Sleeping(0), Talking on telephone(0), Resting(0), Putting away dishes(2), Putting away groceries(2), Putting away laundry(2), Taking out the trash(0), Lawnwork(1), Pet care(0), Home education(0) Going out to school(0), Going out for entertainment(1), Working at computer(0), Going out to exercise(0), Going out for shopping(2), Listening to music(0), and Watching TV(3).					
Activities Not Recognized Better than Random Guess					
Preparing dinner(8), Washing dishes(7), Preparing a snack(14), Going out to work(12), and cleaning(8)					

than the average duration of the activity represented by the feature windows to detect the activities during these "dead intervals" of sensor firings.

The evaluation method with the highest detection accuracies per activity was the "best interval detection". The classes with the highest "best interval detection" accuracies (over 70%) were "bathing", "toileting", "grooming", and "preparing lunch" for the first participant. For the second participant the higher "best interval detection" accuracies (over 50%) were "preparing breakfast", "preparing lunch", "listening to music" and "toileting".

Accuracy vs number of sensors. Since activities such as "going out to work" and "doing laundry" are represented by sensor firings from a single sensor (door and washing machine respectively), it was expected that they would show higher detection accuracies than other activities. However, the sensors were also activated during other activities which decreased their discriminant power. The most activated sensors overall for both subjects were

Table 5. Leave-one-day-out crossvalidation accuracies per class for the multiclass naive Bayes classifier using the best two combination of features for subject two. E stands for the *exist* feature, and BT stands for the *before type* feature.

Multiclass Naive Bayes Classifier for Subject Two					
Activity	No. Examples	E	E+BT	Random Guess	Evaluation
Preparing lunch	20	0.22	0.22	0.10	
Listening to music	18	0.20	0.09	0.10	
Toileting	40	0.20	0.23	0.10	Percentage of Time
Preparing breakfast	18	0.30	0.24	0.10	Activity is Detected
Washing dishes	21	0.05	0.11	0.10	
Watching TV	15	0.04	0.16	0.10	
Preparing lunch	20	0.51	0.48	0.40	
Listening to music	18	0.61	0.44	0.40	
Toileting	40	0.52	0.48	0.40	Activity Detected
Preparing breakfast	18	0.68	0.59	0.40	in Best Interval
Washing dishes	21	0.51	0.54	0.40	
Watching TV	15	0.25	0.52	0.40	
Preparing dinner	14	0.38	0.30	0.24	
Preparing lunch	20	0.48	0.61	0.26	
Listening to music	18	0.66	0.45	0.38	
Toileting	40	0.46	0.43	0.10	Activity Detected
Preparing breakfast	18	0.75	0.65	0.16	at Least Once
Washing dishes	21	0.15	0.28	0.09	
Watching TV	15	0.08	0.45	0.30	
Activities with Less than Six Examples					
Work at home(0), Going out to work(0), Eating(0), Bathing(3), Grooming(3), Dressing(5), Washing hands(0), Sleeping(0), Talking on telephone(4), Resting(0), Preparing a beverage(1), Putting away dishes(3), Putting away groceries(1), Cleaning(3), Doing laundry(0), Putting away laundry(1), Taking out the trash(0), Lawnwork(1), Pet care(0), Home education(2), Going out to school(0), Going out for entertainment(1), Working at computer(5), Going out to exercise(0), and Going out for shopping(3).					
Activities Not Recognized Better than Random Guess					
Preparing dinner(14), Taking medication(14), and Preparing a snack(16).					

the kitchen door, refrigerator, and cabinets. For subject one, the three other most activated sensors were the bathroom sink faucet (165), medicine cabinet (118), and kitchen drawer #84 (74). For subject two the three other most activated sensors were the microwave oven (197), garbage disposal (79), and living room TV (63). The least activated sensors were located in the living room and hallway area.

Accuracy vs sensors installation locations. The state change sensors were not appropriate for installation on some useful objects. For example, sensors were not installed on pans, dishes, chairs, tables and other locations that could improve the recognition accuracy of for preparing dinner. A new version of the tape on sensors in development that uses accelerometry instead of reed switch sensing will permit a 2-3 times increase in the number of sensors that can be installed in a given environment. This may improve recognition, but it also may increase the need for larger training data sets.

Multiclass vs multiple binary classifiers. We have described our implementation of a multiclass naive Bayes classifier. However, we also tested the system with multiple binary classifiers – one per activity. Multiple binary classifiers do not enforce mutual exclusivity, since each classifier is independent of the others. On our dataset, the multiclass and multiple binary classifiers performed with approximately the same accuracy ($\pm 5\%$). However, the multiple binary classifiers may perform better in future studies as more accurate activity labels become available with multi-tasking.

Some of the false positives obtained in this work almost certainly resulted from the fact that a considerable number of short or multi-tasked activities carried out by the subjects were not labelled in our dataset.

Subject one vs Subject two results. Overall, recognition accuracies for the first subject's data were higher than those for the second subject's data. This was mainly for two reasons: (1) the number of sensors that were dislodged, failed and were noisy was higher in the second subject apartment, and (2) the quality of the labels for the first subject was higher because the sensor firings for each activity were easier to label. One possible explanation is that the first subject spent less time at home and the sensor firings were not as complex as those for the second subject.

Improvement over the random guess baseline. The results shown in Tables 4 and 5 represent a significant improvement for some activities over the random guess baseline[3].

Even though the accuracy for some activities is considerably better than chance, it is not as high as expected. The main problems faced were: (1) the quality and number of activity labels, and (2) the small training set of two weeks. It is expected that training sets collected over months, better labels generated by video observation or other methods, and improved versions of the data collection boards and sensors will improve the detection accuracies. Although the results presented here are preliminary, we believe they show promise. Experiments are underway to improve the data collection and annotation process and to acquire substantially more detailed datasets.

9   Subject Experiences

The studies described here have led to some useful qualitative observations about subject reaction to the sensor system and the ESM data collection method. For instance, the participants felt that they became "oblivious" to the presence of the sensors after a few days of the installation. The first subject reported forgetting that the sensors were in her apartment after two days. The second subject was not even sure where some of the sensors were installed. We suspect the acclimation period to more invasive sensors such as cameras would be substantially longer. One of our subjects told us she would not have agreed to the study if it had involved video observations. The second subject would have agreed but would have restricted cameras from the bathroom.

---

[3] The random guess baseline for the "percentage of time" evaluation method is $1/n$, where n is the number of activities. The random guess is $1 - ((n-1)/n)^i$ for the "best interval" and "at least once" methods, where $i = (2\phi)/\triangle t$ for "best interval" and $i = activity\_average\_duration/\triangle t$ for "detected at least once".

Subjects had a more difficult time adjusting to the experience sampling device. They started to find the ESM highly disruptive by the second and third day of the study. This was probably because of the high sampling rate (15 minutes), but even with extremely helpful volunteers it has become clear that tolerance for repetitive sampling of the same activities is low. In short, subjects did not mind "teaching the computer" about new activities but did not enjoy having to tell the computer about doing the same activities repetively. The percent of total ESM prompts responded to was 17% and 24% for the first and second subject respectively.

Finally, even these simple sensors can impact behavior. The first participant reported that that being sensed did cause her to alter some behaviors. For example, she always made sure to wash her hands after using the bathroom. The second subject did not report any changes in behavior due to the sensors.

10   Summary

The work described here is preliminary but demonstrates that ubiquitous, simple sensor devices can be used to recognize activities of daily living from real homes. Unlike prior work in sensor systems for recognizing activities, the system developed in this work was deployed in multiple residential environments with actual occupants. The occupants were not researchers or affiliated with the experimenters. Moreover, the proposed sensing system presents an alternative to sensors that are perceived by most people as invasive such as cameras and microphones. Finally the system can be easily retrofitted in existing home environments with no major modifications or damage.

References

1. Elite Care's Oatfield Estates. http://www.elite-care.com/oatfield-tech.html.
2. T. Barger, M. Alwan, S. Kell, B. Turner, S. Wood, and A. Naidu. Objective remote assessment of activities of daily living: Analysis of meal preparation patterns. Poster presentation, Medical Automation Research Center, University of Virginia Health System, 2002.
3. P. Clark and T. Niblett. The CN2 induction algorithm. *Machine Learning*, 3(4):261–283, 1989.
4. S. K. Das, D. J. Cook, A. Bhattacharya, E. O. Heierman, and T.Y. Lin. The role of prediction algorithms in the MavHome smart home architecture. In IEEE Wireless Communications, editor, *IEEE Press*, volume 9, pages 77–84, 2002.
5. P. Domingos and M. Pazzani. Beyond independence: Conditions for the optimality of a simple bayesian classifier. In L. Saitta, editor, *Proceedings of the Thirteenth International Conference on Machine Learning*, pages 194–202. Morgan Kauffman, 1996.
6. J. Friedman. On bias, variance, 0/1 - loss, and the curse-of-dimensionality. *Data mining and knowledge engineering*, 1:55–77, 1997.
7. S. Hollar. *COTS Dust*. Ph.D. thesis, University of California, Berkeley, 2001.
8. S.S. Intille and A.F. Bobick. Recognizing planned, multi-person action. *Computer Vision and Image Understanding (1077-3142)*, 81(3):414–445, 2001.

9. S.S. Intille, E. Munguia Tapia, J. Rondoni, J. Beaudin, C. Kukla, S. Agarwal, L. Bao, and K. Larson. Tools for studying behavior and technology in natural settings. In A. K. Dey, A. Schmidt, and J. F. McCarthy, editors, *Proceedings of UBICOMP 2003*, volume LNCS 2864. Springer, 2003.

10. S.S. Intille, J. Rondoni, C. Kukla, I. Anacona, and L. Bao. A context-aware experience sampling tool. In *Proceedings of the Conference on Human Factors and Computing Systems: Extended Abstracts*. ACM Press, 2003.

11. J. Himberg J. Mantyjarvi and T. Seppanen. Recognizing human motion with multiple acceleration sensors. *IEEE International Conference on Systems, Man, and Cybernetics.*, pages 747–52, IEEE Press, 2001.

12. G.H. John and P. Langley. Estimating continuous distributions in Bayesian classifiers. In *Proceedings of the Eleventh Conference on Uncertainty in Artificial Intelligence*. San Mateo, AAAI Press, 1995.

13. J.M. Kahn, R.H. Katz, and K.S.J. Pister. Mobile networking for Smart Dust. In *ACM/IEEE International Conference on Mobile Computing and Networking (MobiCom 99)*, pages 271–278. 1999.

14. O. Kasten and M. Langheinrich. First experiences with Bluetooth in the Smart-Its distributed sensor network. In *Workshop on Ubiquitous Computing and Communications, PACT*. 2001.

15. H. Kautz, O. Etziono, D. Fox, and D. Weld. Foundations of assisted cognition systems. Technical report CSE-02-AC-01, University of Washington, Department of Computer Science and Engineering, 2003.

16. P. Langley, W. Iba, and K. Thompson. An analysis of Bayesian classifiers. In *Proceedings of the Tenth National Conference on Artificial Intelligence*, pages 223–228. AAAI Press, San Jose, CA, 1992.

17. M.P. Lawton and E.M. Brody. Assessment of older people: self-maintaining and instrumental activities of daily living. *Gerontologist*, 9:179–186, 1969.

18. S.W Lee and K. Mase. Activity and location recognition using wearable sensors. *IEEE Pervasive Computing*, 1(3):24–32, 2002.

19. M. Makikawa and H. Iizumi. Development of an ambulatory physical activity monitoring device and its application for categorization of actions in daily life. In *MEDINFO*, pages 747–750. 1995.

20. M. Mozer. The Neural Network House: an environment that adapts to its inhabitants. In *Proceedings of the AAAI Spring Symposium on Intelligent Environments*, Technical Report SS-98-02, pages 110–114. AAAI Press, Menlo Park, CA, 1998.

21. B. Ornstein. Care technology: Smart home system for the elderly. In *Proceedings of NIST Pervasive Computing*. 2001.

22. R.J. Orr and G.D. Abowd. The Smart Floor: A mechanism for natural user identification and tracking. In *Proceedings of the 2000 Conference on Human Factors in Computing Systems (CHI 2000)*. ACM Press, 2000.

23. M. Philipose, K.P. Fishkin, D. Fox, H. Kautz, D. Patterson, and M. Perkowitz. Guide: Towards understanding daily life via auto-identification and statistical analysis. In *UbiHealth Workshop*. Ubicomp, 2003.

24. W.A. Rogers, B. Meyer, N. Walker, and A.D. Fisk. Functional limitations to daily living tasks in the aged: a focus groups analysis. *Human Factors*, 40:111–125, 1998.

25. S. Szalai. *The Use of Time. Daily Activities of Urban and Suburban Populations in Twelve Countries*. Mouton, The Hague, 1973. Edited by Alexander Szalai.

Automatic Calibration of Body Worn Acceleration Sensors

Paul Lukowicz[1,2], Holger Junker[2], and Gerhard Tröster[2]

[1] Institute for Computer Systems and Networks, UMIT Innsbruck, Austria
paul.lukowicz@umit.at
[2] Wearable Computing Lab, ETH Zurich, Switzerland
junker,troester@ife.ee.ethz.ch

Abstract. The paper presents a scheme for automatic calibration of body worn acceleration sensors which does not require any user interaction and any knowledge about the position and orientation of the sensors on the body. We describe the theoretical principle behind the method, discuss the main practical implementation concerns, and present experimental validation results.

1 Introduction

The use of body worn acceleration sensors for the recognition of user activities has been proposed and demonstrated by a number of research groups. Such sensors can be easily miniaturized, consume little power while at the same time providing valuable information on the position and motion of body parts. Applications described to date included the distinction between such simple activities as walking, sitting and running [4,2], the recognition of high level actions such as greeting a person [1] or operating a certain home appliance as well as the analysis of motion patterns [7,5,6] (e.g. for medical diagnosis).

A substantial practical problem with many wearable applications of acceleration sensors, is the need for calibration. Each sensor has a characteristic sensitivity s and offset o value, which allows the acceleration in g a to be computed from the measured value x as: $a_g = (x - o)/s$. In general o and s can differ significantly between devices. In addition both parameters are subject to thermal drift, which means that they change over time.

For traditional applications (e.g. in the automotive industry) the problem can be easily solved by embedding the sensor in a circuit containing ROM memory with factory calibration values together with appropriate drift stabilization logic. However in distributed, reconfigurable wearable systems envisioned by our group such a solution might not always be practical. Size and power consumption limitations might require 'naked' sensors to be integrated in the clothing. Using communication lines embedded in the textile such sensors would then be connected to an exchangeable, external processing unit unaware of the calibration values. Furthermore in many experimental systems sensors are provided and used without stored calibration values and stabilization circuits. In both

A. Ferscha and F. Mattern (Eds.): PERVASIVE 2004, LNCS 3001, pp. 176–181, 2004.

cases the sensors might need to be manually re-calibrated each time the system is switched on. Furthermore, a periodic recalibration might be needed to compensate the thermal drift.

To solve the above problem, this paper proposes a scheme that allows the system to automatically recalibrate the sensors. While the scheme uses some ideas that have been put forward by [3] to automatically derive the orientation of body mounted acceleration sensors, to our knowledge this is the first time that a scheme that requires neither user interaction nor information about sensor position is presented for calibration.

2   Calibration Algorithm

The acceleration sensor signal can be decomposed into two components: a static and a dynamic one. The static component corresponds to the earth gravity. Its value for any single sensor axis depends on the angle between the axis and the vertical plane. It has a maximum of 1g for a sensor aligned parallel to the vertical plane and the $arcos(\alpha)$ for an axis at an angle α with that plane. The dynamic component is caused by changes in speed (or rotations) that lead to 'true' acceleration. It is overlaid (added to) the static component and in general the two can not be separated.

In the standard manual calibration two measurements are taken for each axis when it is oriented parallel to the gravity facing down and up.

This gives two equations from which the two unknown parameters (o and s) can be computed. The problem with this approach is, that it requires the user to explicitly assume a series of well defined positions. In addition, the accuracy of the calibration depends on how well the sensors are positioned, which can be difficult in a body mounted system.

2.1   Calibration

The algorithm proposed in the paper overcomes these problems allowing the system to calibrate a 3-axis acceleration sensor without any user interaction and without any requirements on a particular position. It is based on two observations.

First, when the dynamic component is zero, the only acceleration acting on the system is earth gravity of 1g. Thus, the total acceleration on all three axis given by the norm of the 3D vector $a = a_x, a_y, a_z$ must be equal to 1g which is equivalent to

$$(a_x^2 + a_y^2 + a_z^2)^{1/2} = 1 \equiv a_x^2 + a_y^2 + a_z^2 = 1 \tag{1}$$

Second, due to mechanical constraints on human movement, it is extremely unlikely that a constant acceleration can be applied on any single axis for an extended period of time. An 'extended period of time' must be seen in context of the typical speed of human movement, which means that it is at most 1sec. Thus, as pointed out by [1,3], whenever the acceleration on all three axes remains constant for about 1 sec, the system can be assumed to be under the influence of gravity only. In addition, it is known to be in a stationary position with respect to the gravity vector.

Looking at how to average the above for calibration, we consider a three axis measurement (x, y, z) with an uncalibrated sensor that is stationary with respect to the gravity vector. Assuming the offsets and sensitivities on the x, y and z axis to be $(o_x, s_x, o_y, s_y, o_z, s_z)$ respectively, equation (1) becomes:

$$((x - o_x)/s_x)^2 + ((y - o_y)/s_y)^2 + ((z - o_z)/s_z)^2 = 1 \qquad (2)$$

This is a nonlinear equation with 6 unknown variables. As a consequence if the measurement is repeated six times in different position, a nonlinear equation system can be constructed that contains enough information to derive all six parameters. It can be solved using numerical methods such as Newton iteration.

2.2   Practical Concerns

To use the above principle to derive the calibration parameters from a body worn sensor system, three practical issues must be considered: (1) the acquisition of the required six, different stationary positions, (2) the complexity of the solution process and (3) the accuracy of the numerical solution.

Acquiring Positions This is obviously related to the location of the sensors on the body and the user activity. However, in general, every body part is likely to be static quite often and, with few exceptions, it is likely to assume a different position each time. The main exception are the hips. While they tend to remain stationary for large time periods, only small variations in the position are observed. To a lesser degree this is also true for the upper legs. For most other typical placements (arms, lower legs, head) we have observed sufficient, different stationary positions.

Complexity While numerical solutions to nonlinear equation systems can be very computationally intensive, the small size of the equation system means that even if a large number of iterations is needed, the solution time is negligible. Solutions involving 100 iterations have been performed using Matlab on a PIII 900MHz system in a matter of seconds.

Accuracy and Averaging The accuracy of the solution is the most serious practical problem. It is determined by a combination of measurement accuracy and solver stability. The former is given by sensor noise the amount of rest motion present in the position that were perceived as static. Both are particularly relevant if some of the positions used for the solutions are close to each other so that the difference between the positions are small with respect to the noise and motion artifacts. In addition, the precision of the perpendicular alignment of sensor axis with respect to each other needs to be considered. If all three axis are part of a single integrated circuit device than this question can be ignored.

However, if single axis devices are mounted on a printed circuit board considerable deviations can be encountered. This deviations mean that equation 1 is not exactly valid with the amount of violation being dependent on the degree misalignment and the position in which a measurement is performed.

Most numerical solution methods are based on an empirical search strategy such as gradient descent in some sort of penalty function. It is well known that depending on the input values such methods can get stuck in local minimum. As a consequence it is to be expected that the solver will not be equally accurate for all measurement. Even the sequence in which the measurement values are inserted into the equations can influence the accuracy of the result.

In summary, the accuracy can be expected to vary for different positions. The variations is partly random (due to solver getting stuck in local minima) and partly dependent on the separation between the positions being large with respect to noise and misalignment related violations of equation 1. This is confirmed by the experimental results presented in the next section. To offset this effect, several computations using different combinations of measurements must be performed and averaged to obtain an accurate results.

3   Experimental Results

To verify the proposed method, four types of tests need to be made (1) a basic soundness of the method, (2) an investigation of the accuracy and solver stability problems and (3) a validation that the of averaging solves the problem of extreme variations and (4) a demonstration that enough valid, stationary segments can be found in real life data.

Basic Validation For basic validation, data was recorded for the standard calibration positions (parallel to gravity) for all three axis. From this data the offset were computed using the standard method as well as by inserting it into equation 1 and solving it with the csolve solver for Matlab. Table 1 shows the results for 5 attempts. As can be seen the computed positions agree with the results obtained using standard calibration.

Table 1. Calibration parameter values obtained using standard calibration method (column 1) and the new proposed method (column 2-6)

Parameter	Standard	Dataset1	Dataset2	Dataset2	Dataset2	Dataset2	Mean	std
X-Sensitivity	754	750.9	747.9	748.0	748.5	747.9	748.6	1.3
X-Offset	1731	1726.0	1725.5	1723.6	1724.3	1722.1	1724.3	1.6
Y-Sensitivity	725	723.1	723.9	721.7	725.9	722.8	723.5	1.5
Y-Offset	1830	1814.0	1812.3	1813.1	1813.1	1814.6	1813.4	0.9
Z-Sensitivity	760	749.9	752.2	749.4	750.2	750.0	750.3	1.1
Y-Offset	1936	1925.3	1922.8	1925.3	1924.4	1923.9	1924.4	1.0

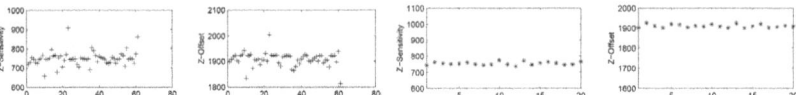

Fig. 1. The results of test calibration for the z-axis. Left two graphs show the results without, the right two with averging.

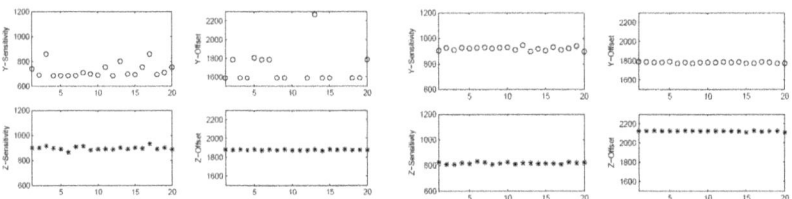

Fig. 2. The results of calibration with averaging over 20 different segment combinations on real life data for the x and z axis with sensor location on the wrist (left) and on the arm (right).

A ccuracy of D i erent Position Sets To investigate the variations of accuracy for different positions a set of stationary positions was extracted from a test data set obtained from wearing the sensor on the wrist for a short period of time. From this test set 83 random combinations of six positions were chosen and used to derive the calibration. The results are shown in figure 1 left. It can be seen that while the majority of values have small variations only and agree with the manual calibration, there are some extreme deviations.

A ccuracy w ith A veraging To prove that the variations can be handled through averaging we have taken 20 averages of 25 random position sets from Fig. 1 left. As can be seen in Fig. 1 right, the variations are reduced to at most 5% with the majority being within 1%.

U sability w ith R eal L ife D ata As a first indication of the value of our scheme in real life situations a subject was asked to wear a system with 3 sensors (wrist, upper arm and upper leg) for about 20 min. before and during a coffee break at our lab. During this time he has done some work on his computer, read a newspaper, ate some cornflakes, and walked around. To ensure realistic data the subject was unaware of the aim of the experiment. Based on our experience with artificial data we have defined 'stationary' to mean segments of 0.5 or 1 sec length during which the standard deviation of the acceleration values remained below 0.5% on all three axis. In the data 45 (wrist), 48 (arm) and 68 (leg) 0.5 sec segments and 35 (wrist) 37 (arm) and 66 (leg) 1 sec segments were found. Since a stable solution requires the individual data points to be sufficiently different, we have then made sure that for any two segments in the set at lest one axis differed by at least 150 (approx 10%). This left 12, 14, 4 segments for the 0.5 sec and 9, 13 4 segments for the 1 sec case. The large reduction of the legs segments from over 60 to 4 reflects the fact that while the legs where immobile for most of the time while the subject was sitting, they stayed in the same position. As a consequence no solution was found for the leg sensor. For the wrist and the arms solutions were found that were nearly as stable as in the case of artificial data described above (see figure 2). The only less accurate case is the z-axis on the wrist, for which the values in the experiment did not have siffcient variation.

4   Conclusion

We have introduced a method that allows body worn acceleration sensors to be calibrated without any explicit interaction with the user and without any prior knowledge about the orientation and location of the sensors. This is particularly interesting for wearable systems where sensors are an integral part of clothing constituting a system that is dynamically reconfigured each time the user puts on a different outfit. The experimental results presented in the paper suggest that solver accuracy and noise problems can be overcome through averaging and that the method is viable for practical applications. Since solver stability has been the major source of inaccuracy, additional, significant improvements can be expected from the use of more advanced numerical methods.

References

1. N. Kern, B. Schiele, H. Junker, P. Lukowicz, and G. Tröster. Wearable sensing to annotate meeting recordings. In *Proceedings Sixth International Symposium on Wearable Computers ISWC 2002*, 2002.
2. J. Mantyjarvi, J. Himberg, and T. Seppanen. Recognizing human motion with multiple acceleration sensors. In *2001 IEEE International Conference on Systems, Man and Cybernetics*, volume 3494, pages 747–752, 2001.
3. D. Mizell. Using gravity to estimate accelerometer orientation. In *Proceedings Seventh International Symposium on Wearable Computers IS WC 2002*, 2003.
4. C. Randell and H. Muller. Context awareness by analysing accelerometer data. In *Digest of Papers. Fourth International Symposium on Wearable Computers.*, pages 175–176, 2000.
5. M. Sekine, T. Tamura, T. Fujimoto, and Y. Fukui. Classification of walking pattern using acceleration waveform in elderly people. *Engineering in Medicine and Biology Society*, 2:1356 – 1359, Jul 2000.
6. M. Sekine, T. Tamura, M. Ogawa, T. Togawa, and Y. Fukui. Classification of acceleration waveform in a continuous walking record. In H. K. Chang and Y. T. Zhang, editors, *Proceedings of the 20th Annual International Conference of the IEEE Engineering in Medicine and Biology Society*, volume 3, pages 1523–1526, 1998.
7. T. Tamura, Y. Abe, M. Sekine, T. Fujimoto, Y. Higashi, and M Sekimoto. Evaluation of gait parameters by the knee accelerations. *BMES/EMBS Conference, 1999. Proceedings of the First Joint*, 2:828, Oct 1999. not useful.

Reject-Optional LVQ-Based Two-Level Classifier to Improve Reliability in Footstep Identification*

Jaakko Suutala, Susanna Pirttikangas, Jukka Riekki, and Juha Röning

Intelligent Systems Group, Infotech Oulu
90014 University of Oulu, Finland
E-mail: Jaakko.Suutala@ee.oulu.fi

Abstract. This paper reports experiments of recognizing walkers based on measurements with a pressure-sensitive EMFi-floor. Identification is based on a two-level classifier system. The first level performs Learning Vector Quantization (LVQ) with a reject option to identify or to reject a single footstep. The second level classifies or rejects a sequence of three consecutive identified footsteps based on the knowledge from the first level. The system was able to reduce classification error compared to a single footstep classifier without a reject option. The results show a 90% overall success rate with a 20% rejection rate, identifying eleven walkers, which can be considered very reliable.

1 Introduction

In this paper, experiments on recognizing walkers on a pressure-sensitive floor are described. Automatic recognition of occupants leads to personal profiling and enables smooth interaction between the environment and the occupant. It facilitates the building of intelligent environments that learn and react to the occupants' behaviour [1].

The idea of using footsteps to identify persons is not new. In [2] and [3] the identification based on small area sensors, measuring the *ground reaction force*, were reported. In our earlier experiments, Hidden Markov Models (initial study of three persons' footsteps) [4] and Learning Vector Quantization [5] (for eleven walkers) [6] were used in the identification of single footsteps, measured by utilizing the pressure-sensitive ElectroMechanical Film [7] (EMFi), as described in [6]. In both cases, the overall classification rate was 78%.

In this paper, a two-level classifier is introduced. It is based on the 0-reject classifier with a reject option developed in [8]. These authors base their classification and rejection on single input, while we use three consecutive input samples (footsteps) to make the final decision. When a person walks into the room, it takes less than three seconds to record three footsteps on the floor. This method improved the identification reliability, which is essential in building smart living room scenarios based on a personal profile.

2 The Reject-Optional Learning Vector Quantization

In a complex classification problem, such as footstep identification, it is useful to reject samples that can not be classified reliably to any of the known classes. The aim is to reject

---

* This work was funded by TEKES and Academy of Finland

A. Ferscha and F. Mattern (Eds.): PERVASIVE 2004, LNCS 3001, pp. 182–187, 2004.

the highest possible percentage of samples that would otherwise be misclassified. As a side effect, some correctly classified input samples are rejected, although they would be correctly classified without the reject option. In this section, the term 0-reject classifier is used to describe a classifier without a reject option.

To achieve the criteria for rejection, two different thresholds must be determined: one for detecting samples lying in an overlapping region, and one for the samples significantly different from the trained class boundaries. In our application, the overlap corresponds to a situation where two persons have similar features in their footsteps or the measurements are noisy. If the input sample is far from the class boundaries, a previously unknown (to the system) person is walking on the floor. To achieve the thresholds, reliability evaluators are calculated from the training data. This evaluator is derived from the properties of the 0-reject classifier. In this paper, an LVQ-based reliability evaluator and the main ideas for determining optimal thresholds are presented. More details can be found in [8].

The reject option can be adaptively defined for the given application domain. This is done by assigning a cost coefficient to the misclassified, rejected and correctly classified samples. Optimal thresholds can be computed using an effectiveness function for given cost values. The effectiveness function P is determined in a form

$$P = C_c(R_c - R_c^0) - C_e(R_e - R_e^0) - C_r R_r, \tag{1}$$

where C_c, C_e, and C_r are the costs for correctly classified, incorrectly classified and rejected samples. R_c^0 and R_e^0 are the percentages of correctly and incorrectly classified samples for a given threshold σ. R_c, R_e and R_r present the percentages of correctly classified, misclassified and rejected samples after the introduction of the reject option. The effectiveness function (1) needs to satisfy $C_e > C_r$.

The output vector of a trained LVQ classifier is the distance between the input sample and the closest prototype vector in a codebook, and is named as O_{WIN}. Moreover, a trained LVQ codebook presents the class boundaries of given training set and can be used to compute the values of reliability evaluators. The first reliability evaluator Ψ_a is defined as

$$\Psi_a = \begin{cases} 1 - \frac{O_{WIN}}{O_{max}}, & \text{if } O_{WIN} \leq O_{max} \\ 0, & \text{otherwise}, \end{cases} \tag{2}$$

where O_{max} is the highest value of O_{WIN} in the training set. This evaluator is used to eliminate samples significantly different from the trained codebook vectors. The second reliability evaluator Ψ_b is

$$\Psi_b = 1 - \frac{O_{WIN}}{O_{2WIN}}, \tag{3}$$

where O_{2WIN} is the distance between the input sample and the second winner prototype vector. This criterion is for rejecting the input samples belonging to an overlapping region.

The optimal value of reject threshold $\underline{\sigma}$ is obtained from the training set. The maximum of the effectiveness function can be found from the derivative for $P(\sigma)$ ([8]) as follows,

$$C_N D_e(\sigma) - D_c(\sigma) = 0, \tag{4}$$

where $D_c(\Psi_a)$ and $D_e(\Psi_b)$ are occurrence densities, and $C_N = (C_e - C_r)/(C_r + C_c)$ is normalized cost. The occurrence densities can be estimated using Eq. (2) and Eq. (3) for every training sample.

The following training algorithm presents the determination of the optimal threshold values $\underline{\sigma}_a$ and $\underline{\sigma}_b$ of fixed cost coefficients.

1. The training set is classified with a 0-reject classifier and then split into the subsets S_c of correctly classified samples and the subset S_e of misclassified samples.
2. The values of the reliability evaluators Ψ_a and Ψ_b are determined for each sample in the sets S_c and S_e. Then, the occurrence density functions $D_c(\Psi_a)$, $D_c(\Psi_b)$, $D_e(\Psi_a)$, and $D_e(\Psi_b)$ are calculated.
3. The values of σ_a and σ_b satisfying (4) are calculated.
4. The values of σ_a and σ_b from Step 3 maximizing (1) are chosen as rejection thresholds.

Figure 1 shows the typical occurrence densities for the footstep data. Samples below the given threshold σ_b are rejected.

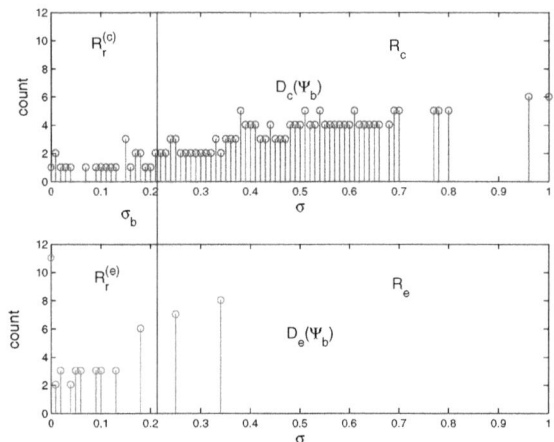

Fig. 1. The example occurrence densities $D_c(\Psi_b)$ and $D_e(\Psi_b)$ for correctly and incorrectly classified footstep data using a reliability evaluator Ψ_b. $R_r^{(c)}$ and $R_r^{(e)}$ are the percentages of rejected samples for a given threshold σ_b. R_c and R_e are the accepted samples of correctly and incorrectly classified samples by the 0-reject classifier

3   Two-level Identification System

By combining decisions of multiple classifiers, we usually get more accurate results compared to the decision of one classifier. The combinations of classifiers are typically multi-level systems where several similar or different classifiers are used to to make a joint decision concerning the input patterns [9]. For example, an application using multiple independent LVQ classifiers to recognize handwritten digits can be found in [10].

In our system, three consecutive footsteps from one person are used in the identification. The architecture of the classifier is shown in Figure 2. The first level consists of two different reliability evaluators, Ψ_a and Ψ_b, as presented in section 2. Level 1 rejects footsteps if Ψ_a is below σ_a, or if Ψ_b is below σ_b. The decision at the second level is based on the knowledge of the classification results at level 1, and the final decisions are defined as

- REJECT
  1. If a majority of the three samples are rejected by Ψ_a.
  2. If one of the three samples is rejected by Ψ_a and another one is rejected by Ψ_b.
  3. If all samples are classified to different classes.
- ACCEPT
  1. If a majority of the samples are classified to the same class.
  2. If two of the samples are rejected by Ψ_b and one is classified to one of the classes.

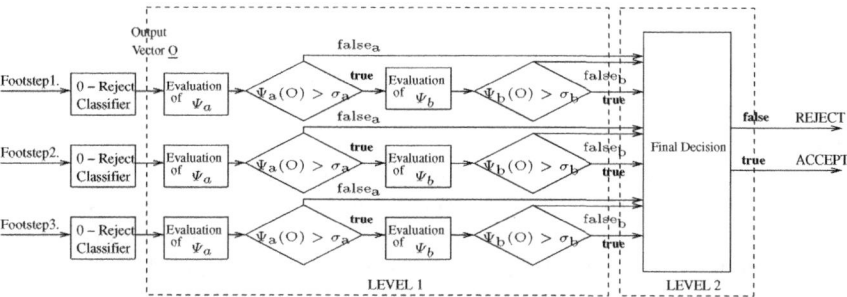

Fig. 2. Two-level identification system, which consists of two different reliability evaluators, Ψ_a and Ψ_b. Level 1 rejects footsteps if Ψ_a is below σ_a, or if Ψ_b is below σ_b. Level 2 rejects or accepts three steps pre-classified at the first level. At the second level, samples can be accepted if a majority of them belong to the same class or if two of them are rejected by σ_b (from the overlapping region) at the level 1

4   Experimental Results

The collected test data consist of the measurements of 11 persons walking on the pressure-sensitive floor, stepping on one particular stripe (containing about 40 footsteps / person). The data collection and processing, as well as the feature selection, are described more detailed in [6]. The rejection threshold determination algorithm and two-level identification system were implemented in the MATLAB technical language.

The two-level classifier system was tested with 10 different, randomly selected groupings for the data. The LVQ codebook consisted of 7 prototype vectors for each class.

Table 1. The results of 10 randomly chosen data sets of eleven persons' footsteps. The cost coefficients of misclassification C_e and rejection C_r were selected as (2,1),(4,3),(6,5),(10,9),(15,14) and (-) (no rejection on the first level), keeping C_c equal to one. The first column presents the cost values of the best results. The second and third column consist of the best total recognition and reject rates of three consecutive footstep. The fourth and fifth columns present total recognition and reject rates using single footsteps, and the last column shows the recognition rates of a single 0-reject LVQ classifier

Test set	(C_e, C_r)	Recog. rate 3 footsteps	Reject rate 3 footsteps	Recog. rate 1 footstep	Reject rate 1 footstep	0-reject classifier 1 footstep
1.	(-)	93.2	14.3	66.4	0.0	66.4
2.	(6,5)	84.8	21.4	67.8	10.7	67.2
3.	(15,14)	87.9	14.3	74.2	22.1	66.4
4.	(15,14)	88.6	9.5	75.7	21.4	70.2
5.	(-)	87.9	16.7	62.6	0.0	62.6
6.	(15,14)	100	21.4	75.4	14.5	70.2
7.	(-)	86.4	11.2	67.2	0.0	67.2
8.	(15,14	90.2	11.9	70.9	1.5	69.5
9.	(6,5)	95.5	30.9	69.2	10.7	68.0
10.	(15,14)	75.0	28.6	59.8	11.5	58.0
Average		89.0%	18.0%	68.9%	9.2%	66.6%

The initial codebook training was similar to [6]. The reject thresholds were determined for each data set using different pairs of cost coefficients for misclassification C_e and rejection C_r.

Different cost coefficients were tested to find the optimal relation between the recognition and reject rates. The relations between C_e and C_r were chosen to be small, as it is essential to keep the percentages of rejection rate quite low. When the relation was raised, it turned out that a majority of the samples were rejected. Using $(C_e, C_r) = (2,1)$, for example, the average reject rate was 65% due to the huge overlap between the different classes.

The best results from these 10 randomly generated data sets are shown in Table 1. The results show that the two-level reject-optional LVQ for three footsteps was able to improve identification compared to the single footstep classification. The average rate shows 89.0% recognition by the two-level classifier with a 18.0% reject rate, while the single-step classifier gives only 68.9% and 66.6% average rates with and without a reject option, respectively. The most general cost coefficients C_e and C_r of the best results were 15 and 14. The typical rejection thresholds for using these costs were $\sigma_a = 0.5$ and $\sigma_b = 0.06$.

5 Conclusions

In this paper, experiments on identifying persons based on their successive footsteps on an EMFi floor were reported. A two-level identification system was developed for decision-making. The system utilizes a reject option for single footsteps and makes a decision on identification based on three consecutive footsteps. Footsteps can be recorded when a person enters the smart living room within three seconds at best.

The reject option and the two-level classifier were able to raise the recognition rates compared to the single-footstep classification. The overall success rate was 89%, and the rejection rate was 18%. It is essential to make the identification as reliable as possible, as the room is supposed to react to the users identity and presence. The number of testees in this experiment was eleven, but in a smart living room scenario is considered, probably fewer occupants are about to use the space. Then, the overall recognition rate might be even better.

Naturally, this research aims at real-time learning and identification. The reject option provides a possibility to enable adaptiveness. The reason for rejecting an identification can be determined based on the algorithm. The thresholds calculated (σ_a, σ_b) explicitly indicate if the person is unknown to the system, or if the rejection is based on noisy measurements. If identification is rejected based on σ_a, the system can interpret the occupant as a new person and start retraining the classifier automatically based on the new footsteps.

References

1. I. A. Essa. Ubiquitous sensing for smart and aware environments: Technologies towards the building of an aware home. *IEEE Personal Communications*, October 2000. Special issue on networking the physical world.
2. M.D. Addlesee, A. Jones, F. Livesey, and F. Samaria. ORL active floor. *IEEE Personal Communications*, 4(5):35–41, October 1997.
3. R.J. Orr and G.D. Abowd. The smart floor: A mechanism for natural user identification and tracking. In *Proc. 2000 Conf. Human Factors in Computing Systems (CHI 2000)*, New York, 2000. ACM Press.
4. S. Pirttikangas, J. Suutala, J. Riekki, and J. Röning. Footstep identification from pressure signals using Hidden Markov Models. In *Proc. Finnish Signal Processing Symposium (FIN-SIG'03)*, pages 124–128, May 2003.
5. T. Kohonen. *Self-Organizing Maps*. Springer-Verlag Berlin Heidelberg New York, 1997.
6. S. Pirttikangas, J. Suutala, J. Riekki, and J. Röning. Learning vector quantization in footstep identification. In M.H. Hamza, editor, *Proc. 3rd IASTED International Conference on Artificial Intelligence and Applications (AIA 2003)*, pages 413–417, Benalmadena, Spain, September 8-10 2003. IASTED, ACTA Press.
7. M. Paajanen, J. Lekkala, and K. Kirjavainen. Electromechanical film (EMFi) - a new multi-purpose electret material. *Sensors and actuators A*, 84(1-2), August 2000.
8. C. De Stefano, C. Sansone, and M. Vento. To reject or not to reject: That is the question - an answer in case of neural classifiers. *IEEE Transactions on Systems, Man, and Cybernetics-part c: Applications and Reviews*, 30(1):84–94, February 2000.
9. J. Kittler, M. Hatef, R.P.W. Duin, and J. Matas. On combining classifiers. *IEEE Transactions on Pattern Analysis and Machine Intelligence*, 20(3):226–239, March 1998.
10. T. K. Ho. Recognition of handwritten digits by combining independent learning vector quantization. In *Proceedings of the Second International Conference on Document Analysis and Recognition*, pages 818–821, October 1993.

Issues with RFID Usage in Ubiquitous Computing Applications

Christian Floerkemeier and Matthias Lampe

Institute for Pervasive Computing
Department of Computer Science
ETH Zurich, Switzerland
{floerkem|lampe}@inf.ethz.ch

Abstract. Radio Frequency Identification (RFID) has recently received a lot of attention as an augmentation technology in the ubiquitous computing domain. In this paper we present various sources of error in passive RFID systems, which can make the reliable operation of RFID augmented applications a challenge. To illustrate these sources of error, we equipped playing cards with RFID tags and measured the performance of the RFID system during the different stages of a typical card game. The paper also shows how appropriate system design can help to deal with the imperfections associated with RFID.

1 Introduction

In his famous article Mark Weiser describes a vision of ubiquitous computing in which technology is seamlessly integrated into the environment and provides useful services to humans in their everyday lives [8]. The potential of Radio Frequency Identification (RFID) tags to contribute to the realization of this vision has been demonstrated by many researchers over the years. Examples include prototypes such as the Magic Medicine Cabinet by Wan [6], the augmentation of desktop items by Want et al. [7] and smart shelves by Decker et al. [1]. These prototypes show that RFID technology has many benefits over other identification technologies because it does not require line-of-sight alignment, multiple tags can be identified almost simultaneously, and the tags do not destroy the integrity or aesthetics of the original object. Due to the low cost of passive RFID tags and the fact that they operate without a battery, there are, however, also some weaknesses associated with RFID-based object identification. Particularly in a multi-tag and multi-reader configuration, the phenomenon of false negative reads occurs, where a tag that is present is not detected.

The goal of this paper is to illustrate the problem of failed RFID reads with the help of a sample application, identify potential causes of the failed reads, and suggest ways to deal with the resulting uncertainty from an application and system perspective. In particular, we address the problems of collisions on the air interface and tag detuning.

The following section gives an introduction to RFID system components and Section 3 looks at the various causes of false negative reads. Section 4 suggests

A. Ferscha and F. Mattern (Eds.): PERVASIVE 2004, LNCS 3001, pp. 188–193, 2004.

ways to deal with the uncertainty caused by failed RFID reads from an application and system design perspective. Section 5 provides a conclusion.

2 RFID Primer

RFID systems consist of two main components: the RFID tag, which is attached to the object to be identified and serves as the data carrier, and the RFID reader, which can read from and sometimes also write data to the tag. Tags typically consist of a microchip that stores data and a coupling element, such as a coiled antenna, used to communicate via radio frequency communication. The readers usually consist of a radio frequency module, a control unit, and a coupling element to interrogate the tags via radio frequency communication. There is a wide variety of RFID systems available. The reader is referred to the book by Finkenzeller [2] for an in-depth classification of RFID systems.

The experiments presented in this paper use the Philips I-CODE System[1], which operates at 13.56 MHz and which is based on the ISO 15693 standard for RFID. The I-CODE tags obtain their power from the magnetic field generated by the reader through inductive coupling. The magnetic field induces a current in the coupling element of the smart label, which provides the microchip with power. The inductively coupled RFID system consequently behaves much like loosely coupled transformers. The ISO 15693 protocol employs a variant of slotted Aloha for access to the shared communication medium, known as framed Aloha [5].

3 Failed RFID Tag Reads and Their Causes

Failure to detect tags that are present in the read range of a reader can be due to a variety of causes including collisions on the air interface, tag detuning, tag misalignment, and metal and water in the vicinity of the RFID system. To illustrate the failed tag reads caused by some of these phenomena we equipped playing cards with RFID tags similar to prior work by Römer [3]. Although other scenarios could have been used to demonstrate some of the challenges involved in the use of RFID, we believe that the playing card scenario is an appropriate example because it aptly demonstrates the most common causes of failed tag reads. The pictures in Figure 1 show the RFID tags on the back of the playing cards with the RFID antenna of the I-CODE System in the background.

To illustrate the different causes of false negative reads, we carried out measurements in a number of configurations that typically occur during a card game. It is evident from Figure 1 and Figure 2 that on average we do not detect all of the 10 tags present in a single frame in any of the arrangements considered and that the arrangement of the cards with respect to the reader and with respect to each other has a strong influence on the read performance. In the following subsections we address the individual causes of the failed reads.

---

[1] http://www.semiconductors.philips.com/markets/identification/products/icode/

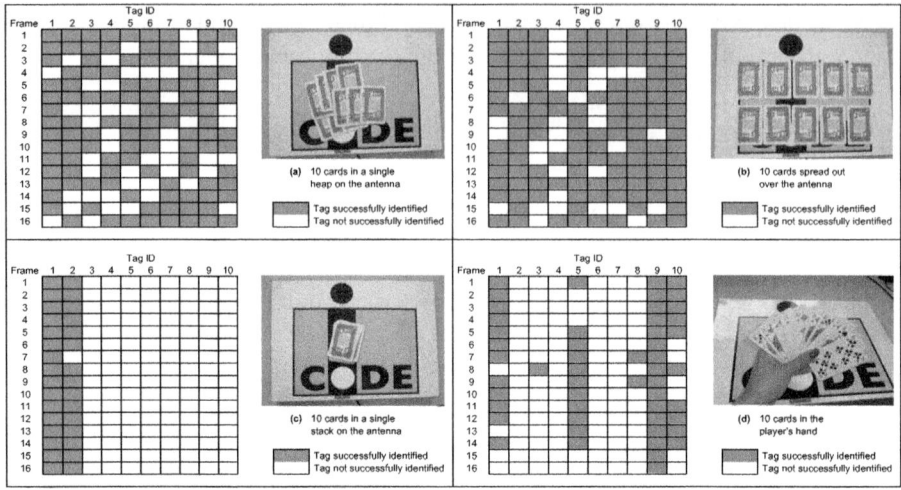

Fig. 1. Four different arrangements of 10 playing cards equipped with RFID tags: (a) in a heap on the antenna, (b) spread out over the antenna, (c) stacked on top of each other, and (d) in the player's hands. The patterns to the left of the images show a snapshot of the data captured by the reader. A dark field indicates a successful detection of a tag in a frame; a light field indicates a failed detection. These measurements were carried out with 32 time slots per frame and a frame rate of 5 Hz.

3.1   Tag Collisions

As mentioned earlier, the system used for our tests employs an anti-collision algorithm based on framed Aloha. In most circumstances, tags transmitting their ID in the same time slot cannot be detected. Exceptions to this rule are due to the capture effect [9], where the reader manages to detect the data sent by one of the tags correctly, although multiple tags respond in the same time slot.

In a stochastic anti-collision algorithm there is hence always a chance that a tag is not detected for at least the duration of a single frame, if more than a single tag is present. Obviously, the probability of collisions decreases with the number of available time slots (see Figure 2) and increases with the number of electronic tags present. Wieselthier et al. [9] developed expressions for the probability of successful transmission in framed Aloha with no capture. We used their analysis to calculate the expected number of tags detected per frame (see bars labelled "prediction" in Figure 2). The slight discrepancy between the prediction and the observed data for the two scenarios that perform well is believed to be due to the capture effect, which is not considered in our estimate. The relatively good match between the prediction and the observation indicates that all negative reads are due to collisions on the air interface in these cases.

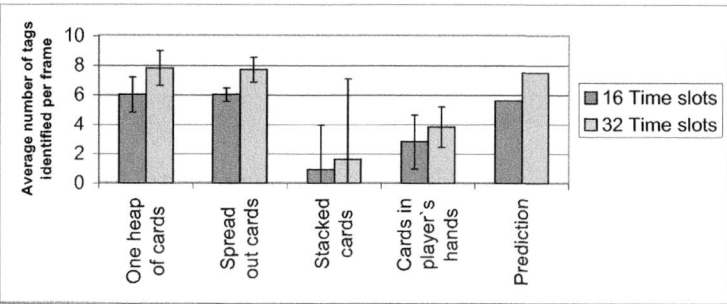

Fig. 2. Average number of tags identified per frame vs. the different card arrangement for frames with 16 and 32 time slots. The prediction bars represent the expected number of tags successfully identified out of the 10 tags present. This calculation is based on the analysis of framed Aloha with no capture.

In the next subsection, we explain why the detection rate for cards organized in a single stack and cards held in the player's hand is much lower than one would expect from the tag collision analysis.

3.2 Tag Detuning

In inductively coupled RFID systems the voltage induced in the antenna coil of the tag by the magnetic field is used to power the microchip. Finkenzeller [2] describes how tag manufacturers create a parallel resonance circuit by adding a capacitor in parallel to the antenna coil so that the resonance frequency of the resonance circuit is tuned to the operating frequency of the RFID system. At resonance, the induced voltage produced across the tuned tag will thus be significantly enhanced compared to frequencies outside the resonant bandwidth resulting in an increased read range.

As a resonant application, the tag is, however, vulnerable to environmental detuning effects which can also cause a significant reduction in reading distance. Clusters of RFID tags in close proximity to each other, for example, exhibit significant detuning effects caused by their mutual inductances. Undesirable changes in the tag's parasitic capacitance and effective inductance can also be caused by metal and different dielectric mediums in the vicinity, e.g. a hand holding the tag [2]. The shift in resonance frequency away from the operating frequency results in the tag receiving less energy from the reader field and hence a decrease in reading distance.

Tag detuning due to other tags in very close proximity is thus also the cause for the low read rates we witnessed for the cards organized in a stack (see Figure 2 for details). The relatively low read rate in the player's hand is believed to be a result of the combination of tag detuning by the player's hand, by the proximity of the tags, and due to the increased distance from the antenna.

3.3   Other Sources of Error

Other causes of failed reads include the presence of metal in the tag vicinity, since it distorts the magnetic flux, thus weakening the energy coupling to the tag. If tags are directly attached to a metal surface, they can often not be detected at all. Similar to tag detuning, metal in the vicinity of the reader antenna results in a read range reduction because the antenna is detuned. For example, in our experiments we witnessed a strong read range variation when we placed the reader antenna on a table that was supported by a metal frame.

The misalignment of the tags with the magnetic field of the reader coil can also lead to failed reads. Maximum power transfer occurs when the tag coil plane is perpendicular to the magnetic field lines. As the label is rotated with respect to the field lines, the coupling is reduced until the tag is no longer detected.

4   Implications for System and Application Design

To reduce the uncertainty that arises from the tag collisions, one could opt for RFID systems in which the time it takes to detect the tags is reduced. Since regulations on the 915 MHz band offer significantly more bandwidth in the communication from the reader to the tag than do the regulations on the 13.56 MHz band, an RFID system operating in the UHF band can detect tags much faster. The large bandwidth under US regulations also permits the use of deterministic anti-collision algorithms such as variants of the binary tree-walking scheme, where the reader traverses a tree of all possible identification numbers [4]. Unfortunately, the large bandwidth in the UHF band is currently not available worldwide, although there are proposals to increase the allocated bandwidth in Europe [2]. There are also RFID systems that exhibit a superior performance at 13.56 MHz, but they rely on more expensive tag and reader designs, e.g. involving the use of multiple frequency channels for the tag to reader communication.

To reduce the tag detuning that occurs when the tags are placed in a stack, we experimented with smaller tags that we placed in random locations on the playing cards. This reduced the tag detuning significantly, but the read range was also decreased due to the smaller labels. Alternatively, there are also specialized tags and readers available that are tuned for stack reading. Redundant tags placed at different orientations on the object to be identified are especially effective at reducing failed reads caused by a misalignment of the tag with the magnetic field of the reader antenna, as at least one of the tags should always be correctly aligned.

From an application design perspective, it is possible to use certain application-specific constraints. This might include group constraints, where a certain group of tags is known to always move together. The presence of a tag that was not detected can now be inferred by the detection of another tag that belongs to the same group. This group constraint would be particularly useful in cases where low cost RFID tags are used in conjunction with tags that are less likely to be affected by the issues presented above, e.g. active tags that are battery powered. Alternatively, the use of additional identification means that augment the RFID system could be considered, e.g. computer vision techniques or weighing scales.

The most radical approach to improve the performance of an RFID supported application is to modify the application itself so that a minimum number of tags are present in the read range simultaneously and the tags which are present are far apart from each other. The tag antennas should also be aligned with the magnetic field of the reader antenna, and metal and water should ideally not be present in the vicinity of the system. Unfortunately, this approach usually requires the active involvement of the user, e.g. by telling him not to place the cards in a stack. It also significantly reduces the appeal of RFID technology as a non-obtrusive identification technology.

5 Conclusion

Radio Frequency Identification (RFID) technology is known to be well-suited to linking the physical and virtual world. Using playing cards augmented with RFID tags as an example, we highlighted some of the issues that arise when multiple tags are present in the read range simultaneously. The paper also addresses the implications of those weaknesses from an application and system design perspective.

References

1. Christian Decker, Uwe Kubach, and Michael Beigl. Revealing the retail black box by interaction sensing. In *Proceedings of the ICDCS 2003, Providence, Rhode Island,* 2003.
2. Klaus Finkenzeller. *RFID Handbook: Radio-Frequency Identification Fundamentals and Applications.* John Wiley & Sons, 2000.
3. Kay Römer and Svetlana Domnitcheva. Smart playing cards: A ubiquitous computing game. *Journal for Personal and Ubiquitous Computing (PUC),* 6, 2002.
4. Sanjay E. Sarma, Stephen A. Weis, and Daniel W. Engels. RFID Systems and Security and Privacy Implications. In *Workshop on Cryptographic Hardware and Embedded Systems,* pages 454–470. Lecture Notes in Computer Science, 2002.
5. H. Vogt. Efficient object identification with passive RFID tags. In F. Mattern and M. Naghshineh, editors, *International Conference on Pervasive Computing,* volume 2414 of *Lecture Notes in Computer Science,* pages 98–113, Zurich, August 2002. Springer-Verlag.
6. D. Wan. Magic medicine cabinet: A situated portal for consumer healthcare. In *Proceedings of the International Symposium on Handheld and Ubiquitous Computing, Karlsruhe, Germany,* 1999.
7. R. Want, K.O. Fishkin, A. Gujar, and B.L. Harrison. Bridging physical and virtual worlds with electronic tags. In *Proc. of ACM SIGCHI,* pages 370–377, May 1999.
8. M. Weiser. The computer of the 21st century. *Scientific American,* pages 94–100, September 1991.
9. Jeffrey E. Wieselthier, Anthony Ephremides, and Larry A. Michaels. An exact analysis and performance evaluation of framed aloha with capture. *IEEE Transactions on Communications,* COM-37(2):125–137, 1989.

A Fault-Tolerant Key-Distribution Scheme for Securing Wireless Ad Hoc Networks

Arno Wacker, Timo Heiber, Holger Cermann, and Pedro José Marrón

Institute for Parallel and Distributed Systems (IPVS)
Universität Stuttgart, Stuttgart, Germany
{wacker,heiber,cermanhr,marron}@informatik.uni-stuttgart.de

Abstract. We propose a novel solution for securing wireless ad-hoc networks. Our goal is to provide secure key exchange in the presence of device failures and denial-of-service attacks. The proposed solution relies solely on symmetric cryptography and therefore is applicable for highly resource-limited devices. In order to avoid a single point of trust, no master device or base station is used. We achieve this by enhancing our previously published approach with redundancy and algorithms for recovery on device failures.

1   Introduction

As the vision of Pervasive Computing becomes reality, many daily life devices will have computational power and wireless communication capabilities that allow the formation of ad-hoc networks. One scenario for such wireless ad-hoc networks is home automation. Here a private home is equipped with a multitude of sensors and actuators to enhance the lifestyle of individuals. For instance, the heating is turned on automatically when the owner of the house comes home; the light is switched on in rooms where motion is detected, etc. Security is a crucial factor for such systems as they introduce many new ways to invade an individual's personal life. For example, a thief could gather information about when somebody is at home before breaking into the house. Encryption is an elementary technique for securing communications. Encryption schemes, however, require keys to be exchanged before secret communications can take place. Our goal is to provide a secure key exchange scheme for wireless ad-hoc networks. As we show below, this is a challenging task in such environments. Our approach is suitable for resource-limited devices like sensors. Devices can exchange keys without referring to a central authority, thus avoiding a single point of trust. Unique keys are exchanged between device-pairs providing authenticity. Even when a device is subverted by an attacker, the key exchange for the remainder of the network remains functional. In [1] we have presented a key-distribution scheme that guarantees the secrecy of a key exchange as long as there are less than s subverted devices, where s can be chosen according to the actual security requirements. Device failures and denial-of-service attacks are not handled by this approach. In this paper, we extend our previous work to additionally cope with less than r device failures or denial-of-service attacks. This parameter can also be chosen according to the actual requirements.

The remainder of this paper is organized as follows: In Sec. 2 we describe the system model. Our previously published key-distribution scheme is introduced in Sec. 3. A first

A. Ferscha and F. Mattern (Eds.): PERVASIVE 2004, LNCS 3001, pp. 194–212, 2004.

extension to this scheme, based on redundancy, is presented in Sec. 4. Thereafter in Sec. 5, we incrementally design a scheme able to recover from multiple device failures even in the presence of attackers. Finally, we give an overview of existing approaches in wireless ad hoc and sensor networks in Sec. 6 and conclude the paper with a short summary and future work.

2   System Model

Our network consists of independent devices, each with its own processor and memory, communicating over a wireless channel. The channel itself is insecure, i.e. anyone can listen and send to the channel. We assume a non-partitioned network, that is, communication between two arbitrary devices is always possible[1]. The number of devices is not predetermined or constrained in any way, as it may change due to the introduction of new devices to the network or device failures. The devices of such a network have to be inexpensive, and therefore they will only have limited resources. We assume that an attacker will subvert a number of devices since this is hard to prevent [2]. Considering these properties, we require that:

– The key distribution scheme must be decentralized — it still must be functional even when some devices are subverted.
– Symmetric cryptography is used in order to deal with resource limited devices — on such devices asymmetric cryptography is problematic [3].

We assume the existence of the following three types of attackers:

1. The *eavesdropping attacker (Eve)*. This attacker is only interested in learning about secrets of other devices, e.g. newly established keys. The objective of the attacker is to eavesdrop on communication between devices.
2. The *fail-stop denial-of-service attacker (fsDoS)*. This attacker silently halts all functions of a device. This class includes device failures (e.g. due to low power).
3. The *byzantine denial-of-service attacker (bDoS)*. This attacker may act arbitrary in order to prevent the correct execution of the key-distribution protocols. This class of attacker subsumes the fsDoS attacker class.

We have thoroughly analyzed Eve in [1]. In this work we analyze the other two types of attackers and provide countermeasures. Note that we do not consider DoS on the physical layer (e.g. jamming the frequency).

3   Basic Key-Distribution Scheme

The work presented in this paper is based on the scheme proposed and analyzed in [1]. We refer to this scheme as the basic key-distribution scheme (bKDS). In this section, we give a short overview of bKDS. Its overall objective is to establish a shared key between

---

[1] See Sec. 4.3 for further discussion of this assumption

each pair of devices in the network. bKDS is robust against eavesdropping (Eve), but not against device failures (fsDoS) or denial-of-service (bDoS).

In order to set up the wireless ad-hoc network (and introduce new devices to it), bKDS requires an initial set of keys to be exchanged between devices via a secure channel. Following [4], this can e.g. be achieved by using physical contact between two devices to establish a unique shared key. However, it is impractical to establish physical contact between each pair of devices in the network, and the number of keys is limited by the memory of the devices. Hence, it is necessary to limit the number of physically exchanged keys and establish additional shared keys between devices on demand.

In the next subsection we define the formal representation of our network, which will be used throughout the rest of this paper. Subsections 3.2 through 3.4 describe the original algorithms while their properties and restrictions are discussed in 3.5.

3.1   Network Model

We represent our network as an undirected graph $G = (V, E)$, where V is the set of devices in the network, and E represents the set of shared keys between devices where $\{v_1, v_2\} \in E$ iff the nodes v_1 and v_2 share a symmetric key. We will use the term *device* to indicate the physical device and the term *node* to indicate the representation of that device in the graph.

3.2   Network Setup

Whenever a new device is added to the network, a certain number of keys needs to be exchanged. We require that each device that is added to the network shares a securely exchanged key with at least s devices that are already part of the network and refer to s as the *security level* of the network. The actual procedure is given in Alg. 1, using the formal representation of the network.

A graph constructed with Alg. 1 with less than $(s + 1)$ nodes will be fully connected. For each additional node introduced, s new edges from the new node to previously existing nodes will be added. This algorithm guarantees the existence of s node-disjoint paths between two arbitrary nodes in the graph. Proof of this and further properties of the algorithm can be found in [1]. The security of the scheme presented in the next subsection is dependent on this property.

3.3   Establishing a New Shared Key

Two devices that do not share a key yet can establish a new one using the following approach: To establish an l-bit key k, a device randomly generates s l-bit shares $k_1, \ldots k_s$, and sends them over s device-disjoint paths (i.e. paths that do not share common devices) to the destination device (Fig. 1(a)). On each link of a path, the key share is encrypted with the existing shared key for this link. The final key k is then calculated by $k = k_1 \oplus k_2 \oplus \ldots \oplus k_s$, where $\oplus$ is the bitwise XOR operation. This approach is also used in [5,6,7].

It should be clear that without having access to all key shares, an Eve-type attacker does not stand a chance to recover the key. If it can be assured that the key shares

---

Algorithm 1 Introducing a new node to the graph

1: Given a graph $G = (V, E)$ with $n = |V|$ with nodes $v_i \in V$ and a new node v_{n+1}
2: $V := V \cup \{v_{n+1}\}$;
3: **if** $s \geq n$ **then**
4:     // device corresponding to v_{n+1} creates n new keys
5:     **for** $i = 1$ to n **do**
6:         $E = E \cup \{v_{n+1}, v_i\}$; // establish a new key through physical contact
7:     **end for**
8: **else**
9:     $V' :=$ random subset of $V - \{v_{n+1}\}$ with $|V'| = s$;
10:     // device corresponding to v_{n+1} creates s new keys
11:     **for** $i = 1$ to s **do**
12:         $E = E \cup \{v_{n+1}, v_i\}$ with $v_i \in V'$; // establish a new key through physical contact
13:     **end for**
14: **end if**

---

are communicated over s node-disjoint paths of the network graph, Eve will need to subvert at least s nodes (one on each path) to compromise the newly established key. Our construction per Alg. 1 guarantees this property.

3.4   Controlled Removal of Devices

Removing a device from the network can destroy the s-connected property of the corresponding graph. In [1], we also proposed an algorithm for removing a device in a controlled shutdown, i.e. a device has the time to announce its impending departure from the network and make all necessary arrangements. Algorithms 2 and 3 describe this procedure for removing a device from the network.

The presented solution is based on pretending that the device that is to be removed had never been there in the first place and replacing existing shared keys accordingly. If the device is still present when this procedure is performed, keys can be replaced

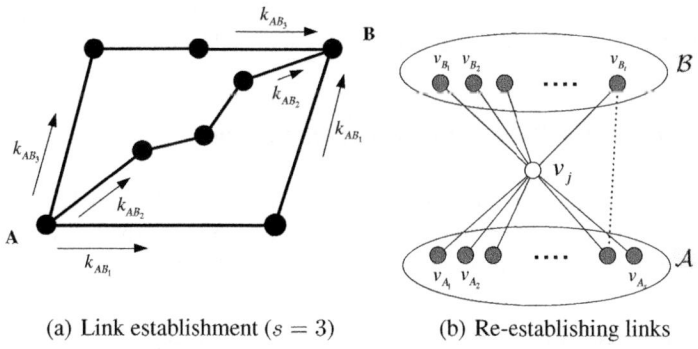

(a) Link establishment ($s = 3$)   (b) Re-establishing links

Fig. 1.

Algorithm 2 Controlled removal from the network

1: *FirstDevices* := first s known devices (set $\mathcal{A}$);
2: *MyDeviceList* := the set of devices for which we have a shared key;
3: **for all** *Device* $\in$ *MyDeviceList* **do**
4:     sendto(*Device*, LeavingIntention{*FirstDevices*});
5: **end for**
6: wait for all ACKs;

Algorithm 3 Action of a neighboring device

1: onReceiveLeavingIntention(*Sender, DeviceList*);
2: *MyDeviceList* := the set of devices for which we have a shared key;
3: *Candidates* := *DeviceList* − *MyDeviceList*;
4: **if** *Candidates* $\neq \emptyset$ **then**
5:     *Device* := pick randomly one from *Candidates*;
6:     establish a new shared key with *Device*;
7:     update *MyDeviceList* by **replacing** *Sender* with *Device*;
8: **else**
9:     *MyDeviceList* := *MyDeviceList* − {*Sender*};
10: **end if**
11: sendto(*Sender*,"ACK");

---

automatically, using the procedure described in Sec. 3.3, thus re-establishing the s-connected property of the underlying graph.

To see why these algorithms work, let the graph under consideration be $G = (V, E)$ with $V = \{v_1, v_2, \ldots v_n\}$ where a node v_i was the i-th node added to the graph according to our construction. Let the node that is to be removed from the graph be v_j. Then we can define two sets, $\mathcal{A}$ and $\mathcal{B}$ (see Fig. 1(b)). The set $\mathcal{A}$ contains all s nodes to which v_j had its first s edges during the construction, i.e. the devices, v_j established the first s keys (*FirstDevices* in Alg. 2). The set $\mathcal{B}$ contains all nodes which established an edge to v_j during their insertion into the graph, i.e. the devices being inserted *after* v_j. When v_j leaves the network, all nodes in set $\mathcal{B}$ need to establish a new edge to some node in set $\mathcal{A}$ which did not exist before. Doing this will result in the same situation as if the nodes in set $\mathcal{B}$ were introduced into the graph when node v_j was not a node of the graph. Thus, the s-connected property of the graph is again guaranteed by the construction algorithm.

3.5   Properties of the Basic Key-Distribution Scheme

bKDS guarantees the *secrecy* of a newly established key as long as Eve controls less than s devices. However, the s-connected property and therefore new key establishment can only be guaranteed in the absence of DoS attacker devices:

A device subverted by a fsDoS attacker will not execute any algorithms anymore. Thus it will not perform the procedure for controlled removal, which might destroy the s-connected property of the underlying graph.

A device subverted by a bDoS attacker will interact with the protocols with the deliberate intention to destroy the s-connected property, i.e. preventing the establishment of new keys. Imagine the following situations:

1. *During key establishment:* The device might forward an altered key-share (see Sec. 3.3). As a result, the two participating devices will not share a common key. Thus, establishing a new key cannot be guaranteed anymore.

2. *Initiating the procedure for controlled removal:* The device might initiate the removal protocol, but send a false list of *FirstDevices* to its neighbors. As a result, the neighbors would make connections to the wrong devices, possibly destroying the s-connected property.

3. *Responding to the procedure for controlled removal:* The device might decide not to respond in a correct manner when the device gets a "LeavingIntention" message. This also might result in the destruction of the s-connected property of the underlying graph.

In order to become robust against fsDoS or even bDoS, these issues have to be addressed. Therefore in the remainder of this paper, we enhance bKDS with redundancy and recovery mechanisms to cope with these attacker types.

4   Basic Redundancy for Coping with DoS Devices

A direct solution for dealing with DoS is the introduction of redundancy: In order to deal with r bDoS attacker devices we need to establish a $(s + r)$-connected graph. We refer to r as the *redundancy level* of our network. Thus, between any pair of nodes, there exist $(s + r)$ node-disjoint paths, of which s paths are needed in order to establish a new key.

4.1   Properties of the Basic Redundancy Approach

Having a $(s + r)$-connected graph, provides room to tolerate up to r bDoS devices. Consider the following situations where bDoS attacker devices can interfere:

Attacks on the key establishment protocol. In order to establish a new key, a device chooses s paths (out of a total of $(s + r)$) to the destination and invokes the key establishment protocol as described in Sec. 3.3. After the key establishment has succeeded, both parties have to check whether they really share the same key. This can be done with a challenge/response protocol using the newly established key. In case the key-establishment fails (or the key is invalid), the device chooses another set of s paths to the destination device, until it eventually establishes a functional key.

Attacks with respect to Alg. 2. When a device removes itself, it is expected to execute Alg. 2. Three cases are possible:

1. The leaving device has failed (e.g. low power), or it has been subverted by a DoS attacker, i.e. it is part of the fsDoS class. In this case it will not initiate the removal protocol. Thus the network will not be repaired, i.e. the $(s + r)$-connected property will not be restored, leaving the graph only $(s + r - 1)$-connected in the worst case. For an analysis see Sec. 4.2.

2. The device is subverted by a bDoS attacker and therefore tries to prevent new key establishments in the network. It does so by providing a false list of devices (*FirstDevices*) when sending its "LeavingIntention" (see Alg. 2). This leads to a false reconstruction of the graph, which also in worst case decreases the connectivity of the graph by one.

3. The leaving device is supposed to wait until all neighboring devices have acknowledged (see Alg. 2) before it can eventually leave the network. Due to failure of a neighboring device, however, this message may never be received. In order not to wait forever, we must introduce a timeout mechanism. For the underlying graph, this means that one of the neighbors did not execute the removal protocol properly, therefore the original connectivity of the graph cannot be guaranteed anymore. However, also this situation would as a result decrease the connectivity of the graph by one.

Attacks with respect to Alg. 3. When a device receives a "LeavingIntention" from a neighboring device, it must act in order to repair the $(s+r)$-connected property of the graph. It must (if necessary) establish new keys in order to bypass the leaving device and when done, send an acknowledgement to the leaving device. Two situations are possible:

1. The device does not send the necessary acknowledgement at the end of Alg. 3, since it is subverted by an fsDoS or even bDoS attacker. When subverted by a bDoS attacker it may establish its new keys to arbitrarily chosen devices. This situation is already described above leading to the decrease of the connectivity by one.

2. The device is subverted by a bDoS attacker, and therefore does not repair the network as it should (i.e. establish keys to a device from the *FirstDevices* list), but sends the acknowledgement message anyway. Also here, in the worst case, the connectivity of the graph is reduced by one.

Therefore we can conclude that given an $(s + r)$-connected network, the network will stay at least s-connected as long as the attacker (fsDoS or bDoS) does not subvert more than r devices.

4.2 Analysis

From a formal point of view, all attacks described in the previous section can be divided into two categories:

1. The removal of a node, without "repairing" this section of the graph, i.e. constructing new edges in order to bypass the removed node.

2. The removal of a node while constructing additional edges to some arbitrary nodes, which do not bypass the removed node.

Clearly, the additional construction of edges in a k-connected graph cannot decrease the connectivity. In the worst case, however, the removal of nodes can decrease the connectivity by the number of removed nodes. In order to show this we need the following definition from graph theory:

Definition 1 (k-connected graph). *A graph $G = (V, E)$ is said to be k-connected iff for any set $W \subseteq V$ with $|W| < k$, the subgraph induced by $V - W$ is still connected.*

Theorem 1. *In a k-connected graph $G = (V, E)$, the removal of t nodes (and all corresponding edges), will result in a graph, which will be at least $(k - t)$-connected.*

Proof. By contradiction. Assume that the graph G' induced by $V - Y$ with $Y \subseteq V$ and $|Y| = t$ is not at least $(k - t)$-connected. Then it suffices to remove h nodes from G', with $h < (k - t)$ in order to disconnect G'. But this means we disconnected G by removing $t + h$ nodes with $t + h < k$, which is a contradiction to G being k-connected since the removal of less than k nodes cannot disconnect G.

In this section, we have seen that an $(s + r)$-connected graph can tolerate up to r bDoS devices while retaining a security level of s. However, with the Basic Redundancy approach the connectivity of the graph will monotonously decrease with every additional devices being in fsDoS or bDoS. In Sec. 5 we will present an approach that allows the graph to recover from device failures, i.e. devices from the fsDoS class.

4.3 Coping with Denial of Service at the Routing Layer

Up to now, we have assumed that devices can always communicate with each other. However, dealing with denial of service attacks also requires dealing with attacks on the communication system.

As mentioned earlier, countermeasures against attacks on the physical layer are beyond the scope of this paper. What remains to be considered, though, are routing attacks, i.e. attacks on layer 3 of the ISO/OSI model.

If all devices are in transmission range of each other, our protocols can be built directly on layer 2 of the ISO/OSI model. There is no need for routing, so the problem is solved trivially.

If routing is used for communication, we need to consider the *communication graph* of the network. The communication graph is a graph whose nodes represent the devices and whose edges describe a direct communication link between two devices. In order to fully prevent denial-of-service with up to r devices exhibiting bDoS behavior, we need to build an $(r + 1)$-connected communication graph. Similar to Alg. 1, it is sufficient if every device is in the communication range of at least $(r + 1)$ other devices when adding it to the network. With an $(r + 1)$-connected graph and an appropriate routing protocol (e.g. flooding), a message will always arrive at the destination.

For the rest of this paper we assume that we either have a $(r + 1)$-connected communication graph or that the devices are in range of each other — thus, as stated in Sec. 2, communication between two arbitrary devices is always guaranteed.

5 Recovery from Device Failure

While the number of attacker devices is a parameter which depends on the security aspects of the networks, the number of probable failures (fsDoS) is not as easy to predetermine. Therefore we propose an enhancement which enables the network to recover

---

Algorithm 4 Detecting a failed device

1: onDeviceFailureDetect(*FailedDevice*);
2: **if** *FailedDevice* ∈ *FirstDevices* **then**
3:    // we are in set $\mathcal{B}$, therefore we must act
4:    broadcast(FailingNotification(*FailedDevice*));
5: **end if**

Algorithm 5 Checking upon a *FailingNotification*

1: onReceiveFailingNotification(*Sender, FailedDevice*);
2: **if** LinkPartnerKeyPosition(*FailedDevice*) $\leq (s + r)$ **and** FailedDevice not responding **then**
3:    // we are in *FirstDevices* (set $\mathcal{A}$) of *FailedDevice* and the device really failed
4:    broadcast(HealingNotification(*FailedDevice*));
5: **end if**

---

from fsDoS attacks, thus making the initially needed connectivity independent of the number of probable device failures.

We introduce a new parameter c which holds the maximum number of device failures (fsDoS) occurring *concurrently* while the recovery algorithm can still recover the connectivity of the underlying graph.

We are going to build our final algorithms incrementally. In Sec. 5.1, we show the basic principle for dealing with device failures. We then modify this approach in Sec. 5.2 to rule out some DoS attacks. Our final algorithm, which handles DoS and also multiple failures at the same time, is presented in 5.3.

5.1 Basic Approach for Recovery from Device Failures

When a device fails, it is not able to perform the controlled removal protocol (as described in Sec. 3.4). Even worse, in the basic removal approach, the departing device still has to be in place while executing the protocol, since the underlying graph does not provide any redundancy (it is exactly s-connected).

We will now introduce the *passive* removal protocol. This protocol is based on the controlled protocol, but in contrast it is initiated by a neighboring device. Note that by having an underlying $(s + r)$-connected graph, any removal protocol can be executed without the aid of the device which is leaving or has just failed. The basic idea behind the passive removal protocol is, that if some device in the network detects that another device has failed, it tries to recover this part of the underlying graph. In order to do so, it first has to gather the information about which nodes are in the set $\mathcal{A}$ (see Sec. 3.4) of the failed device. It does so by broadcasting a request (Alg. 4) in order to get a answer from any device being in the *FirstDevices* set of the *FailedDevice* (Alg. 5). After receiving a "HealingNotification" from any device, which it does *not* share a key with, it can create a new key to this device to substitute for the failed one. Since all messages are sent using broadcasts, other devices — having the failed device in their *FirstDevices* set — can also make use of the sent "HealingNotification", even if they did not send a "FailingNotification" themselves. Algorithms 4, 5 and 6 describe our approach in detail.

Algorithm 6 Action of a healing device

1: onReceiveHealingNotification(*Sender, FailedDevice*);
2: **if** *FailedDevice* not responding **and** *Sender* ∈ *FirstDevices* **then**
3:    establish a new shared key with *Sender*;
4:    update *MyDeviceList* by **replacing** *Sender* with *FailedDevice*;
5: **end if**

There is still an open question in the above algorithm: How does a device know if it belongs to the set *FirstDevices* of another device? Clearly, asking the device in question is not possible, since it failed and therefore will not be able to respond to any requests. This information can be saved together with a new key, when integrating a device into the network. Keys are stored in the order in which they where established, making it easy to determine the set *FirstDevices* for a certain device. Additionally, in order to determine whether a device is among the set *FirstDevices* of another device, for each key we store the position of this key at the other device. For device x this information is *LinkPartnerKeyPosition(x)*. When this position changes, also the remote stored information has to be updated. Note that this still scales with respect to the network size — it just increases the needed storage for a key on the device by a constant amount. The basic key distribution scales with on average $2s$ keys per device.

To see why these algorithms work, consider the following: In 3.4 we stated that on failure of a node v_j any node $v_{B_i} \in B$ has to replace its missing edge to v_j with an edge to a node $v_{A_i} \in A$ (see Fig. 1(b)). In the above algorithms, a node in the set B will detect the failure of the corresponding device and will therefore send a "FailingNotification". All nodes $v_{A_i} \in A$ will receive and answer to this request with a "HealingNotification", thereby giving all nodes $v_{B_i} \in B$ the information needed to replace the missing edge.

As the above algorithms are based on the controlled removal protocol, they provide the same properties, i.e. after a device failure it is ensured that the underlying graph has still the same connectivity as before. However, this will not work under all circumstances:

If a bDoS device is present the following can happen: In case of device failure, some other device will detect this failure and per Alg. 4 will ask for the set of *FirstDevices* (set A) of the failed device. The attacker, who wants to prevent the network from establishing new keys, will announce himself as part of this set. In this case, other devices will establish links to the attacker in order to repair the network, possibly decreasing the connectivity of the underlying graph by one. Therefore doing this $r+1$ times (on $r+1$ device failures), the connectivity of the underlying graph may drop below s, although there was just a single bDoS attacker device present.

Multiple failures at the same time also pose a problem: Imagine two devices failing, where one device is in the *FirstDevices* set of the other. Not all required devices (those of set A) will respond to a "FailingNotification" and therefore it cannot be guaranteed that every node from set B will find a suitable node from set A to replace the "dead" key. Note that a fundamental property for the correctness of the protocol is the fact that the size of set A is equal to the connectivity of G — see [1] for details.

There are several more possibilities for a single DoS attacker device. All these attacks are possible due to the fact that the complete set *FirstDevices* is only known to one device and are lost if this device fails.

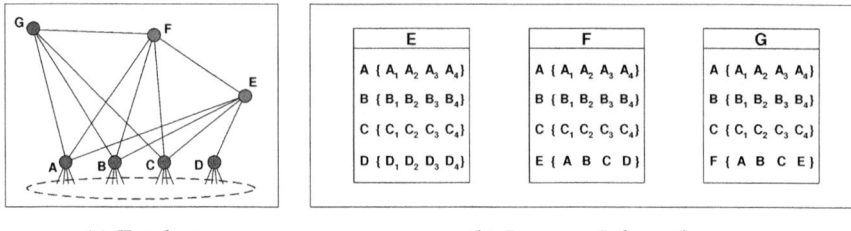

(a) Topology   (b) Recovery Information

Fig. 2. Recovery Information

5.2 Recovery with Replicated Data in Set B

As pointed out in the last section, the major weakness in algorithms 4-6 is that the *FirstDevices* set of the failed device is not known to other devices. Therefore, an attacker can forge this information. To overcome this, we need to replicate this information to all devices that may need it. By Sec. 3.4 these are the nodes which are in the set B of a node. As an example consider Fig. 2. The *FirstDevices* set of device A is $\{A_1, A_2, A_3, A_4\}$ (analogously for B, C, D) — in Fig. 2(a) the area below A, B, C and D. Fig. 2(b) shows the information stored on nodes E, F and G. For instance, node E is in the set B of device A, B, C and D, therefore it stores the *FirstDevices* sets for these nodes (first column). The *FirstDevices* set of node E is $\{A, B, C, D\}$. This information is replicated only on node F since F is in the set B of node E (column 2, line 4).

The replication of this list is done during the initialization phase of a device. Whenever a new device is introduced into the network, a number of keys are shared with other devices. Immediately after these keys are shared, the device knows its own *FirstDevices* set, and can replicate this information to every *new* device in *MyDeviceList* — this corresponds to the set B of the underlying graph.

Having stored this information, every device can locally execute the function get-FirstDeviceList(*Device*), which delivers the corresponding list for each device in its own *FirstDevices* set. That way, failures can be repaired based only on local information.

Note that by executing Alg. 7 on every device, there is no need for additional communication anymore — Alg. 7 replaces the original algorithms for passive removal. The

---

Algorithm 7 Action on a device failure on every node

1: onDeviceFailureDetect(*FailedDevice*);
2: *FailedDeviceFirstDevices* := getFirstDeviceList(*FailedDevice*)
3: *Candidates* := *FailedDeviceFirstDevices* − *MyDeviceList*;
4: **if** *Candidates* ≠ ∅ **then**
5:     *Device* := pick randomly one from *Candidates*;
6:     establish a new shared key with *Device*;
7:     update *MyDeviceList* by **replacing** *FailedDevice* with *Device*;
8: **end if**

---

controlled removal algorithm can still be used as an optimization but due to the additional data, an attacker device cannot cheat about any list sent.

Properties of the Approach with Replication in set $\mathcal{B}$. Our modified approach uses additional memory on each device. For a $z := s + r$-connected graph, we need on average $2z \cdot s_k$ bytes memory storage for the keys on each device, where s_k is the size of a key in bytes. On top of that, the following amount of storage is needed for the *FirstDevices* sets: Each set contains z device ids. If s_i is the size of a device id in bytes, we need $z \cdot s_i$ bytes for a single set, of which z sets are stored per device. Thus the additional memory requirements for storing the replicated lists are $z^2 \cdot s_i$ bytes per device. The total required memory per device is $2z \cdot s_k + z^2 \cdot s_i$ bytes, which is not dependent on the network size n. This means that it scales with respect to the network size n, but the memory overhead with respect to the security parameters is $\mathcal{O}((s+r)^2)$.

Alg. 7 prevents some of the DoS attacks described in Sec. 5.1 The reasons why this algorithm can recover from non-simultaneous failures remain the same as in Sec. 5.1. The problem of multiple device failures, however, is still unresolved: Alg. 7 cannot recover when multiple devices which share a key fail simultaneously. The reasons are the same as in Sec. 5.1. Therefore, although this approach prevents some DoS attacks, it is still vulnerable and does not tolerate arbitrary device failures.

5.3 Recovery with Replicated Data in Set $\mathcal{A}$

In the last section we have proposed to replicate the *FirstDevices* set on all devices which are in the set $\mathcal{B}$ of a particular node. We showed that this does not prevent all weaknesses described in 5.1. The main reason is that the size of set $\mathcal{B}$ is not predetermined in any way. Thus we instead propose to replicate the *FirstDevices* set of a device on all devices which are part of the *FirstDevices* set, i.e. in set $\mathcal{A}$. This set always contains z elements when the underlying graph is z-connected.

In this scheme we set $r := d + c + m$ with

$$m := \begin{cases} d + 1 - s : s < d + 1 \\ 0 \quad\quad : otherwise \end{cases}, \tag{1}$$

where c stands for the maximum number of *concurrent* device failures and d for the maximum tolerated number of bDoS attacker devices. Thus the graph G will be z-connected with

$$z := s + r = \begin{cases} s + d + c : s < d + 1 \\ 2d + c + 1 : otherwise \end{cases}. \tag{2}$$

Since $s + d + c \geq 2d + c + 1$, G will be at least $(2d + c + 1)$-connected. The basic idea behind this approach is that having the *FirstDevices* set replicated $(2d + c + 1)$-times, even when there are d subverted devices, acting as bDoS attacker devices and additionally c concurrent failures in the network, there are still a majority of $(d + 1)$ devices with the replicated information available, which will cooperate and therefore enable the algorithms to proceed.

Algorithm 8 Detecting a failed device

1: onDeviceFailureDetect(*FailedDevice*);
2: **if** *FailedDevice* ∈ *FirstDevices* **then**
3:     // we are in set B of the failed device
4:     *FirstFailedDevice* := *FailedDevice*; // remember the device we started the recovery with
5:     **execute** recoverDeviceFailure(*FailedDevice*);
6: **end if**

Algorithm 9 Broadcasting the Recovery Request

1: **function** recoverDeviceFailure(*FailedDevice*);
2: *DeviceToRecover* := *FailedDevice*;
3: *VoterSet* := ∅; // empty set
4: *FirstDevicesLists* := []; // empty list
5: broadcast(RecoveryRequest(*FailedDevice*));

Thus, when building an z-connected graph as above, we tolerate s eavesdropping devices, d bDoS attacker devices and an arbitrary number of failures, of which at most c failures can occur concurrently.

A device failure will eventually be detected by a neighboring device. This device will request the *FirstDevices* set with a broadcast. A device which holds a replica of this set will first establish a new key with the requesting device (if necessary) and then send the set to the requesting device over this secure channel. Note that with at most d bDoS attacker devices (and c concurrent failures), in the worst case the requesting device will still receive $(2d+c+1)-c-d = d+1$ sets from non-subverted devices. When $(d+1)$ times the same *FirstDevices* set has been submitted by different senders, the receiving device can be sure that this is the correct set and use it for determining the device to which it must create a new key in order to recover from the failure. If this device is not available either (i.e. it has also failed or it is a DoS attacker device), the scheme is applied recursively. Algorithms 8 through 11 describe this behavior in detail.

When a device detects the failure of another device belonging to its *FirstDevices* set, it saves the corresponding device id in the variable *FirstFailedDevice* (see Alg. 8). Then the recovery is started by executing the function recoverDeviceFailure() (Alg. 9). This function will be called recursively in case of multiple device failures.

Consider Fig. 3. The parameters used in this example are $s = 2$, $d = 1$ and $c = 1$, which results in $r = 2$.

In Fig. 3 devices F and E have just failed simultaneously. Note that since bDoS behaviour subsumes fsDoS behaviour, it is possible to tolerate these failures. The failure of F is eventually detected by G, which needs to replace the key to F with a key to some device in the *FirstDevices* set of F. In order to gather this information, it broadcasts the request for *FirstDevices* of F (Fig. 3(a)).

A device receiving "RecoveryRequest(*FailedDevice*)" must first check if it belongs to the *FirstDevices* set of the failed device. This is done using the function LinkPartnerKeyPosition introduced in Sec. 5.1. Note that the information needed for LinkPartnerKeyPosition can be gathered from the locally replicated *FirstDevices* sets — there is no need to store additional key positions as proposed in Sec. 5.1.

Algorithm 10 Actions upon a receiving a RecoveryRequest

1: onReceiveRecoveryRequest(*Sender*, *FailedDevice*);
2: **if** LinkPartnerKeyPosition(*FailedDevice*) $\leq (s + r)$ **then**
3:     // we are in set $\mathcal{A}$ of *FailedDevice*, therefore we hold a replica of the *FirstDevices*-set
4:     **if** *Sender* $\notin$ *MyDeviceList* **then**
5:         // we don't share a key with the requesting device, thus we have to establish one
6:         establish a new shared key with *Sender*;
7:         // send the replicated *FirstDevices* of *FailedDevice* to the sender
8:         sendto(*Sender*, RecoveryAnswer(*FailedDevice*,*FirstDevices*(*FailedDevice*)));
9:         release shared key with *Sender*;
10:     **else**
11:         // send the replicated *FirstDevices* of *FailedDevice* to the sender
12:         sendto(*Sender*, RecoveryAnswer(*FailedDevice*,*FirstDevices*(*FailedDevice*)));
13:     **end if**
14: **end if**

In the next step, every device in the *FirstDevices* set of the failed device will send the local copy of this set to the requesting device. This is done over a secure connection, establishing a new key with the requesting device if necessary. Fig. 3(b) shows devices A, B and C sending this information to G — device E should also send this information, but since it failed it will not do so. Device G collects the responses, i.e. 3 times the *FirstDevices* set of device F: $\{A, B, C, E\}$. A response is counted only once per responding device.

Device G, having received the same set at least twice ($d + 1 = 2$), tries to recover from the failure of device F by establishing a new key to device E (Fig. 3(c)). However, E has also failed, thus the key establishment will fail. Device G will therefore recursively try to recover from the failure of E by broadcasting the request for *FirstDevices* of E (Fig. 3(d)). In this case devices A, B, C and D will answer, providing G with the needed information (Fig. 3(e)). After the second identical *FirstDevices* set ($d + 1 = 2$), G can be confident that it has the correct set, namely $\{A, B, C, D\}$, and will choose D — the only one with which G does not share a key yet. In order to recover the connectivity of the underlying graph, device G will now replace the key for device F with the newly established one for device D (Fig. 3(f)).

Properties of the Approach. Just as in the previous approach, additional memory is required. The additional memory requirements are the same as in Sec. 5.2, since we also replicate the *FirstDevices* set z times, with $z := s + r$. However, during the execution of the recovery, additional temporary memory is needed to store the received sets. A maximum of z additional sets will be received, thus the temporary overhead is $z \cdot z \cdot s_i = z^2 s_i$. This approach again is independent of the network size n, but has quadratic overhead with respect to the security parameters s and r.

This final approach, however, can cope with up to c concurrent device failures and uncooperative devices, i.e. DoS attackers. Building the graph with a redundancy of $r = c + d$ gives room to tolerate d attacker devices and c concurrent failures. This is due to the fact that even in the worst case, with d DoS attackers and c failing devices, there will be $d + 1$ devices left which will cooperate. After a successful execution of the

---

Algorithm 11 Actions due to requested answers

1: onReceiveRecoveryAnswer(*Sender, FailedDevice,FailedDeviceFirstDevices*);
2: **if** *FailedDevice = DeviceToRecover* **and** *Sender ∉ VoterSet* **then**
3:      // first answer from this sender
4:      *VoterSet := VoterSet ∪ Sender*; // just one answer per sender
5:      *FirstDevicesLists := FirstDevicesLists +(FailedDeviceFirstDevices)*;
6:      *Candidates :=* select a *FirstDevicesLists*-entry, if included at least $(d + 1)$ times, else $\emptyset$;
7:      *Candidates := Candidates - FirstDevices*;
8:      **if** *Candidates ≠ $\emptyset$* **then**
9:          *Device :=* pick randomly one from *Candidates*;
10:         **if** establish a new shared key with *Device* **then**
11:             update *MyDeviceList* by **replacing** *FirstFailedDevice* with *Device*;
12:             *DeviceToRecover :=* none;
13:         **else**
14:             **execute** recoverDeviceFailure(*Device*);
15:         **end if**
16:     **end if**
17: **end if**

---

recovery algorithms, the connectivity of the graph will be restored, i.e. the graph will again be $(s + r)$-connected. To see why this approach works, consider the possibilities a DoS attacker device has:

Denying to forward key-shares or send altered key-shares. This cannot harm, since there are always redundant paths available.

Refusing to establish a new key with some device. The attacker device will be considered as a failed device, and therefore removed from the network. Due to the recovery mechanism, the connectivity will be restored without this device. Thus only the attacker device itself is affected.

Incorrectly initiating a recovery process. Three cases are possible:

1. The attacker device initiates the recovery process for a non-existing device. On receiving the broadcast every device will locally determine that it is not part of the corresponding *FirstDevices* set. Therefore no further action will be taken.
2. The attacker device initiates the recovery process for an existing device which is not part of its *FirstDevices* set. With this action the attacker gains information about the network topology, i.e. the *FirstDevices* set of the announced Failed-Device and it causes some unnecessary communication. The connectivity of the underlying graph will not be decreased by such messages, since every device has to initiate the recovery process on its own.
3. The attacker device initiates the recovery process, but forges its own id — this is possible since group communication (i.e. broadcast) is not authenticated. This attack yields the same communication overhead as described before. Devices receiving answers (the real devices to which the forged id belongs) will discard these messages, as long as they did not broadcast the same request.

Incorrectly responding to a recovery request. There are two cases possible: The responding device might not answer at all, or it replies with a bogus *FirstDevices* set.

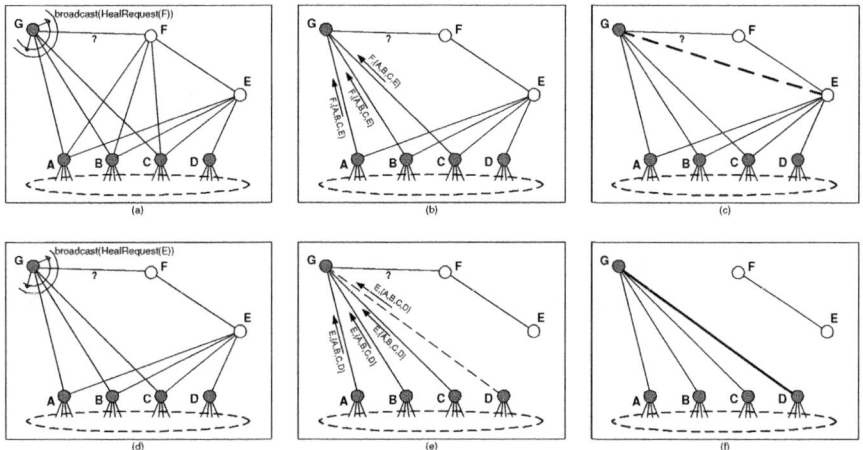

Fig. 3. Recovery from multiple device failures, ($s = 2, r = 2$)

Both cases will not do harm since there will be enough correct answers (at least $d + 1$).

Incorrect processing of recovery messages. Upon a failure, an attacker device might decide to announce this failure correctly, thus gathering the needed information about the *FirstDevices* set, but behave arbitrarily afterwards. This decreases the connectivity of the graph by at most one. However, this happens per attacker device and not per failed device and is therefore covered by the introduced redundancy.

The extensions to our algorithms guarantee the secrecy of the newly established keys, as long as there are less than s eavesdropping attackers in the network, and also guarantee the key establishment functionality of the network as long as there are at most r DoS attacker devices present. Devices belonging to the fsDoS class are fully covered in these algorithms, i.e. the number of fsDoS devices is only limited by the network size n and such failures can always be recovered from as long as there are less than c simultaneous failures. In general, in order to deal with g attackers of *any kind* while recovering up to c simultaneous failures, we have $s := g + 1$ and $d := g$, hence we need to initially establish a z-connected graph with $z := 2d + c + 1 = 2g + c + 1$ (since $s \not< d + 1$, see (2) in Sec. 5.3).

6   Related Work

A number of solutions for key-distribution in wireless ad hoc or sensor networks have been proposed. Most of these approaches do not address the issue of easy addition or removal of devices.

The straightforward solution is to use certificates issued by a central certification authority. A central server thereby stores and signs the public keys of the individual devices. The major problems with this approach are the limited resources of the devices and the need for the central (potentially vulnerable) authority itself.

Since a centralized authoritative device is a single point of failure, asymmetric decentralized approaches have been proposed. Such designs can be achieved by distributing the certification authority over the participating devices [8,6].

Asymmetric approaches share a common problem: If small sensor devices are used — even with the use of supporting high performance nodes [9] — asymmetric cryptography is often not possible due to delay and energy constraints [10,3].

Perrig et al. [11] proposed SPINS, a centralized approach using symmetric cryptography. They address device failures due to energy loss or destruction. If a device fails, the base station will remove all references on other devices. The establishment of new session keys is done solely by the base station. Therefore, device failure cannot affect this functionality or communication. However, the failure of the base station will disable the whole key-distribution. Also, the base station is a single point of trust, thus the hostile take over of the base station will subvert the whole network at once.

To avoid these problems, a few approaches using symmetric cryptography without the use of a base station have been proposed. Secure Pebblenets [12] represent a simple approach proposed by Basagni et al. using only one key — known by all devices — for the whole network. A link creation is not required since all messages are encrypted with the same key, independent of the receiver. As a drawback, no device to device authenticity can be achieved, and additionally, the tampering of one single device reveals the data sent by all devices to the attacker.

Eschenauer and Gligor [13] proposed a solution using a random key predistribution. Before deployment, every device is supplied with a probabilistic set of keys from a key pool. After deployment, the devices try to establish connections by finding a commonly shared key or by creating a new key through a secure path including other devices. Since their approach is probabilistic, no clear assumptions about network connectivity can be made afterwards. After the initial network establishment, every device creates a secure link to all of his neighbors in communication range, so the connectivity of the network will not be threatened by single device failures. The removal of subverted devices is handled by a centralized revocation scheme using a trusted base station.

Extensions have been introduced by Chan et al. [5]. They present three different approaches. In one of their approaches, a set of secure and authenticated links can be established after deployment of the sensors. Due to the random predistribution, two arbitrary devices might not be able to establish a secure link without relying on other devices, since only some *randomly chosen* nodes can communicate directly and authenticated with each other. Due to this fact, real authenticated communication between arbitrary devices is not always possible. A device failure decreases the probability that two devices may communicate.

The fourth decentralized symmetric approach by Zhu et al. [14] also uses a initially distributed set of random keys. Additionally to [5], Zhu et al. propose a pairwise key establishment protocol using *multiple* paths. This way, splitting of a pairwise key over multiple untrusted paths, as proposed in [7], can be used to improve attacker resistance. Due to the random key predistribution, the actual existence of different paths in the network is not assured in any way. It follows that — in case of device failures — no presumption about the real number of the device disjunct paths is possible.

Previous work on fault-tolerance has been done for instance in [15]. The focus of this work is the interplay of network connectivity and secure communication in a general way. The issue of easy addition or removal of devices is not within the scope of these approaches.

7 Conclusion and Future Work

In this paper, we presented a fault-tolerant approach for key distribution in wireless ad-hoc networks based on symmetric cryptography. In contrast to the related work, we have proposed a parametrized algorithm which *guarantees* the ability for an arbitrary pair of devices to exchange a key in a secure fashion, provided that the number of subverted devices remains below certain threshold parameters. We proposed a recovery algorithm which ensures the key-exchange functionality in case of device failures.

We have implemented our approach on the Atmel ATMega16 microcontroller. This implementation will be used to conduct a performance evaluation of our approach.

Furthermore, we are working on new mechanisms to discover the device-disjoint paths. Additional work includes key revocation schemes combined with intrusion detection mechanisms.

References

1. Wacker, A., Heiber, T., Cermann, H.: A key-distribution scheme for wireless home automation networks. In: Proceedings of IEEE CCNC 2004, Las Vegas, Nevada, USA, IEEE Communications Society, IEEE (2004)
2. Anderson, R., Kuhn, M.: Tamper resistance - a cautionary note. In: Proceedings of the Second Usenix Workshop on Electronic Commerce. (1996) 1–11
3. Brown, M., Cheung, D.: PGP in constrained wireless devices. In: Proceedings of the 9th USENIX Security Symposium. (2000)
4. Stajano, F., Anderson, R.: The resurrecting duckling: Security issues for ad-hoc wireless networks. In: 7th International Workshop on Security Protocols. LNCS (1999) 172–194
5. Chan, H., Perrig, A., Song, D.: Random key predistribution schemes for sensor networks. In: IEEE Symposium on Security and Privacy. (2003)
6. Zhou, L., Haas, Z.J.: Securing ad hoc networks. IEEE Network 13 (1999) 24–30
7. Gong, L.: Increasing availability and security of an authentication service. IEEE Journal on Selected Areas in Communications 11 (1993) 657–662
8. Hubaux, J.P., Buttyan, L., Capkun, S.: The quest for security in mobile ad hoc networks. In: Proceeding of the ACM Symposium on Mobile Ad Hoc Networking and Computing (MobiHOC). (2001) 146–155
9. Modadugu, N., Boneh, D., Kim, M.: Generating RSA keys on a handheld using an untrusted server. In: Cryptographer's Track RSA Conference. (2000)
10. Carman, D., Kruus, P., Matt, B.: Constraints and approaches for distributed sensor network security. Technical Report #00-010, NAI Labs (2000)
11. Perrig, A., Szewczyk, R., Wen, V., Culler, D.E., Tygar, J.D.: SPINS: Security protocols for sensor networks. In: Mobile Computing and Networking. (2001) 189–199

12. Basagni, S., Herrin, K., Bruschi, D., Rosti, E.: Secure pebblenets. In: Proceedings of the ACM Symposium on Mobile Ad Hoc Networking and Computing. (2001) 156–163
13. Eschenauer, L., Gligor, V.D.: A key-management scheme for distributed sensor networks. In: Proceedings of the 9th ACM Conference on Computer and Communication Security (CCS-02). (2002) 41–47
14. Zhu, S., Xu, S., Setia, S., Jajodia, S.: Establishing pair-wise keys for secure communication in ad hoc networks: A probabilistic approach. Technical Report ISE-TR-03-01, George Mason University (2003)
15. Dolev, D., Dwork, C., Waarts, O., Yung, M.: Perfectly secure message transmission. J. ACM **40** (1993) 17–47

ProxNet: Secure Dynamic Wireless Connection by Proximity Sensing

Jun Rekimoto, Takashi Miyaki, and Michimune Kohno

Interaction Laboratory, Sony Computer Science Laboratories, Inc.
3-14-13 Higashigotanda, Shinagawa-ku, Tokyo 141-0022 Japan
http://www.csl.sony.co.jp/person/rekimoto.html

Abstract. This paper describes a method for establishing ad hoc and infrastructure-mode wireless network connections based on physical proximity. Users can easily establish secure wireless connections between two digital devices by putting them in close proximity to each other and pressing the connection button. The devices "identify" each other by measuring each other's signal strength. We designed a set of protocols to support secure connections between digital devices by using a proximity communication mode to exchange session keys. We also introduce a "dummy point" that is analogous to a wireless access point but handles proximity-mode communication. The dummy point represents physical locations of digital devices and supports context-sensitive network communications.

1 Introduction

Digital appliances based on wireless networks are becoming increasingly popular, and the operation of such appliances differs from the operation of conventional devices based on wired networks. Users of wireless appliances frequently change network connections depending on their needs and typically should be able to perform the following operations:

- exchange digital pictures between two digital cameras by dynamically creating an ad hoc wireless connection between the two cameras. They should be able to see the screens of both devices to interactively decide which pictures to copy from one device to the other. They also need to transmit files securely so unauthorized users cannot intercept their files.
- connect a digital device to the Internet in a public place in order to share files with other users. This file sharing should be protected from other users at the same hotspot.
- use wireless mobile game devices to dynamically establish ad hoc wireless connections in order to play games on the network.
- dynamically connect a PDA to a projector (for example, during presentations) without knowing the IP address and login password of the projector.
- upload pictures from a cellular phone to a public wall display.

To do these things, users must be able to establish and change network connections quickly. We therefore need to

(1) develop a method to identify target devices. Users should be able to identify devices they want to connect to without explicitly specifying them by IP addresses or other symbolic references.

A. Ferscha and F. Mattern (Eds.): PERVASIVE 2004, LNCS 3001, pp. 213–218, 2004.

(2) provide communication security. Established wireless connections must be secure in the sense that users should be able to connect to the correct devices and that communication on these connections should be protected to prevent the interception of files by a third party.

While traditional networking largely relies on specifying destination devices by (IP) addresses and using passwords for authentication, the management of nearby wireless devices can be based on the physical relationship between them. The establishment of network connections, for example, can be based on the proximity of devices. Devices can also be connected via a wireless connection by pointing to a target device (e.g., a TV set) with a mobile device (e.g., a PDA). In both cases, once a wireless connection has been established, device location is no longer restricted (i.e., the devices do not have to remain in proximity and the user does not need to keep pointing at the target device).

The sensors and user operations required in some previously developed systems based on this idea are outlined in Figure 1 [7,6]. These systems establish wireless connections by using communication devices with a limited range, such as an RFID tag/reader, or directional communication methods such as infrared beaming. These devices and methods, however, are not always suitable for all mobile devices. In this paper we describe a simpler method that uses standard wireless cards and establishes wireless connections by measuring the signal strength of received packets.

Sensor Type	User Operation	Required Sensors / Devices
RFID[7]'@	put devices in proximity	RFID tag / reader
Infrared[7]	pointing to device A with device B	infrared transmitter-receiver
SyncTap[6]	Synchronous user operations	buttons etc.
	(e.g., synchronized pressing and releasing of buttons)	

Fig. 1. User operations and sensors for establishing wireless connections between devices.

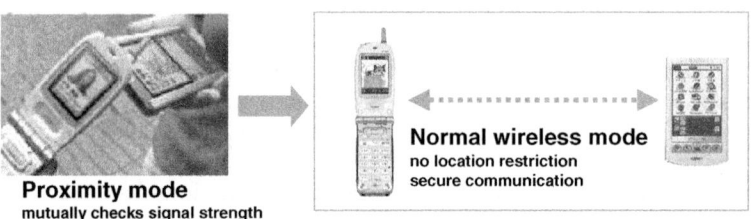

Fig. 2. Establishing a wireless connection by measuring the signal strength

2 ProxNet: Establishing Wireless Connections Based on Proximity

Figure 2 shows the idea behind our proposed method. Unlike other location-sensing methods based on wireless signal strength [1], our method determines only the relative distance between devices. The proximity of two devices can be determined because the strength of a signal one receives from the other is inversely proportional of the square of the distance between them.

Our method has two communication modes: a "proximity mode" in which both devices check the signal strength of incoming wireless packets, and a normal wireless mode. When users want to establish a network connection between two wireless devices, they put these devices close to each other and press the "connection" button on one of these devices. The first device then transmits a wireless packet containing the initiator's address and a dynamically generated public key. The second device receives this packet, and it acknowledges receipt of the packet only when the signal strength of this packet is greater than a predefined threshold. The acknowledging packet contains the responder's address and a one-time session key that will be used to establish a wireless connection. Because this acknowledging packet is encrypted by the public key sent from the first device, only the first device can access the information in it. The first device then checks the signal strength of the acknowledging packet and establishes a normal wireless connection if it is greater than the threshold..

We implemented this method by using an Intersil PRISM2 802.11b wireless chip with our modified version of the "Host AP" Linux wireless driver [4]. The Host AP device driver supports a "monitor" mode that allows applications to monitor wireless packets transmitted by other devices. This mode also supports the monitoring of signal and noise levels (RSSI information) on each packet. While the original monitor mode is receive-only and does not support packet transmission, our modified driver allows the monitor mode to both receive and transmit packets. This modified driver makes it possible for devices listen to transmissions from other devices, recognize connection requests, check signal strength, and then use the normal wireless mode.

Because this device driver modification does not require any hardware changes, it can potentially be used with a variety of wireless devices without requiring special hardware or sensors. It should also be possible to apply this method to wireless networks other than 802.11, such as Bluetooth.

3   Communication Protocols

The actual protocols are shown in Figure 3. There are two modes for establishing network connections: the ad hoc mode, in which the devices establish wireless connections themselves, and the infrastructure mode, in which the devices communicate through an access point.

Ad hoc Mode. To establish an ad hoc mode wireless connection, two devices must share the same network identifier (e.g., ESSID) and a session key (e.g., WEP key). This information is exchanged in the proximity mode (Figure 3-a). In this mode, the signal strength of the received packets is measured on both sides. Once the information needed for establishing an ad hoc connection has been exchanged, the two devices start normal wireless communication. The WEP key and ESSID are dynamically generated for each session, and users do not need to manually enter them.

Infrastructure Mode and the "Dummy point". An operation similar to the one in the ad hoc mode could be used in the infrastructure mode: users could put mobile devices close to an access point and exchange the necessary information. This operation is not always practical, however, because access points are often located in physically inaccessible places (e.g., near the ceiling).

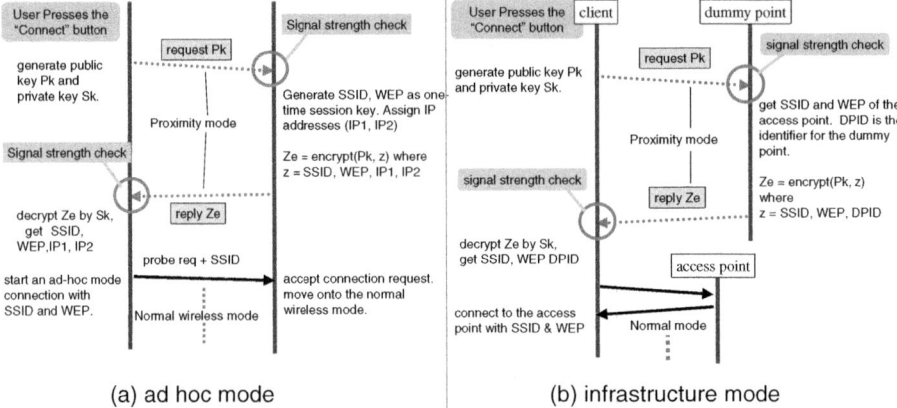

(a) ad hoc mode   (b) infrastructure mode

Fig. 3. ProxNet protocols for ad hoc and infrastructure-mode wireless connections.

We therefore introduce a "dummy point," which is a wireless device that supports only proximity-based communication. For example, a dummy point can be placed on a meeting table to enable users to connect their mobile devices (e.g., wireless notebook PCs) to the access point in that room. Users do that by putting their mobile devices close to the dummy point, and the dummy point responds to the request packets from the mobile devices by sending information needed to establish an infrastructure-mode wireless connection (e.g., the ESSID and the WEP key for the access point). Then the mobile devices connect to the access point (Figure 3-b).

The physical form of the dummy point can vary, and a hand-held dummy point (Figure 4, right) can be used. At an Internet cafe, for example, a manager can put a dummy point like this close to a notebook PC in order to "invite" it into the cafe's wireless environment.

The dummy point can also support context-sensitive communication. We can place more than one dummy point in any environment, and each dummy point can represent a location or a corresponding object. When a user starts a wireless connection by using a dummy point placed next to a printer, for example, the default output device would be automatically set to that printer. Similarly, a wall-sized public display can have a corresponding dummy point. People can upload data by establishing a wireless connection with the corresponding dummy point.

Fig. 4. Dummy point examples: (left: dummy-point modules based on a single board Linux. right: a handheld dummy point)

4   Applications

Sharing files by using an ad hoc wireless connection (Figure 5, left). Two mobile devices start an ad hoc- mode network connection and exchange files. Users first establish a connection by putting two devices close to each other and pressing the connection button on one of them. Once the connection has been established, users do not need to stay close to each other. By sharing the screens of their devices, users can interactively decide which files to transmit. This cannot be done in file transmission using physical media such as memory cards.

Setting up a network game session (Figure 5, middle). A network game can be set up by putting two devices close to each other. Users can also create connections dynamically during a game session. For example, a user can start a game in a single-user mode but then share this game with other players on the network.

Transferring information between mobile devices and public wall displays (Figure 5, right). People working in the same office can share information by uploading it to a public wall-mounted display. They approach this display and put their mobile devices close to the dummy point attached to it. Then they can upload digital picture files onto the public display or get files from the display. Similarly, users making a presentation can display presentation materials from a mobile device by first approaching a dummy point placed near the presentation screen. This method can also be used with digital devices without user interfaces. A file server or a network router, for example, can be controlled from mobile devices in this way.

5   Related Work

Pick-and-Drop [5] and mediaBlocks [10] are systems using physical manipulation to support digital data transfer among various devices. While these systems were designed primarily for "one-shot" data transfer using physical objects, our method supports continuous wireless network connections between devices. Our method also supports secure communication.

Several other systems use range-limited communication channels, such as those provided by infrared beaming or used by radio-frequency identification (RFID) tag reader/writes, for associating wireless devices [9,7,8]. Our method uses wireless communication in two ways – one is for normal wireless communication and the other is for sensing signal strength (i.e., for measuring a range) .

Fig. 5. Applications of ProxNet networking.

The measuring of signal strength has been used for network security management in various applications, such as wireless keys for automobiles or login keys for computers. In these applications, however, only the signal strength of the key is measured. Our method supports a two-way measuring of signal strength for better security.

Another way to recognize spontaneous device associations is by using synchronicity [3,6,2]. For example, if two devices were tied together and moved together, the movements of both devices would be quite similar. Devices with acceleration sensors would be able to exchange acceleration information to find matching devices [3]. Similarly, if a user performs synchronized inputs, such as pressing buttons on two devices, this information can be used for finding device pairs [6]. These synchronicity approaches and our proximity-sensing approach are complementary and can be used in combination. For example, the selection of candidate devices to be associated can first be based on signal strength, and then an actual pair can be determined by synchronicity.

6   Conclusion

We developed a method for establishing secure wireless connections between mobile devices by measuring the strength of signals between them. The method supports ad hoc and infrastructure modes, and the same operations can be performed in both modes. This method uses normal 802.11 wireless chipsets without additional hardware and is applicable to a wide range of wireless devices.

References

1. P. Bahl and V. Padmanabhan. RADAR: An In-Building RFBasedUser Location and Tracking System. In *IEEE Computer and Communications Societies (INFOCOM 2000)*, pp. 775–784, 2000.
2. Ken Hinckley. Synchronous gestures for multiple users and computers. In *ACM User Interface Software and Technologies (UIST 2003)*, pp. 149–158, 2003.
3. L. Holmquist, F. Mattern, B. Schiele, P. Alahuhta, M. Beigl, and H. Gellersen. Smart-Its Friends: A Technique for Users to Easily Establish Connections between Smart Artefacts. In *Ubicomp 2001*, pp. 116–122. Springer-Verlag, 2001.
4. Jouni Malinen. Host ap driver for intersil prism2/2.5/3. http://hostap.epitest.fi/.
5. Jun Rekimoto. Pick-and-Drop: A Direct Manipulation Technique for Multiple Computer Environments. In *Proceedings of UIST'97*, pp. 31–39, October 1997.
6. Jun Rekimoto. Synctap: An interaction technique for mobile networking. In *Proc. of MOBILE HCI 2003*, 2003.
7. Jun Rekimoto, Yuji Ayatsuka, Michimune Kohno, and Haruo Oba. Proximal Interactions : A direct manipulation technique for wireless networking. In *Proc. of INTERACT 2003*, 2003.
8. Colin Swindells, Kori M. Inkpen, John C. Dill, and Melanie Tory. That one there! pointing to establish device identity. In *Symposium on User Interface Software and Technology (UIST'02)*, pp. 151–160, 2002.
9. Peter Tandler, Thorsten Prante, Christian Muller-Tomfelde, Norbert Streitz, and Ralf Steinmetz. Connectables: dynamic coupling of displays for the flexible creation of shared workspaces. In *Proceedings of the 14th annual ACM symposium on User interface software and technology (UIST 2001)*, pp. 11–20, 2001.
10. Brygg Ullmer, Hiroshi Ishii, and Dylan Glas. mediaBlocks: Physical containers, transports, and controls for online media. In *SIGGRAPH'98 Proceedings*, pp. 379–386, 1998.

Tackling Security and Privacy Issues
in Radio Frequency Identification Devices

Dirk Henrici and Paul Müller

University of Kaiserslautern, Department of Computer Science, PO Box 3049
67653 Kaiserslautern, Germany
{henrici,pmueller}@informatik.uni-kl.de

Abstract. This paper introduces shortly into the security and privacy issues of RFID systems and presents a simple approach to greatly enhance location privacy by changing traceable identifiers securely on every read attempt. The scheme gets by with only a single, unreliable message exchange. By employing one-way hash functions the scheme is safe from many security threats. It is intended for use in item identification but is useful in other applications as well.

1 Introduction and Survey of Approaches for Enhancing Privacy

Radio Frequency Identification (RFID) enables many applications: By attaching tags to products an automated inventory can be easily maintained. Tags also allow customers to pay and checkout automatically by pushing a loaded trolley past a reader. Postal services can equip shipped goods with tags for tracking purposes and other application fields like libraries, toll collect, bank notes, and many others emerge as well.

To enable automated inventory and supply chain management, each tag needs to contain a unique identifier to enable item tracking. Since products equipped with RFID tags containing unique identifiers enable tracking of persons by the tags they carry [1/2], RFID devices recently gained unexpected attention. Announcements of established companies like Wal-Mart, Benetton, Michelin, and Gillette to deploy RFID tags in their products [e.g. 3/4] raised immense privacy concerns ranging from complaints up to boycotts [5/6]. In the popular press even "cradle-to-grave surveillance" scenarios are stated [7]. Also, many threats are conceivable that intend to compromise security in RFID systems: Physical attacks, traffic analysis, eavesdropping, counterfeiting, spoofing, and denial of service. In consequence, security and privacy in RFID systems are important aspects that need particular attention.

But tags need to be comparatively simple devices to keep cost per item low. Therefore, the cheapest ones only contain a small amount (e.g. 96 bits) of read-only storage. Extended ones have larger memory, read-write storage, integrated sensors, or more gates for computation purposes. Only the most complex ones can have enough capabilities for creating "good" random numbers, symmetric or even asymmetric cryptography but are too expensive for the mass market.

A. Ferscha and F. Mattern (Eds.): PERVASIVE 2004, LNCS 3001, pp. 219–224, 2004.

It is obvious that security/privacy and cost are at odds. Thus, a suitable tradeoff is required providing the desired features at optimal cost. The level of privacy required depends on the application but should in any case comply with a minimum being agreed upon to satisfy consumer demand.

The most obvious solution is restricting the functional range of the tags by removing the serial number and keeping only manufacturer and product type intact or even a complete "killing" of tags at checkout – but it is not satisfactory [8]. In the first scenario, tracking of people is still possible examining the constellation of products; both prohibit legitimate applications as well as illegitimate and do not render corporate espionage impossible as long as the tags are active. Further, mechanisms to prevent or detect unauthorized identifier removal or killing of tags are required so that the schemes are not as simple as they look like.

For operations that require exchange of sensitive information a tag can be equipped with a physical contact channel to bypass the vulnerable wireless interface. Drawbacks are inconvenience and increased tag size and cost.

Many approaches like "Blinded Tree-Walking" take advantage of the backward channel being weaker than the forward channel (with passive tags) to counteract eavesdropping. Such schemes take care that sensitive information is never broadcast over the forward channel. Note that eavesdropping at the backward channel, which is an important issue as well, is not prevented.

A more complex proposal is the "Hash Lock"-approach counteracting unauthorized reads: A tag does not reveal its information unless the reader has sent the right key being the preimage to the hash value sent by the tag. The scheme requires implementing cryptographic hash functions on the tag and managing keys on the backend [9]. This is regarded as economic for the near future [10]. Unfortunately, the scheme offers data privacy but no location privacy since the tag can be uniquely identified by its hash value. Another drawback is that the key is sent in plain text over the forward channel which can be eavesdropped easily from a large distance.

The extended scheme called "Randomized Hash Lock" [9] ensures location privacy but is not scalable for a huge number of tags since many hash-operations must be performed at the back-end and it additionally relies on the implementation of a random number generator in the tags to randomize tag responses. Such devices need sources for physical randomness so that the implementation is rather complex and expensive.

"Blocker Tags" [8] aim at blocking readers to provide consumer privacy. This is done by interfering with the anti-collision algorithm used to singularize tags. The tags itself remain unaltered. The approach is imaginative and applicable for some scenarios but is neither free of drawbacks nor a fully satisfactory solution as the problem itself is not dealt with.

Other approaches are based on re-encryption to cause a ciphertext to change its appearance. Such operations normally require the tag of being capable of strong cryptography. But for use in banknotes that need a proof of authenticity but should not contain traceable identifiers, in [11] the calculation is moved outside the tag. But the scheme relies on optical information printed on the banknotes for validation and detection of forgery. This check must be performed manually being a great disadvantage of the approach.

Using strong cryptography like symmetric or even better asymmetric encryption is capable of solving most security and privacy issues. For maximum security, no long-term secrets like master keys, private keys, or other sensitive information may be stored on a tag. The development done for smart cards applies to RFID technology accordingly [10/12]. Unfortunately, strong cryptography is costly to implement and offers aggressors many opportunities for attacks [13].

2 High-Level Description of "Privacy by Hash-Based ID Variation"

Based on the design implications derivated from the conceivable threats and the limited tag resources we developed a simple hash based scheme that greatly enhances security and privacy in RFID systems without restricting legitimate applications or relying on complex cryptography.

The scheme is based on the same prerequisites as the "Hash Lock", i.e. a one-way hash function and key management at the backend, but is safe from eavesdropping and cannot be compromised by spoofing or replay attacks. To enhance location privacy, the general idea is to change the identifier of a tag on every read attempt in a secure manner, i.e. a tag changes its identity on every query.

Independent from this identifier variation, moving data storage into the backend is recommended. With this, access to data no longer needs to be controlled by the tag thus reducing complexity and cost. Further, physical attacks can be of no use for getting valuable data or other stored secrets: Only short-term values are kept in a tag.

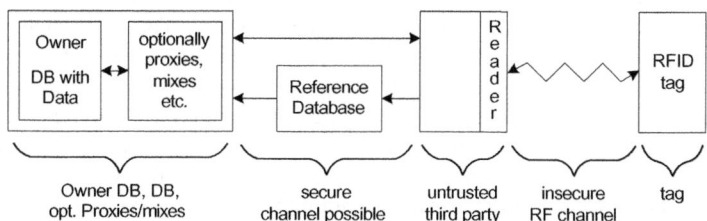

Fig. 1. Involved system components

The system is set-up according to figure 1. The set-up is quite straight forward except for the reference database before the actual back-end database. The reference database is needed for the following purpose: When the reader/third party queries the tag it needs to know who to contact for getting the data associated with the tag. This information may not come directly from the tag, because a URL or similar data could be exploited for tracking the tag. Thus, the reference database is used to conceal the owner. It should favorably be operated by the manufacturer of the tag or an independent organization. Note that there should only be a few number of reference databases so that an individual cannot be tracked by seldom constellations of tags pointing to certain reference databases.

The scheme itself relies on a two-message data exchange. With this message exchange the tag identifies itself to a database in the backend and the identifier of the tag is changed. This is done in a secure way while maintaining the desired location privacy by not sending traceable values:

- the scheme is resilient to message loss or service interruption
- an attacker cannot change the identifier of the tag since he does not have the information to fake the necessary authentication

- the scheme cannot be compromised by replay attacks or spoofing since transaction numbers and shared secrets are used
- the two messages do not expose traceable information and have no content that would allow conclusion of the tag identity or its owner by an attacker (for instance, transaction numbers are never sent in plain text)

A complete description of the protocol and an example for its operation can be found in [14].

2.1 Advantages of the Scheme

In return for that only limited resources used, the scheme achieves a high level of location privacy and is save from many security threats.

Neither at the tag nor at the database entity complex processing is required. The tag only needs to perform hashing and exclusive-or-operations. The database entity additionally must provide a single random number which is used to calculate the new tag identifier.

Messages are only regarded as valid if the correct transaction numbers were used when creating them. Therewith, an attacker does not have the necessary information to compromise the scheme.

As mentioned above, the scheme is resilient to loss of messages – due to transmission errors or as well ones being provoked by an attacker. This is accomplished by storing two sets of information in the database entity. The set which was used to identify the tag in a read attempt is always preserved; the second set of information is then set to the newly calculated tag identity. This way it is ensured that the tag never looses its connection to the corresponding information at the database entity.

In contrast to most other approaches, the scheme does not rely on a trustful reader or third party respectively. The reader itself or the third party who operates the reader only works as a kind of proxy for the communication with the tag.

It is recommended that no user data is stored on the tags. This saves costs and access control for this data is no longer required at the tag. Access control to data and changing ownership and other properties of a tag should be moved to the backend where plenty of computing power and the feasibility of a certificate management is available inexpensively [15]. For getting data, the third party might be required to authenticate itself to the database entity, and data retrieval might be restricted depending on access restrictions set for the third party requesting the data. For other applications, communication between third party and database entity could be made anonymous by means of a mix network [16] or a similar technology. Further, in the backend, security systems and access control schemes can be changed easily according to current requirements.

The hash-function that is employed can be changed easily. With small extensions in the databases, several hash-functions can be used in parallel in the system.

Collisions of ID values that are a problem in many schemes in which IDs are calculated randomly can be easily coped with: Duplicate IDs are detected when attempting to update the reference database and a new ID is requested in this case.

2.2 Drawbacks of the Scheme and Ideas for Resolution

For gaining maximum location privacy, the number of reference databases must be kept as low as possible, ideally one. Fortunately, the operations that have to be performed at the reference database are not complex and can be implemented efficiently. Load sharing can be performed for instance by splitting the range of values for the primary index among several servers for each reference database. Nevertheless, if the scheme is used extensively, traffic to and from the reference databases might become a problem and a well thought-out distributed database is required. Also security for the reference databases is a very important issue since the system relies on their appropriate operation.

Tracking an individual cannot be prevented employing the proposed scheme if traffic analysis (counting the number of items carried etc.) is used. But apart from that location privacy is enhanced considerably since no static, traceable IDs exist any more.

Mimicking of a tag is restrainedly possible: An attacker could act as a reader and record the tag's answers of the queries. Later, the attacker could replay the answers in same sequence to the queries of a legitimate reader to make it appear as if the real tag was still there albeit long taken away. Anyhow, such operation can be detected by suspicious transaction numbers and does not compromise the scheme itself. To eliminate the possibility of the attack completely, the reader could send a timestamp along with its query that must be included in calculating the authenticating hash value in the first message at the tag.

Using the proposed scheme, the tag needs to stay online until the reader gets the reply message from the database entity. This lowers the rate in which tags can be read. But by extension of the protocol this issue can be solved as well.

Loss, interception or blocking of the reply message results in preventing the tag ID from being changed but has no other implications: The tag can use its old ID in the next request which will still work. Errors in message transfer can be detected afterwards comparing transaction numbers. Suspicious values attract attention and counteractive measures can be taken. By extending the scheme employing a third message, changing of the ID can be ensured.

3 Conclusion

Vast deployment will make RFID a pervasive technology. Thus, security and privacy are import issues to be considered. Many location-aware applications emerge; systems are capable of tracking all of our movements and recording anything [17]. This is done in the name of convenience or due to economic reasons. Legislation must ensure that privacy of the individual is still protected; researchers must develop the required techniques.

An overview of current approaches for enhancement of privacy has been given above and their capability of handling the various threats RFID technology faces has been explained. Afterwards we proposed a hash-based scheme with a high inherent security rendering it a useful technique for all kinds of applications where static identifiers are used currently and location privacy is an issue.

The main benefit of the proposed scheme is its simplicity: It only requires implementation of a hash function in the tag and data management at the backend and does not rely on random numbers generated by the tag, strong symmetric or even asymmetric encryption. It offers a high degree of location privacy and is resistant to many forms of attacks like eavesdropping. Further, only a single message exchange is required, the communications channel need not be reliable, the reader/third party need not be trusted, and no long-term secrets need to be stored in tags.

Most of the denoted drawbacks have already been solved and the solutions will be published in subsequent publications.

References

1. Black, J.: Playing Tag with Shoppers Anonymity, Business Week online, 2003, available at http://www.businessweek.com/technology/content/jul2003/tc20030721_8408_tc073.htm
2. Garfinkel, S.: An RFID Bill of Rights, Technology Review, 2002, available at http://www.technologyreview.com/articles/garfinkel1002.asp
3. Crane, J.: Benetton Clothing to Carry Tiny Tracking Transmitters, Associated Press, 2003
4. RFID Journal: Gillette to Buy 500 Million EPC Tags, 2002; Michelin Embeds RFID Tags in Tires, 2003; available at http:// www.rfidjournal.com
5. Consumer Group Calls for Immediate Worldwide Boycott of Benetton., Website, available at http://www.boycottbenetton.org/PR_030313a.html
6. http://www.stoprfid.org/, Website
7. McCullagh, D.: RFID tags: Big Brother in small packages, CNET, 2003, available at http://news.com.com/ 2010-1069-980325.html
8. Juels, A. et al.: The Blocker Tag: Selective Blocking of RFID Tags for Consumer Privacy, 10th ACM Conference on Computer and Communications Security, 2003
9. Weis, S. et al.: Security and Privacy Aspects of Low-Cost Radio Frequency Identification Systems, First International Conference on Security in Pervasive Computing (SPC), 2003
10. Weis, S.: Security and Privacy in Radio-Frequency Identification Devices, Massachusetts Institute of Technology, 2003
11. Juels, A.; Pappu, R.: Squealing Euros: Privacy Protection in RFID-Enabled Banknotes, Financial Cryptography, 2002
12. Abadi, M. et al.: Authentication and Delegation with Smart-cards; Theoretical Aspects of Computer Software, pp. 326-345, 1991
13. Weingart, S.: Physical Security Devices for Computer Subsystems: A Survey of Attacks and Defenses, CHES 2000, volume 1965, pages 302-317, Springer LNCS, 2000
14. Henrici, D.; Müller, P.: Hash-based Enhancement of Location Privacy for Radio-Frequency Identification Devices using Varying Identifiers, PerSec'04 at IEEE PerCom, 2004
15. Sarma, S. et al.: RFID Systems and Security and Privacy Implications, Workshop on Cryptographic Hardware and Embedded Systems, pages 454-470, LNCS, 2002
16. Chaum, D.: Untraceable Electronic Mail, Return Addresses, and Digital Pseudonyms, Communications of the ACM, vol. 24(2), pp. 24-88, 1981
17. Beresford, A. R.; Stajano, F.: Location Privacy in Pervasive Computing; IEEE Pervasive Computing, January-March 2003, pp. 46-55, 2003

Towards Wearable Autonomous Microsystems

Nagendra B. Bharatula, Stijn Ossevoort, Mathias Stäger, and Gerhard Tröster

Wearable Computing Laboratory
ETH Zürich, Switzerland
{bharatula, ossevoort, staeger, troester}@ife.ee.ethz.ch
http://www.wearable.ethz.ch

Abstract. This paper presents our work towards a wearable autonomous microsystem for context recognition. The design process needs to take into account the properties of a wearable environment in terms of sensor placement for data extraction, energy harvesting, comfort and easy integration into clothes and accessories. We suggest to encapsulate the system in an embroidery or a button. The study of a microsystem consisting of a light sensor, a microphone, an accelerometer, a microprocessor and a RF transceiver shows that it is feasible to integrate such a system in a button-like form of 12 mm diameter and 4 mm thickness. We discuss packaging and assembly aspects of such a system. Additionally, we argue that a solar cell on top of the button – together with a lithium polymer battery as energy storage – is capable to power the system even for a user who works predominantly indoors.

1 Introduction

Ubiquitous computing [1] or Wearable computing [2] in combination with an intelligent environment are often cited to change our future significantly by supporting its user and enhancing everyday life. Applications range from tourist guides [3], health monitoring [4] to remembrance agents [5].

A system that can perform all these tasks needs to be context aware. Context awareness can be described as the ability of a system to model and recognize what the user is doing and what happens around him, and to use this information to automatically adjust its configuration and functionality [6].

While some of the context information can be retrieved from sensors not carried by the user (e.g. beacons in the environment for location information), many applications require sensors to be placed on the user's body or in his clothes (e.g. accelerometers for motion recognition or sensors to monitor body function like pulse and blood pressure).

In this paper we show how recent advances in CMOS, RF frontend design and microsystems packaging technologies can be used to combine sensors, computing elements, power generation and ultra low power wireless transceiver to develop autonomous microsystems suitable for embedding in clothes. We will discuss the feasibility of a wearable sensor button. Moreover, We will address power generation and power consumption issues to show that even with a few

A. Ferscha and F. Mattern (Eds.): PERVASIVE 2004, LNCS 3001, pp. 225–237, 2004.

microwatts of power available, such a small system can be built for useful context recognition. Furthermore, we will discuss wearability and fashion aspects. As an example, we will show how a sensor node with three different sensors (light sensor, accelerometer and microphone), processor, RF transceiver, solar cell and battery can be packaged as a button-like autonomous microsystem.

Related Work. Some projects like the Smart-Its [7], the Active Badge [8], the Smart Badge [9] or the TEA Device [10] project explore the possibility of wearable systems for context recognition but don't focus on small and ultra low-power hardware. Till today few research groups tried to develop autonomous microsystems which are of several cubic millimeter in total size. The 'Smart Dust' project [11,12] at Berkeley and the 'eGrains' project [13] at IZM Fraunhofer are focussed on building the system in a volume of a few cubic millimeters. For the operation of such small electronic devices often power sources of a few microwatts to a milliwatt are sufficient. Attempts to use human body heat based on the seebeck effect in the form of micro thermoelectric generators [14,15] or vibration based energy generation [16,17,18] are elsewhere reported. These power sources based on MEMS (MicroElecto Mechanical System) technology are advantageous in terms of cost, performance and batch fabrication.

Paper Contributions and Organization. The main aspect which sets us aside from the work done by other groups in the field of autonomous microsystems is our focus on wearable, autonomous systems. The paper addresses the main issues in the development of an autonomous sensor node with respect to wearable systems. This means that the sensor node, if rightly placed, is allowed to have a volume of a cubic centimeter and can exploit the human body as a power source.

However, the means of communication between nodes is limited. Due to a missing line-of-sight, optical communication is out of question. Instead, wireless RF communication will be discussed.

The remaining paper is organized as follows: Chapter 2 will show which components are required for such a system. In Chapter 3 we will discuss the human body in terms of its possibility to act as a power source. We will discuss the location of the sensor nodes on the human body with respect to wearability, context recognition and power generation. In Chapter 4 an example system for system validation will be presented. We will show how the intended system can be packaged to fulfill the energy and size requirements. Finally, Chapter 5 will give a conclusion and an outlook.

2   Overall System Layout

A context aware system needs to gather sufficiently detailed personal information about the wearer and his/her environment. One widely pursued approach to build context aware systems is via arrays of simple ultra low power sensors distributed over the user's body [19]. The sensors provide context information to a central wearable computer. Examples of sensors that have been shown to provide useful context information are accelerometers for motion analysis and gesture recognition as in [20,21], galvanic skin response sensors, blood volume pulse sensors and respiration sensors for recognizing the user's state as in [22] or a collection of other sensors like light, humidity, temperature and atmospheric pressure sensors for various context information as in [19]. Even sound has proven to be a useful source of information that can be handled with low power resources [23,24].

The chief requirements for such sensor nodes can be summarized as follows:

- **Low Power Consumption:** Since context monitoring has to be performed continuously during the whole day it only makes sense if it can be accomplished with minimal power consumption. Ideally, the sensor nodes should be fully autonomous, operating for months or years on a miniature battery or even better, extracting the energy from the environment.
- **Small Size:** The sensor nodes should be small in size (a few mm^3 to one cm^3) to ensure good wearability and possible integration with clothes and accessories.
- **Wireless Transmission:** To guarantee flexibility, the sensor nodes should transmit their results wirelessly to a central wearable computer. The wireless connection is essentially important if the sensors are located in different pieces of clothing that can not be connected by wires (e.g. shoes, trousers, jackets or glasses).
- **Local Processing Capabilities:** The sensor nodes may need to perform certain amount of local processing, to reduce the amount of data that needs to be transmitted because wireless transmission consumes more energy than computation.

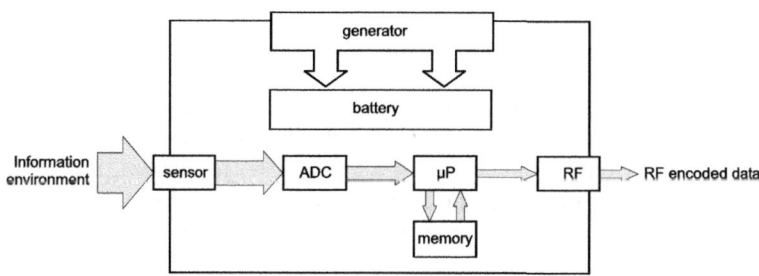

Fig. 1. Overall system architecture of an autonomous microsystem

From the above requirements a rough system architecture as shown in Fig. 1 can be derived. A sensor node that represents an autonomous microsystem contains the sensors itself, an AD converter, a computational unit (microprocessor μP, DSP, or custom made ASIC), a small memory, a RF transceiver including an antenna, a power generator and an energy storage device, for example a solar cell and a battery.

3   The Human Body Aspect: Placement of Sensor Nodes

When designing a wearable autonomous microsystem one has to take into account the limitations imposed by the human body. One key aspect of such a system is wearability. Wearability [25] is described as the interaction between the human body and the wearable object. As optimal examples of wearability can be found amongst our clothes, we propose to integrate autonomous microsystems in our daily clothing. Since clothes are produced according to well established production methods, it would be best to create autonomous microsystems that can be seamlessly integrated into the existing production processes (see Section 3.3). Each sensor node could be part of a sensor network linked via data lines woven into the fabric [26], however if its placement is arbitrary, the system could best communicate via a RF wireless system.

Autonomous microsystems can not just be placed anywhere on our clothing, each position offers different (im)possibilities. Some places diminish our comfort, some places do not give adequate sensor feedback while some places can not be reached from the production point of view. In the following sections we will discuss these placement issues, that pose constraints to our system.

3.1   Data Extraction Possibilities

The type of sensors and its optimal position highly depend on the tasks that need to be recognized. Generally the placement of sensors is simply defined by the type (e.g. light sensors on top surfaces, acceleration sensors for each limb).

3.2   Energy Harvesting Possibilities

A true autonomous microsystem should generate its own energy extracted from the environment. This means, both energy directly derived from our body and energy from our environment are useful to power the system. Starner [27] mentions energy from breathing, body heat, blood transport and various types of motion. Because we focus on a system that can be integrated to our clothing, we are limited to motion and body heat. The possibility to use motion and body heat is highly dependent on the placement of the system. The dark shades in Fig. 2 indicate where most of the energy from motion (left) and heat (right) can be extracted to generate power.

Kinetic generators can be best placed on body parts that are subject to high acceleration or move frequently, preferably limbs. To harvest energy from heat, one would think that the torso is the best place since its temperature is kept constant. However, our clothing prevents a temperature gradient between the

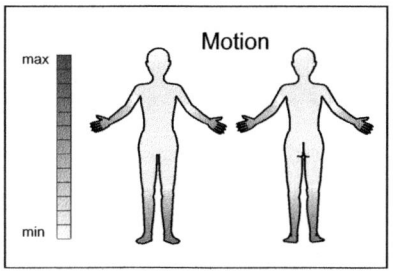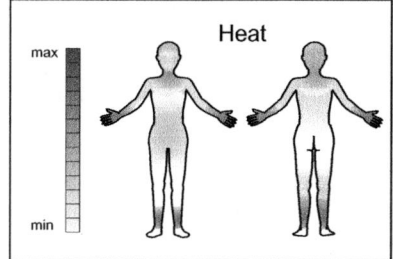

Fig. 2. Places where energy from motion and heat can be extracted

sensor and its surrounding. So, only body surfaces exposed to air can be used for harvesting energy from body heat. Therefore thermoelectric generators need to be placed in close contact to the body, if possible integrated with clothes that are usually worn in a single layer.

An important source of energy that is available from our environment is solar energy. Solar cells offer a greater degree of freedom, they can be placed at all points that are frequently exposed to light. Another useful source of energy could be electro-magnetic radiation. The energy could be provided by inductive coupling by a coil integrated in the fabric, placed underneath the microsystem. Tab. 1 gives a comparison of these options.

Table 1. A comparison of power generators useful for wearable microsystems

Power Generator	Power	Remarks
Solar cell (outdoors) [28]	$150\mu W/mm^2$	direct sun
	$1.5\mu W/mm^2$	cloudy day
Solar cell (indoors) [28]	$5.7\mu W/mm^2$	desk lamp
	$0.06\mu W/mm^2$	standard desk
Motion (shoe generator) [29]	1-10mW	piezoelectric
	50-250mW	electro-magnetic (rotary)
Motion (inertial generator) [18]	$200\mu W$	for legs when walking
	$50\mu W$	for torso when walking
Thermoelectric generator [30]	$0.2\mu W/(cm^2 K)$	

3.3 Comfort and Integration

We realize that, although autonomous microsystems can be very small, their placement is still subject to comfort [31]. Areas of our body that are often in contact with various objects (feet, back, bottom, underside of arms and hands) should be free from any electronic system unless the system can be seamlessly integrated within the thickness of the fabric. Elsewhere, the microsystem could be an element of embroidery, integrated in a tuck, part of a label or encapsulated in a button, as shown in Fig. 3, to allow easy integration in the production process.

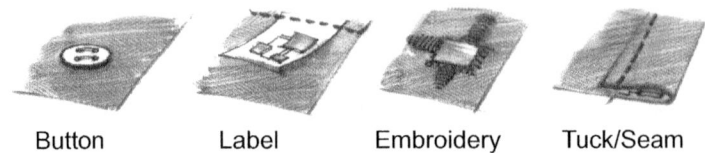

Button   Label   Embroidery   Tuck/Seam

Fig. 3. Possible placement of microsystems in a fabric

4   Feasibility Study

To show that it is feasible to build a wearable autonomous microsystem, we will consider one specific example. This study uses latest research results in CMOS, RF frontend and MEMS technology to create an ultra-low power microsystem in the size and shape of a button.

Our system consists of three different sensors: a differential MEMS microphone [32], a photo diode as a light sensor (SFH 3410 from Opto Semiconductors or PDCA12-68 from Optospeed Ticino SA), and an accelerometer (3-axis accelerometers from [33] or [34], or micro accelerometers such as [35]). Signal processing is handled by a low power microprocessor (CoolRisc 88 or 816 core [36]. To AD convert the analog signal of the photo diode and the accelerometer an ultra-low power successive approximation ADC similar to the one in [37] is choosen. A MEMS microphone similar to [38] with an ADC can be used. A memory in the range of 16-32 KByte supports the microprocessor. An ultra-low power direct conversion RF transceiver similar to [39], operating in the ISM band at 868 MHz provides the communication capabilities to the microsystem. The RF transceiver also needs some external components. A solar cell and a lithium polymer battery as an energy storage completes the system. In the following section packaging and assembly of the system in the form of a button is discussed. Size and energy consumption will be addressed, before analyzing the power budget for a typical scenario of an office worker.

4.1   Packaging and Assembly

The packaging and assembly of different modules (communication, memory, power and sensor) places a restriction on the selection of standard solutions. This is due to the differences in the characteristic behavior and technology used in the development of individual modules. Packaging is one of the important factors which determines the overall size and total cost of the microsystem. The sensors chosen for our system place additional constraints to the packaging technology. Solar cells, the photo diode and the microphone need to be on the surface of the system. The RF chip should be shielded from the other chips. Antenna placement turns out to be difficult. We suggest a helix antenna embedded in the side walls of the plastic housing. A different solution would be a patch antenna beneath the solar cell.

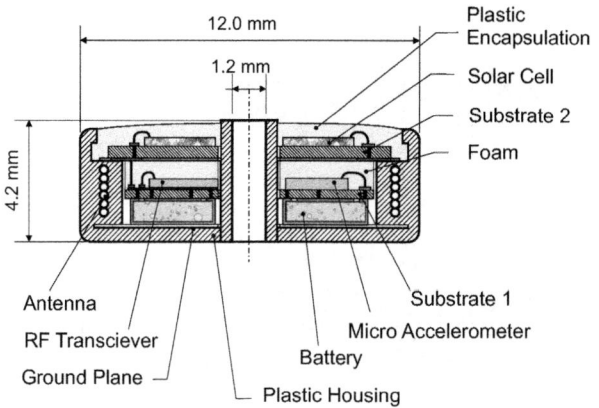

Fig. 4. Button-sized microsystem: side view

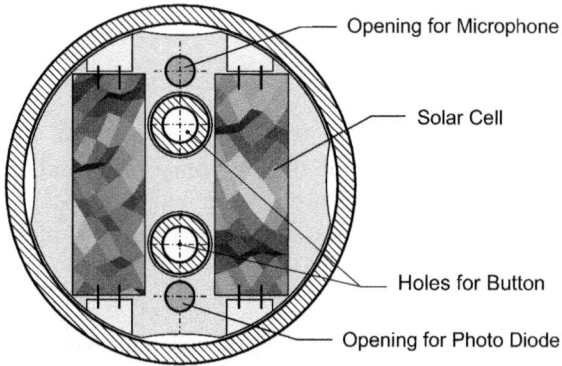

Fig. 5. Button-sized microsystem: top view, top level (substrate 2)

Fig. 4, 5 and 6 show the packaged and assembled system. The sequential buildup starts with wirebonding the accelerometer, the microphone, the RF transceiver and the microprocessor to a FR-4 or liquid crystal polymer (LCP) substrate 1 by copper wires (see Fig 6). This setup is connected to the battery compartment with vias. The thin film solarcells and the photodiode need to be wirebonded to a separate substrate 2 (see Fig 5). The two packaged parts are attached to each other. This entire part can be placed in a injection moulded plastic housing, which is designed to provide sufficient mechanical stability, stiffness and easy assembly. The side walls of the housing encapsulate the helix antenna. The whole system is sealed off by a translucent resin.

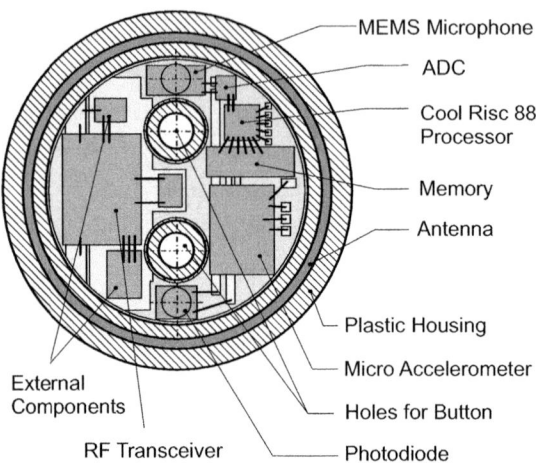

Fig. 6. Button-sized microsystem: top view, bottom level (substrate 1)

4.2   Size of the System

The overall size of the system is limited by its components and the used packaging and assembly technique. Bare dies are recommended in order to reduce the size of the system [40]. The area requirements of the individual components are summarized in Tab. 2.

Table 2. Area requirements of the individual components

Component	Area [mm^2]
ADC (active die area)	0.06
CoolRisc 88 processor	1.5
Memory	2.5-3
MEMS microphone (baredie)	1.69 - 2.0
Micro accelerometer	6.6
Photo diode	1.5
RF Transceiver	10.5
External components	4
2 Solarcells	34-38

The analog and digital parts, including the RF components altogether occupy an estimated area of 28 mm^2. The total area of substrate 1 measures 44 mm^2 (after deducting 6.3 mm^2 for the 2 holes), which leaves enough remaining space for the wirebonding. The available space on substrate 2 is 72 mm^2 which allows the two solar cells to occupy a maximum area of 37 mm^2 After carefully considering the dimensions of battery(s) and the plastic housing with integrated antenna we believe that it is possible to design this system in a button-like form with 12 mm diameter and 4.2 mm thickness.

4.3 Power Consumption of the System

Tab. 3 shows the power consumption of the major components when continuous operation is assumed. For many applications it is possible to reduce the duty cycle whereby reducing the power consumption. The value for wireless data transmission at a rate of 1kBit/s includes analog RF frontend and baseband processing in the microprocessor. The power consumption value of the accelerometer, the microphone and the light sensor include signal conditioning and analog to digital conversion. Tab. 3 also lists the sampling frequencies f_s for the sensors. These values have proven to be sufficient to create accurate results for context recognition [5,19,23]. The Cool Risc 88 processor is running at 20 MHz. The power consumption values for the microprocessor include memory operations.

Table 3. Power consumption for continuous operation

Functional Unit	f_s [Hz]	Power [μW]
Accelerometer	50	150–200
MEMS microphone	5000	50–80
Light sensor	2	50–100
Microprocessor	(20 MHz)	140–170
Data transmission	(1 kBit/s)	100–150
Total		490–700

In total the energy consumption is in the range of 490-700μW. We conclude that such a microsystem can be realized with a power consumption of 1 mW or less.

4.4 Energy Production

Our proposed system uses solar energy to collect, process and forward sensory data to a central wearable computational device. Under continuous lit circumstances solar cells can provide between 0.06 to 150 μW/mm^2 (see Tab. 1). However, mounted on a wearable device, lighting conditions vary strongly depending on the user's activities, his environment and the placement of the solar cell. For any solar cell applied to clothing it would be more appropriate to consider the energy harvested during a daily activity. To illustrate this, we estimated the total amount of energy a solar cell could harvest in a scarce lit environment like an office. We created a scenario of a north European office worker and estimated the amount of solar power this person could collect during 24 hours (see Fig. 7a). Our fictive person wakes up before 7.00h and walks to his office between 8.00h

a

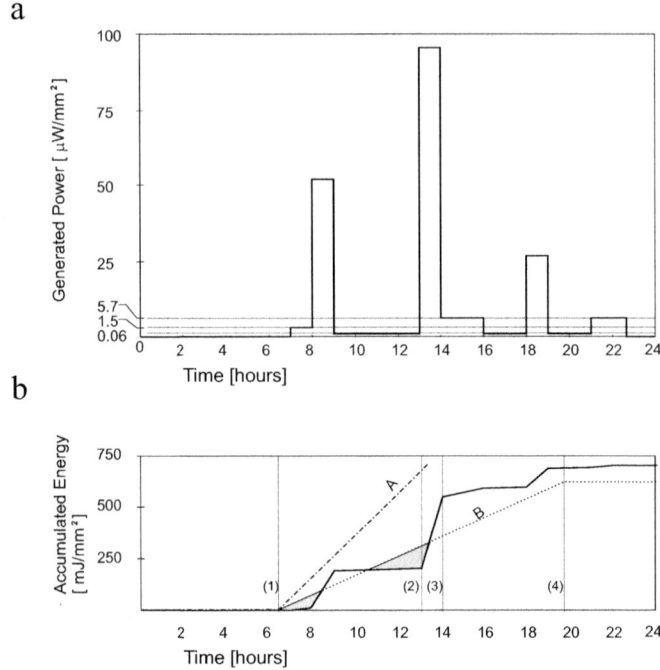

b

Fig. 7. Power and accumulated energy collected with a microsystem worn by an office worker

and 9.00h. Between 13.00h and 14.00h he is outdoor for lunch and has a meeting in a brighter space between 14.00h and 16.00h. Between 18.00h and 19:00h he walks back home. The difference between indoor and outdoor activity is clearly visible in Fig. 7a.

The continuous line in Fig. 7b shows the accumulated energy derived by integrating the values of Fig. 7a over time. The final amount of energy ($707 \ \mathrm{mJ/mm^2}$) collected within 24 hours would be adequate to provide our system (solar cell area of 37 mm^2) continuously with $37 \cdot (707/(24 \cdot 60 \cdot 60))\mathrm{mJ/sec} = 300\mu\mathrm{W}$.

Assuming that our system is deactivated during night time, it will only consume energy between 6.20h and 20.00h. If our system would consume about 1 mW (line A) it would largely exceed the collected energy which fails the system to be autonomous. However, if our system would consume 0.5 mW during it's on-time it could be autonomous (line B). In this case:

(1) At 6.20h the system activates. Because no solar energy is available, the system has to use previously stored energy from the battery.
(2) At 13.10h the difference between the collected and used energy reaches a minimum value of $-125 \ \mathrm{mJ/mm^2}$.
(3) At 14.05h the difference between the collected and used energy reaches a maximum value of about $200 \ \mathrm{mJ/mm^2}$.

(4) At 20.00h the system deactivates. The surplus of collected energy is stored in the battery.

To overlap the periods of energy shortage or surplus the system should include a small battery. If we would like the system to run according to this scenario, the minimum storage capacity of the battery should be $200\text{mJ/mm}^2 \cdot 37\text{mm}^2 = 7.4\text{J}$, which is well feasible with a small lithium polymer battery.

We conclude, that if our proposed system with a power consumption of 0.5 to 0.7 mW (see Section 4.3) is running on duty cycle below 70% the solar cells and the battery are well capable to power the system in the given scenario.

5   Conclusions

This paper has shown that the design process for a wearable autonomous microsystem needs to take into account aspects of the human body. While the placement of the microsystems can pose problems in terms of comfort and wearability, it can also offer possibilities in terms of energy harvesting through motion or body heat. The wearable environment also offers the possibility to integrate the microsystem into clothes and accessories. We suggest to encapsulate the system in an embroidery or a button.

A study of a microsystem consisting of a light sensor, a microphone, an accelerometer, microprocessor and a RF transceiver shows that it is feasible to integrate such a system in a button-like form of 12mm diameter and 4mm thickness. In continuous operation the system consumes less then $700\mu\text{W}$. We argue that if the system is running on a 70% duty cycle, which is acceptable for context recognition tasks, it can be made autonomous: A solar cell of 37mm^2 size on top of the button – together with a lithium polymer battery as energy storage – is well capable to power the system even for a user who works predominantly indoors. However, here we presented an idealized scenario and effects such as occlusion of solarcells, various indoor environments and light absorbing interiors were not considered.

Therefore our future work is focussed on extending the scenarios mentioned in section 4.4 to various users with different daily activities in order to verify the implications on the energy production. We are also investigating the use of alternative power sources, such as thermoelectric and inertial generators, to support the solar cell. Furthermore, in our future work we will address issues of power consumption, reliability, alternative packaging designs and fabrication of a prototype.

References

1. Weiser, M.: Hot topics: Ubiquitous computing. IEEE Computer **26** (1993) 71–72
2. Billinghurst, M., Starner, T.: Wearable devices: new ways to manage information. IEEE Computer **32** (1999) 57–64
3. Randell, C., Muller, H.L.: The well mannered wearable computer. Personal and Ubiquitous Computing **6** (2002) 31–36

4. Lukowicz, P., Anliker, U., Ward, J., Tröster, G., Hirt, E., Neufelt, C.: AMON: A wearable medical computer for high risk patients. In: ISWC 2002: Proceedings of the 6th International Symposium on Wearable Computers. (2002) 133–134
5. Kern, N., Schiele, B., Junker, H., Lukowicz, P., Tröster, G.: Wearable sensing to annotate meeting recordings. In: ISWC 2002: Proceedings of the 6th International Symposium on Wearable Computers. (2002) 186–193
6. Abowd, D., Dey, A., Orr, R., Brotherton, J.: Context-awareness in wearable and ubiquitous computing. Virtual Reality **3** (1998) 200–211
7. Holmquist, L., Gellersen, H., Kortuem, G., Schmidt, A., Strohbach, M., Antifakos, S., Michahelles, F., Schiele, B., Beigl, M., Maze, R.: Building intelligent environments with smart-its. IEEE Computer Graphics and Applications **24** (2004) 56–64
8. Want, R., Hopper, A., Falcao, V., Gibbons, J.: The active badge location system. ACM Transactions on Information Systems **10** (1992) 91–102
9. Maguire, G.Q., Smith, M.T., Beadle, H.W.P.: Smartbadges: A wearable computer and communication system. In: The 6th International Workshop on Hardware/Software Co-design. (1998)
10. Homepage of the TEA project. (http://www.teco.edu/tea/)
11. Warneke, B., Last, M., Liebowitz, B., Pister, K.S.J.: Smart Dust: communicating with a cubic-milimeter computer. IEEE Computer Magazine **34** (2001) 44–51
12. Warneke, B.A., Scott, M.D., Leibowitz, B.S., Lixia, Z., Bellew, C.L., Chediak, J.A., Kahn, J.M., Boser, B.E., Pister, K.S.J.: An autonomous 16 mm^3 solar-powered node for distributed wireless sensor networks. In: Proceedings of the 1st IEEE International Conference on Sensors. Volume 2. (2002) 1510–1515
13. Wolf, M.J.: Homepage of Fraunhofer eGrain project. http://www.egrain.org (2004)
14. Bottner, H.: Thermoelectric micro devices: current state, recent developments and future aspects for technological progress and applications. In: Proceedings ICT'02: 21st International Conference on Thermoelectrics. (2002) 511–518
15. Strasser, M., Aigner, R., Franosch, M., Wachutka, G.: Miniaturized thermoelectric generators based on poly-Si and poly-SiGe surface micromachining. Sensors and Actuators, A: Physical **A98-A98** (2002) 535–542
16. Amirtharajah, R., Chandrakasan, A.P.: Self-powered signal processing using vibration-based power generation. IEEE Journal of Solid-State Circuits **33** (1998) 687–695
17. Meninger, S., Mur-Miranda, J.O., Amirtharajah, R., Chandrakasan, A.P., Lang, J.H.: Vibration-to-electric energy conversion. IEEE Transactions on Very Large Scale Integration (VLSI) Systems **9** (2001) 64–76
18. von Büren, T., Lukowicz, P., Tröster, G.: Kinetic energy powered computing - an experimental feasibility study. In: ISWC 2003: Proceedings of the 7th IEEE International Symposium on Wearable Computers. (2003) 22–24
19. Lukowicz, P., Junker, H., Stäger, M., von Büren, T., Tröster, G.: WearNET: A distributed multi-sensor system for context aware wearables. In: UbiComp 2002: Proceedings of the 4th International Conference on Ubiquitous Computing, Springer: Lecture Notes in Computer Science (2002) 361–370
20. Junker, H., Lukowicz, P., Tröster, G.: PadNET: Wearable physical activity detection network. In: ISWC 2003: Proceedings of the 7th IEEE International Symposium on Wearable Computers. (2003) 244–245
21. Brashear, H., Starner, T., Lukowicz, P., Junker, H.: Using multiple sensors for mobile sign language recognition. In: ISWC 2003: Proceedings of the 7th IEEE International Symposium on Wearable Computers. (2003) 45–52

22. Healey, J., Seger, J., Picard, R.: Quantifying driver stress: Developing a system for collecting and processing bio-metric signals in natural situations. In: Proc. of the Rocky Mountian Bio-Engineering Symposium. (1999)

23. Stäger, M., Lukowicz, P., Perera, N., von Büren, T., Tröster, G., Starner, T.: SoundButton: Design of a Low Power Wearable Audio Classification System. In: ISWC 2003: Proceedings of the 7th IEEE International Symposium on Wearable Computers. (2003) 12–17

24. Peltonen, V., Tuomi, J., Klapuri, A., Huopaniemi, J., Sorsa, T.: Computational auditory scene recognition. In: IEEE International Conference on Acoustics, Speech, and Signal Processing. Volume 2. (2002) 1941–1944

25. Gemperle, F., Kasabach, C., Bauer, M., Martin, R.: Design for wearability. In: ISWC 1998: Proceedings of the 2nd International Symposium on Wearable Computers. (1998) 116–122

26. Cottet, D., Grzyb, J., Kirstein, T., Tröster, G.: Electrical characterization of textile transmission lines. IEEE Transactions on Advanced Packaging 26 (2003) 182–190

27. Starner, T.: Human-powered wearable computing. IBM Systems Journal 35 (1996) 618–629

28. Rabaey, J.M., Ammer, M.J., da Silva Jr., J.L., Patel, D., Roundy, S.: PicoRadio supports ad hoc ultra-low power wireless networking. IEEE Computer Magazine 33 (2000) 42–48

29. Kymissis, J., Kendall, C., Paradiso, J., Gershenfeld, N.: Parasitic power harvesting in shoes. In: ISWC 1998: Proceedings of the 2nd International Symposium on Wearable Computers. (1998) 132–139

30. Strasser, M., Aigner, R., Lauterbach, C., Sturm, T.F., Franosch, M., Wachutka, G.: Micromachined CMOS thermoelectric generators as on-chip power supply. In: 12th International Conference on Solid-State Sensors, Actuators and Microsystems. Volume 1. (2003) 45–48

31. Knight, J., Baber, C., Schwirtz, A., Bristow, H.: The comfort assessment of wearable computers. In: ISWC 2002: Proceedings of the 6th International Symposium on Wearable Computers. (2002) 65–72

32. Rombach, P., Müllenborn, M., Klein, U., Rasmussen, K.: The first low voltage, low noise differential silicon microphone, technology development and measurment results. Sensors and Actuators A95 (2002) 196–201

33. Homepage of Brüel and Kjaer: 3 axis accelerometers. (http://www.bksv.com)

34. Homepage of Endevco: 3 axis accelerometers. (http://www.endevco.com)

35. Junseok, C., Kulah, H., Najafi, K.: An in-plane high-sensitivity low-noise micro-g silicon accelerometer. In: The 16th Annual International Conference on Microelectromechanical systems. (2003) 466–469

36. Piguest, C., Masgonty, J.M., Arm, C., Durand, S., Schneider, T., Rampogna, F., Scarnera, C., Iseli, C., Bardyn, J.P., Pache, R., Dijkstra, E.: Low-power design of 8-b embedded coolrisc microcontroller cores. IEEE Journal of Solid State Circuits 32 (1997) 1067–1078

37. Scott, D.M., Boser, B.E., Pister, K.S.J.: An ultralow-energy ADC for Smart Dust. IEEE Journal of Solid State Circuits 38 (2003) 1123–1129

38. Homepage of Sonion: MEMS microphones. (http://www.sonion.com)

39. Porret, A.S., Melly, T., Python, D., Enz, C.C., Vittoz, E.A.: An ultralow-power UHF transceiver integrated in a standard digital CMOS process: architecture and receiver. IEEE Journal of Solid State Circuits 36 (2001) 452–466

40. Becker, K.F., Ghahremani, C., Jung, E., Neumann, A.: Rapid prototyping for advanced system integration. In: The GOOD-DIE newsletter. (2003) 28–32

Ubiquitous Chip: A Rule-Based I/O Control Device for Ubiquitous Computing

Tsutomu Terada[1], Masahiko Tsukamoto[1], Keisuke Hayakawa[2],
Tomoki Yoshihisa[1], Yasue Kishino[1], Atsushi Kashitani[2], and Shojiro Nishio[1]

[1] Graduate School of Information Science and Technology, Osaka University, Japan
[2] Internet System Research Laboratories, NEC Corp., Japan

Abstract. In this paper, we propose a new framework for ubiquitous computing by rule-based, event-driven I/O (input/output) control devices. Our approach is flexible and autonomous because it employs a behavior-description language based on ECA (Event, Condition, Action) rules with simple I/O control functions. We have implemented a prototype ubiquitous device with connectors and several sensors to show the effectiveness of our approach.

1 Introduction

As a result of the development of computer software/hardware technologies, the processing power and storage capacity of personal computers are rapidly increasing. At the same time, technological advances are contributing to the continued miniaturization of computers and component devices, such as microchips, sensors, and wireless modules[5][9][10]. In the near future, these devices will be embedded into almost any artifact and provide various services to support human daily life. This computing style is called ubiquitous computing. In ubiquitous computing environments, we can acquire various services with multiple interconnected computers that are embedded everywhere[4][12][13].

These embedded computers should automatically perform information exchanges and physical actions in response to surrounding circumstances. Although these computers may have low processing power and small memory, they must have the flexibility to change their function dynamically. Consequently, we apply rule-based technologies to describe the behavior of these ubiquitous computers. In this paper, we propose a new style of computing with rule-based I/O (input/output) control devices for constructing ubiquitous computing environments. We call this device the ubiquitous chip. The remainder of this paper is organized as follows. Section 2 outlines rule-based ubiquitous computing, presents several related works, and describes the design of the behavior description language for the proposed devices. Section 3 explains the software/hardware architectures of the ubiquitous chip, and Section 4 describes the application development environments for the ubiquitous chip. Section 5 presents several examples of its application and Section 6 sets forth the conclusion and planned future work.

A. Ferscha and F. Mattern (Eds.): PERVASIVE 2004, LNCS 3001, pp. 238–253, 2004.

2   Rule-Based Ubiquitous Computing

In conventional computing, a user operates systems with input devices such as a mouse and a keyboard, and acquires computational results with output devices such as a display and speakers. On the other hand, since embedded computers are essentially invisible and must work without these conventional input/output devices, they need to exchange information and perform physical actions automatically in response to surrounding circumstances. Here, we define the following three characteristics, which are requirements for ubiquitous computers:

1. **Autonomy**: computers work automatically without human operation
2. **Flexibility**: computers are applied to various purposes
3. **Organic cooperation**: complex behaviors are achieved by organic coordination with multiple computers

Most previous prototyps of ubiquitous computing environments did not completely fulfill these requirements. For example, although the devices of Aware Home Project[1] and the Active Badge system[11] realize important applications for ubiquitous computing, they have been developed for one special purpose and are not intended for reuse in other applications.

On the other hand, there are several projects to construct a common framework and device. Smart-Its[2] is small computing device that consists of two independent boards, a core board that consists mainly of processing and communication hardware and a sensor board containing a separate processing unit, various sensors, and actuators. Motes[3], MICA[7], and U-cube[6] are also small ubiquitous devices which are separated into two units; a core unit and other (sensor) units. These devices have enough flexibility because we can customize the system configurations by changing attached sensors and other devices. However, in these devices, since running programs are closely related to device configurations, we cannot change their functions or attached devices dynamically while the programs are running. Therefore, it is difficult to customize the behaviors of embedded devices in response to the user. Moreover, since applications are developed in a C-like programming language, it is difficult for public users to program or to customize applications. From this point of view, previous devices lack simplicity in programming and the flexibility/autonomy in terms of changing the functionality of a device dynamically.

To construct flexible, scalable, and easy exploitable ubiquitous computing environments, there is a need for a general device and an architecture that fulfills above three requirements. Consequently, to present our new device for ubiquitous computing, we employ a system design philosophy that consists of two characteristics: the separation between the I/O control and the attachments, and the rule-based approach.

As for the former one, the separation helps to achieve a flexible system configuration by changing attached devices (other devices in related works have the same advantage). Moreover, the proposed device plays the role of a network hub that receives signals from multiple devices such as sensors, and sends signals to multiple devices such as actuators. We can develop the device with a low-processing-power chip because the most of primary responsibility of the device is limited to a circuit switching.

In regard to the latter one, we apply the rule-based (event-driven) principle to the behavior description of ubiquitous computers because a person generally comprehends an event in the real world as a causal relation (event and action). We use ECA rules for the event-driven programming language. An ECA rule consists of the following three parts:

EVENT(E): Occurring event
CONDITION(C): Conditions for executing actions
ACTION(A): Operations to be carried out

ECA rules have been used to describe the behaviors of active databases. An active database is a database system that carries out prescribed actions in response to a generated event inside/outside of the database[14]. Using ECA rules, we can achieve the following advantages:

- As a consequence of their simplicity, we can program applications easily and intuitively. Anyone can construct and change applications for various devices embedded everywhere to enrich daily-life.
- We can change a full/part of a program dynamically because applications are described as a set of ECA rules, and each rule is stored independently in the device. This characteristic also enables users to customize behaviors of devices easily in response to users' requests.
- ECA rules in our system are described as a short-bit string. Therefore, we can send ECA rules as a message to devices via the network.
- Since ECA rules can be processed with a step-by-step approach, we can process rules with high efficiency (see 3.2).

2.1   Language Design of an ECA Rule

In conventional active databases, database operations such as SELECT, IN-SERT, DELETE, and UPDATE are considered events in ECA rules. Further, active databases can carry out actions only concerning database operations. Since ubiquitous computers may have little processing power and small memory, we must simplify the language specification of ECA rules while maintaining the ability to fulfill various requirements in ubiquitous computing environments.

Consequently, we decided that ECA rules are almost always used for I/O control. In other words, the on/off states of an attached switch and inputs from a sensor are handled as "input signals." In addition, output operations such as ringing a buzzer and turning on an LED (Light Emitting Diode) are handled as "output signals." In this way, a ubiquitous chip works almost for I/O control, and various devices are attached to the ubiquitous chip, as shown in Figure 1. In this figure, a ubiquitous chip evaluates the input from these sensors and devices, and outputs signals to connected devices. Moreover, a ubiquitous chip features

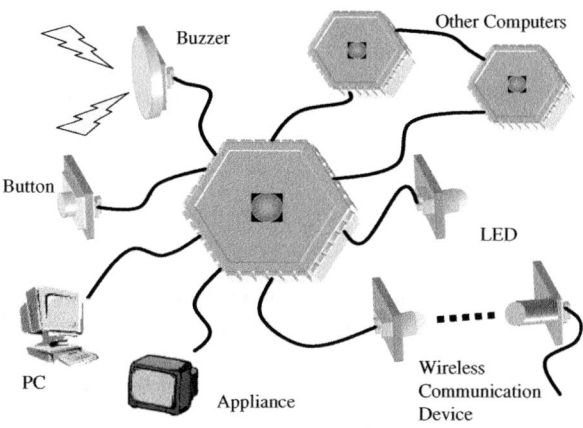

Fig. 1. Ubiquitous chip with connected devices

registers for storing the internal state and includes a flash memory for storing ECA rules. In addition, it has serial ports to communicate with other ubiquitous chips and has multiple timers for flexible timer functions.

Based on this principle, we define the events available in our system as shown in Table 1. There are two types of inputs to a ubiquitous chip: one is packet reception via a serial port, while the other is an input from a sensor via a normal port. When a ubiquitous chip receives a packet, the system generates a RECEIVE event. As for inputs from sensors, the system deals with the port state as conditions to execute actions. Therefore, the system allows ECA rules to be described without any events for executing actions by depending on the port state only.

We can specify the port and internal states in the condition part of the ECA rules. The system allows the description of multiple conditions and executes actions when all of the conditions are satisfied.

Actions provided by the system are shown in Table 2. The OUTPUT action changes the on/off state of each port to control connected devices such as a buzzer and an LED. The OUTPUT_STATE action changes the internal states, while the TIMER_SET action creates a new timer by specifying the interval of the timer and a once/repeat flag. The SEND_MESSAGE action sends a message that has a specific ID via a serial port. Since another ubiquitous chip generates a RECEIVE event on receiving this message, cooperation between ubiquitous chips is achieved using this event and the action. The SEND_COMMAND action sends the control commands shown in Table 3. The HW_CONTROL action controls the hardware. This action controls a general LED (described in Section 3.1), power saving functions, and a relay to switch serial ports. We can describe multiple actions with an ECA rule, and these actions are executed in a sequential order.

Table 1. Events

Name	Contents
RECEIVE	Data reception via the serial port
TIMER	Firing a timer
NONE	Evaluate conditions at all times

Table 2. Actions

Name	Contents
OUTPUT	On/off control of output ports
OUTPUT_STATE	On/off control of state variables
TIMER_SET	Setting a new timer
SEND_MESSAGE	Sending a message
SEND_COMMAND	Sending a control command
HW_CONTROL	Hardware control

Table 3. Commands for the SEND_COMMAND

Name	Contents
ADD_ECA	Adding a new ECA rule
DELETE_ECA	Deleting specific ECA rule(s)
REQUEST_ECA	Requesting a specific ECA rule

2.2 Binary Coding

ECA rules are translated to the binary format according to regulations shown in Figure 2. Basically, an ECA rule consumes four bytes (two bytes for Event and Condition, two bytes for Action). However, since the system allows the description of multiple conditions and multiple actions, we can describe a maximum of two conditions in a rule by switching on a multi-condition flag. We can also describe any number of actions by switching continue flags on. The format of each action varies by type. The second bit of the action discriminates the OUTPUT/OUTPUT_STATE actions from other actions. If the OUTPUT/OUTPUT_STATE actions are selected, the next 14 bits specify the output state. Otherwise, the next three bits specify the type of action and the remaining 11 bits are used to describe the action's content. For example, an ECA rule that states, "When INPUT1 and INPUT5 are ON and INPUT2 is OFF, the system fires a timer five seconds later," is translated as the four-byte binary format "0x91 0x13 0x49 0x15."

3   Prototype of the Ubiquitous Chip

Based on the design described in the previous sections, we developed a prototype device of the ubiquitous chip. In the following sections, we explain the prototype device, focusing on the hardware architecture, the software architecture, and the attachments.

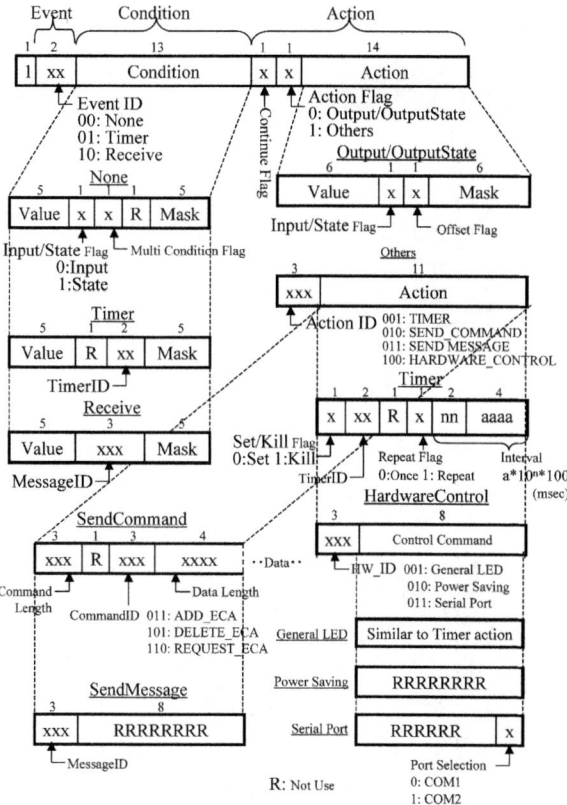

Fig. 2. Binary coding formula

3.1   Hardware Architecture

The prototype device consists of two parts: the core-part (Figure 3, left) and the cloth-part (Figure 3, right). The core-part (34 mm in diameter) contains a microprocessor (PIC16F873), a power on/off switch, and a general LED that indicates the state of the ubiquitous chip. As shown in Figure 4, the core-part has six input ports (IN1-6), 12 output ports (OUT1a-6a, 1b-6b), six power-supply ports (VCC), and two serial ports (COM1-2). The cloth-part (59 mm in diameter) operates as a converter between the core-part and external sensors/devices. The cloth-part houses a Li-ion battery, connectors for attaching sensors/devices, and input/output ports as well as the core-part. Figure 5 shows a connection example between the prototype devices, sensors, and other devices. Table 4 shows the hardware specifications of the ubiquitous chip.

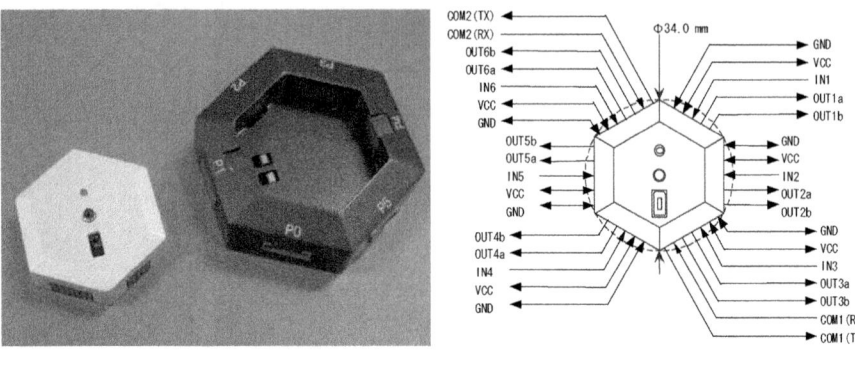

Fig. 3. A prototype device   **Fig. 4.** I/O ports of a prototype device

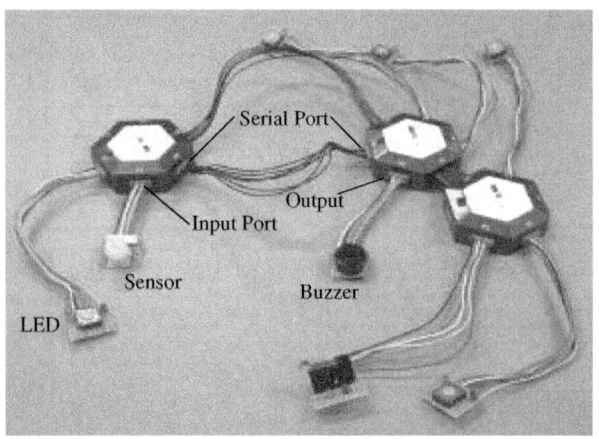

Fig. 5. An example showing connections

Table 4. Hardware specifications

CPU	PIC16F873
Operating voltage	2.9 – 6.0V(3.3V)
Weight	11g (51g includes cloth)
Power resource	300mAh(4.2V)
Program memory	4000 words
RAM	192 bytes
EEP-ROM	128 bytes

3.2 Software Architecture

Our prototype device uses the PIC16F873, which is a programmable RISC (Reduced Instruction Set Computer) type processor. Although this processor is cheap and easily programmed, its memory size is small and processing power is low. Hence, it is necessary to implement a rule-based system to improve efficiency. For example, our prototype stores ECA rules into the rule-base and

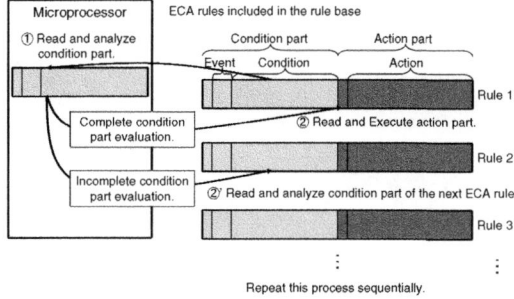

Fig. 6. A process for ECA Rules of the prototype

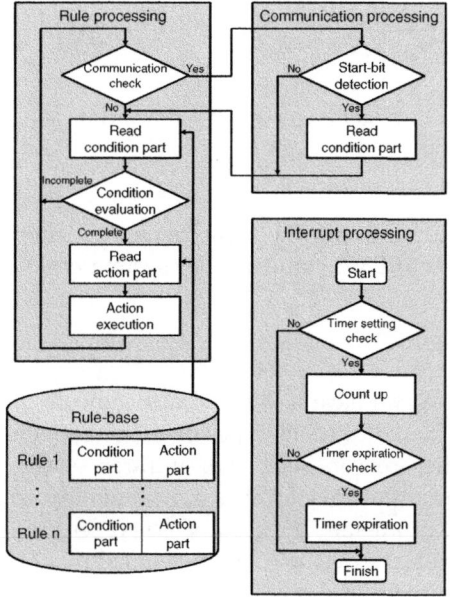

Fig. 7. A block diagram of the prototype

processes the rules every two bytes. The microprocessor reads a condition part of the first rule from the rule base, and if the condition is satisfied, the micro-processor reads the action part; otherwise, it reads the next rule.

This processing formula is illustrated in Figure 6. Therefore, by reading just the necessary data from the rule base, the memory size required for rule pro-cessing is only two bytes.

Figure 7 shows a block diagram of the rule processing in the prototype. The rule-processing part reads and executes rules, and the communication-processing part checks the serial ports. When the system receives data from a serial port,

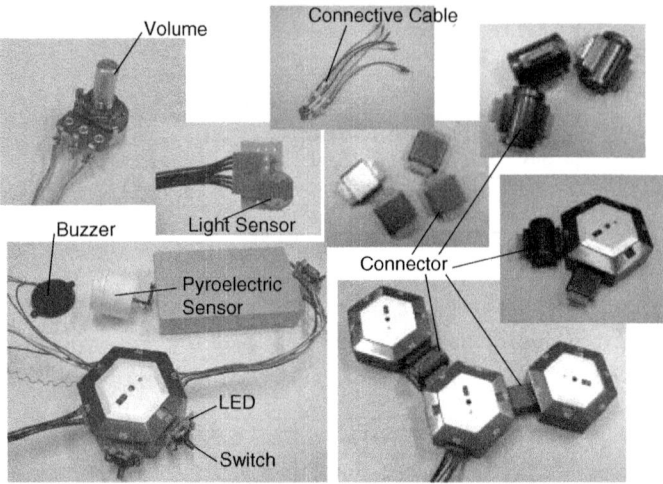

Fig. 8. The developed attachments

it generates a R ECEIVE event. Since the required memory size to process rules is small, even the PIC16F873 can drive the system easily.

3.3 Attachments

As shown in Figure 8, we have developed attachments for the ubiquitous chip. These attachments include various types of connectors for connecting one ubiquitous chip to another, sensors such as an infrared sensor, a pyroelectric sensor, an ultrasonic sensor, ans ultraviolet sensor, as humidity sensor, a light sensor, an acceleration sensor, a pressure sensor, a thermal sensor, a geomagnetic sensor, and an audio sensor. Moreover, we have developed input devices such as various types of buttons, switches, rheostats, and actuators such as a vibrator, buzzers, and motors.

In addition to these attachments, we have developed devices that enhance the cloth-part of the ubiquitous chip, as shown in Figure 9. One is a battery box for AAA rechargeable batteries that have long battery-life and are easy to obtain. The other is a wireless unit that makes the serial port wireless. This device employs the unprocedure communication via RF. Using this device, a ubiquitous chip can communicate other chips in wireless.

4 Development Environment

Since an ECA rule is represented in the binary format, as shown in Figure 2, it is difficult to describe ECA rules directly[8]. Therefore, we implement two development tools: the rule editor and the rule writer. Using these tools on

Fig. 9. The battery box (left) and the wireless unit (right)

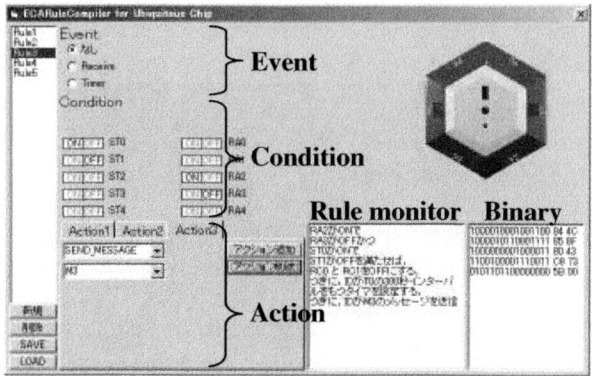

Fig. 10. A screen shot of the Rule Editor

a PC, we can easily describe ECA rules and store them in ubiquitous chips. Moreover, we implement a ubiquitous chip emulator that achieves cooperation between PCs and ubiquitous chips.

4.1   ECA Rule Editor and ECA Rule Writer

Figure 10 shows a screen shot of the ECA rule editor. In this application, we can make ECA rules by specifying events, conditions, and actions graphically with easy mouse operations. When we specify a rule, bit sequences of this rule are displayed on the binary display part and contents of this rule are displayed in natural language on the rule monitor part.

Figure 11 shows a screen shot of the ECA rule writer. This application reads, analyzes, and displays rules from a binary file, or a ubiquitous chip that is connected via a serial port. This application also writes displayed rules into ubiquitous chips via a serial port. We can add, delete, and modify ECA rules on ubiquitous chips freely using this application.

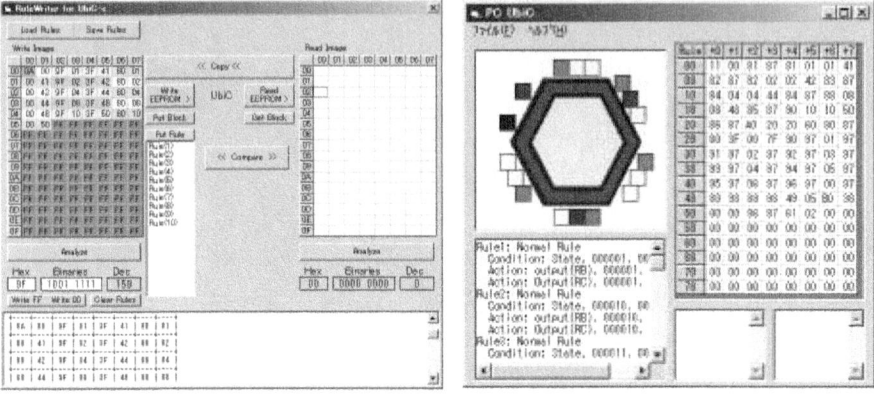

Fig. 11. A screen shot of the Rule Writer   **Fig. 12.** A screen shot of the Emulator

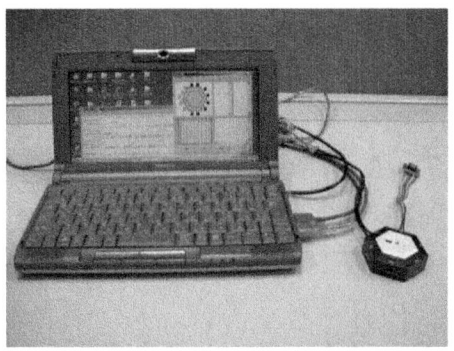

Fig. 13. Example of a connection between a PC and a ubiquitous chip

4.2   Ubiquitous Chip Emulator

We also implement the ubiquitous chip emulator as shown in Figure 12. This emulator is used for simulations of rules/applications. It is also used to achieve cooperation between PCs and ubiquitous chips.

It is possible to use the same rules employed for ubiquitous chips to operate the emulator. The on/off states of the input/output ports on an emulated ubiquitous chip are represented by lighting tones, and we can control the input states by a click operation on each port. Figure 13 shows an example of a connection between a PC and a ubiquitous chip. In this way, a user can collect/utilize information from sensors connected to ubiquitous chips.

Moreover, since the emulator does not have any restriction on memory size, the emulator can not only simulate a ubiquitous chip, but also store ECA rules for another ubiquitous chip and deliver them to connected ubiquitous chips.

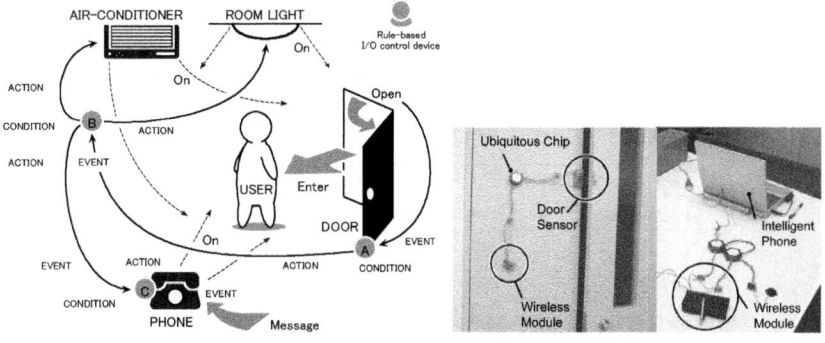

Fig. 14. An example of Application (1)

5   Applications

In this section, we show two examples of applications using ubiquitous chips. The first application is the room automation system illustrated in Figure 14. This application covers the following scenario:

- The user downloads the rule for the room control to his ubiquitous chip by reserving the room on his PC.
- When the user goes to the room and inserts his ubiquitous chip to the slot on the door, the door is automatically unlocked.
- When the door opens, the ubiquitous chip works to customize the room, such as by turning on the room light and the air-conditioner in cooperation with other ubiquitous chips embedded in the room.
- After once shutting the door, the system sounds a buzzer if the door is left open for more than one minute.
- The ubiquitous chip controls the air-conditioner with its temperature sensor.
- When the telephone rings, if there is a person in the room, it plays the message automatically. Otherwise, it records or transfers the message.

Table 5 shows the principals of ECA rules for these services. RULE 1-5 are the downloaded rules for ubiquitous chip A. RULE 6-8 and RULE 9-11 are stored rules in ubiquitous chip B and ubiquitous chip C. RULE 1 detects the door opening and sets a one-minute timer. RULE 2 sounds the buzzer when the timer fires. If the door is closed within one minute, RULE 3 resets the timer. RULE 4 requests the door to unlock and RULE 5 notifies ubiquitous chip B of the entry. RULE 4 is activated when this message arrives. This rule resends the entry message to ubiquitous chip C and turns on the room light. RULE 7 and 8 control the air conditioner using the information from the temperature sensor. RULE 10 and 11 change the behavior of the telephone according to the state of the user's presence.

This type of application produces a "smart" space. We consider that most of the requirements in the smart space are simple, and easy realizable using our

Table 5. The rule set for Example (1)

RULE 1	RULE 2	RULE 3
E:	E: TIMER	E:
C: I1=0, S1=0	C:	C: I1=1, S1=1
A: S1=1, TIMER(1min)	A: O1=1	A: S1=0, TIMER(0), O1=0
RULE 4	RULE 5	RULE 6
E:	E: RECEIVE(M2)	E: RECEIVE(M3)
C: I2=1	C:	C:
A: SEND(M1)	A: SEND(M3)	A: O2=1, SEND(M4), S2=1
RULE 7	RULE 8	RULE 9
E:	E:	E: RECEIVE(M4)
C: I3=1, S2=1	C: I3=0	C:
A: O3=1	A: O3=0	A: S3=1
RULE 10	RULE 11	
E:	E:	
C: I4=1, S3=1	C: I4=1, S3=0	
A: O4=1	A: O4=0	

I1: door sensor   O1: buzzer   S1: door flag   M1: unlock request
I2: key sensor   O2: room light   S2: enter flag (for chip B) M2: unlock complete
I3: heat gauge   O3: air conditioner S3: enter flag (for chip C) M3: enter (to chip B)
I4: phone call   O4: phone   M4: enter (to chip C)

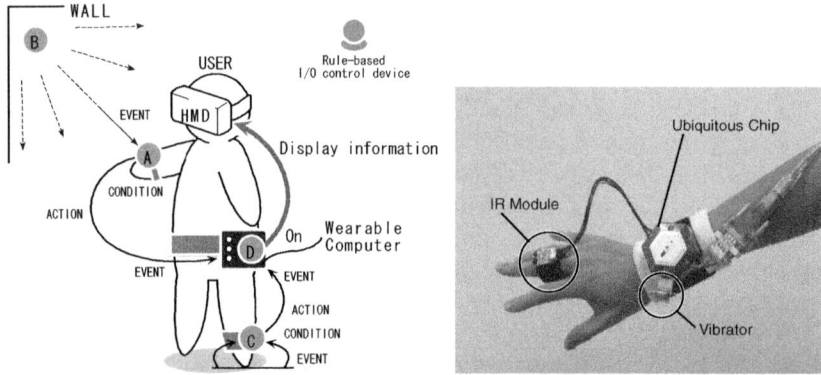

Fig. 15. An example of Application (2)

device. Of course, the flexibility of our device enables the smart space to change its functionality in response to different situations by using the dynamic change of rules.

The second example, illustrated in Figure 15, is the application for wearable computing environments. This application covers the following scenario:

- The user can acquire local information disseminated from a wall when he points his fingers at the wall. However, this acquirement is performed only when he stops walking.
- The user can change the wall light's blinking pattern by pushing a button on his wrist when he points his fingers at the wall light.

Table 6 shows the ECA rules for this application. RULE 1-3 are stored in ubiquitous chip A, while RULE 4, 5, and 6 are stored in ubiquitous chip B, C, and

Table 6. The rule set for Example (2)

RULE 1	RULE 2	RULE 3
E: RECEIVE(M1)	E: RECEIVE(M3)	E:
C:	C:	C: I1=1
A: SEND(M2)	A: O1=1	A: SENDECA

RULE 4	RULE 5	RULE 6
E: TIMER	E:	E: RECEIVE(M2)
C:	C: I2=1, I3=1	C: I4=1
A: SEND(M1)	A: O2=1	A: SEND(M3), SEND(M4)

I1: button	O1: vibrator	M1: hello packet
I2: right foot	O2: notify of stay	M2: detection
I3: left foot		M3: vibration request
I4: connected with O2		M4: detection to wearable PC

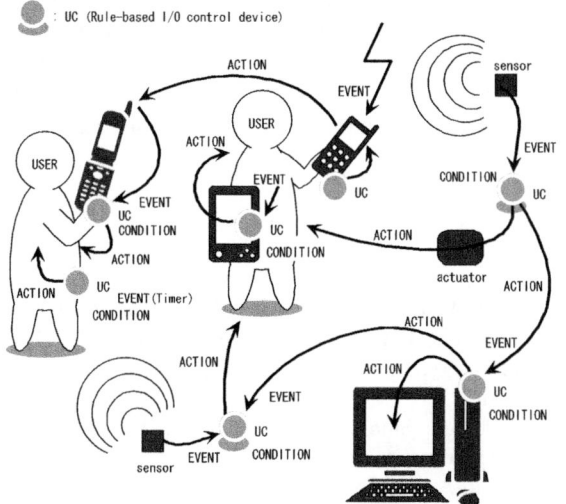

Fig. 16. A ubiquitous computing environment with rule-based I/O control devices

D. RULE 4 disseminates the beacon at a regular interval, and RULE 5 sends the status of the user's feet to ubiquitous chip D. RULE 1 detects the beacon from the wall (ubiquitous chip B) and notifies ubiquitous chip D. If ubiquitous chip D receives the notification and he stops walking, RULE 6 sends the detection of the disseminated beacon to the wearable computer and ubiquitous chip A. RULE 2 then turns on the vibrator. RULE 3 sends the specific ECA rules to ubiquitous chip B via the infrared module when the user pushes the button on his wrist.

In this way, using our devices, we can construct a ubiquitous computing environment that integrates embedded computers, wearable computers, artefacts, and users. Such a system is illustrated in Figure 16.

6   Conclusion

In this paper, we have described the design and implementation of the ubiquitous chip, which is a rule-based I/O control device for ubiquitous computing. The two characteristics of our device, (1) the separation between I/O control and attachments and (2) the rule-based approach, work effectively to construct applications in ubiquitous computing environments. Moreover, we have presented examples of services that use our devices. These applications show the possibility of integration of ubiquitous computing environments and wearable computing environments by using ubiquitous chips.

In the future, we plan to construct ad-hoc networking functions with ubiquitous chips, and various other functions such as analog data processing. Moreover, in its current state of implementation of development environment, it is still rather difficult for users to develop/customize the system behaviors and to develop larger systems composed of hundreds of ubiquitous chips. Therefore, we plan to provide new development tools for end-user programming and larger systems.

References

1. G. D. Abowd, C. G. Atkeson, A. F. Bobick, I. A. Essa, B. Macintyre, E. D. Mynatt and T. E. Starner: The Future Computing Environments Group at the Georgia Institute of Technology, In Proc. of the 2000 Conference on Human Factors in Computing System.
2. M. Beigl, H. Gellersen: Smart-Its: An Embedded Platform for Smart Objects, Smart Objects Conference (sOc) 2003.
3. J. Hill, R. Szewczyk, A. Woo, S. Hollar, D. Culler, and K. Pister: System architecture directions for networked sensors, In Proc. of the 9th International Conference on Architectural Support for Programming Languages and Operating Systems, pp. 93–104, 2000.
4. L. Holmquist, F. Mattern, B. Schiele, et al: Smart-Its Friends: A Technique for Users to Easily Establish Connections between Smart Artefacts, In Proc. of 3rd International Conference on Ubiquitous Computing (UbiComp 2001), pp. 116–122, 2001.
5. J. Kahn, R. Katz and K. Pister: Mobile Networking for Smart Dust, In Proc. of ACM/IEEE International Conference on Mobile Computing and Networking (MobiCom99), pp.271–278, 1999.
6. Y. Kawahara, M. Minami, H. Morikawa, T. Aoyama: Design and Implementation of a Sensor Network Node for Ubiquitous Computing Environment, In Proc. of VTC2003-Fall 2003.
7. MICA, http://www.xbow.com/products/Wireless_Sensor_Networks.htm
8. M. Resnick and S. Ocko: LEGO/Logo: Learning Through and About Design, http://llk.media.mit.edu/papers/1991/11.html.s.
9. P. Saffo: Sensors: The Next Wave of Infotech Innovation, Ten-Year Forecast, Institute for the Future (IFTF), pp. 115–122, 1997.
10. K. Sakamura: TRON: Total Architecture, In Proc. of Architecture Workshop in Japan'84, pp. 41–50, 1984.

11. R. Want, A. Hopper, V. Falcao and J. Gibbons: The Active Badge Location System, ACM Transactions on Information Systems 10(1), pp. 91–102, 1992.
12. M. Weiser: The Computer for the Twenty-first Century, Scientific American, Vol. 265, No. 3, pp. 94–104, 1991.
13. P. Wellner, E. Machay, R. Gold, M. Weiser, et al: Computer-Augmented Environments: Back To The Real World, Communications of the ACM, Vol. 36, No. 7, pp. 24–97, 1993.
14. J. Widom and S. Ceri: Active Database Systems, Morgan Kaufmann Publishers Inc, 1996.

eSeal – A System for Enhanced Electronic Assertion of Authenticity and Integrity

Christian Decker[1], Michael Beigl[1], Albert Krohn[1], Philip Robinson[1], and Uwe Kubach[2]

[1]Telecooperation Office (TecO), University of Karlsruhe
Vincenz-Priessnitz-Strasse 1, 76131 Karlsruhe, Germany
{cdecker,michael,krohn,philip}@teco.edu
[2]SAP AG, Corporate Research
Vincenz-Priessnitz-Strasse 1, 76131 Karlsruhe, Germany
uwe.kubach@sap.com

Abstract. Ensuring authenticity and integrity are important tasks when dealing with goods. While in the past seal wax was used to ensure the integrity, electronic devices are now able to take over this functionality and provide better, more fine grained, more automated and more secure supervision. This paper presents eSeal, a system with a computational device at its core that can be attached to a good, services in the network and a communication protocol. The system is able to control various kinds of integrity settings and to notify authenticated instances about consequent violations of integrity. The system works without infrastructure so that goods can be supervised that are only accessible in certain locations. The paper motivates the eSeal system and its design decisions, lists several types of integrity scenarios, presents the communication protocol and identifies practical conditions for design and implementation. An implementation in a business relevant scenario is presented as a proof of concept.

1 Introduction

It is an important issue to claim and assert the authenticity and integrity of goods, documents or other valued objects in storage or transit. In these times objects of value like documents, deeds, contracts, goods for trade, and other articles, which we collectively refer to as goods, were stored in a container, which in turn was sealed with wax and the imprint of a seal ring (bearing an insignia) or a plumb. This method ensured two important fundamentals of secure and dependable object handling: Authenticity and integrity. The object's authenticity is detectable through the seal ring imprint on the wax and the integrity can be discerned by inspecting for either of the two physical states of the seal - *valid* or *broken*. Nevertheless, modern technology provides advanced methods for violating both integrity and authenticity but also enables us to better protect objects of value.

This paper introduces an electronic seal concept, the eSeal. Like a wax-seal, an eSeal can be applied to physical goods to electronically claim and assert their authenticity and integrity. The eSeal is intended to claim and assert states but not to protect the object itself. However, unlike a wax-seal, an eSeal can detect a larger variety of

A. Ferscha and F. Mattern (Eds.): PERVASIVE 2004, LNCS 3001, pp. 254–268, 2004.

integrity violations – including electronically originated attempts. It can collect context information about this violation including the time and location, can actively monitor and alert and it can perform all these tasks automatically and autonomously. An eSeal can exchange relevant information with other computer systems, maintaining a fine-grained correlation of physical conditions and an interpretation in the information world.

Although the design and concept are generally applicable in many areas, this paper motivates and explains the eSeal concept alongside business applications. In the area of integrity and authenticity supervision, business applications provide an interesting environment with numerous demands for eSeal-related applications. Dwelling in the business application domain also motivates interaction and hence extension of existing information systems through appropriate interfaces to eSeal components.

The paper proceeds to give an analysis about various integrity classes the eSeal can keep track of and explain the eSeal practical considerations, which include particular requirements and constraints of the overall system. This leads into the system design, where the components and their dependencies and tasks are explained and the operational features are outlined. Due to the system constraints and operational features, there were particular security and technical challenges that necessitated further analysis, before practically evaluating the concept through a concrete application. We also discuss related work towards the end of this document.

2 Motivation and Analysis

As a motivating example, see Figure 1, we select a representative logistic scenario to clarify the capability and advantage of an eSeal.

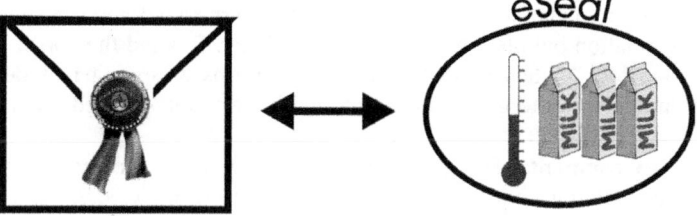

Fig. 1. A traditional seal compared to an eSeal

In this example, a temperature sensitive good like milk is transported and need to be kept in a certain range of temperatures for goods' quality reasons. An eSeal is used to assure the temperature of the goods during transport between two locations. During the transport the eSeal permanently monitors the current temperature of the goods. As long as the temperature is within the acceptable range, the eSeal is considered to be valid, otherwise broken. Once broken it can never be recovered to the valid state like a broken wax seal. When the transported goods arrive at their final destination, the eSeal can report authentically whether the temperature range was held.

Like the traditional wax seal on envelopes the eSeal can protect valuable goods. The simple protection with a wax seal can be matured with the surveillance of additional conditions. The eSeal can provide protection for goods sensitive to for instance temperature and light changes, vibration and radiation.

To go a bit deeper into the eSeal system, we analyzed two aspects of its general problem domain, presented as research questions. First, what are potential breach-of-integrity/authenticity situations and how are they classified? Second, what are practical constraints for an eSeal system design and implementation?

2.1 Integrity Considerations and Classifications

The concept of integrity we want to target with the eSeal system is more than "inviolability", as guaranteed by wax seals. Depending on the object and context, integrity may still be in place even if the object is touched.

The eSeal domain spans over four different integrity classes, which we derived from an analysis of four scenarios in the business areas: storehouse, supply-chain management, office document management and production. These classes are:

- **Conditional Integrity**. This is upheld when the object's physical properties and object state remain unaltered or undamaged. In this case a full access to a sealed object may be allowed, in that the object may be used, but it is forbidden to change the state of the internal – e.g. information – or external – e.g. physical shape – of the object
- **Relational Integrity**. This is similar to the above, but considers the orientation and relation of constituent objects. Integrity is violated when someone adds or removes something from a sealed object collective. Objects may consist of several constituent objects like a palette of goods consist of several goods.
- **Authorization Integrity**. This is the classical wax seal integrity where no unauthorized party is allowed visual or tangible access to the sealed object. Integrity is broken if someone was able to see a defined state e.g. internal information but also the outline of the object. Beyond the "open the container and look" integrity violation, modern forms of spying include x-ray scans and methods to get access to internal information – stored programs and data – of an object.
- **Environmental Integrity**. In this case the object's integrity is violated if its surrounding conditions or context are unfavorable, e.g. that the object is brought into a place where it should not be.

These classes of integrity concerns must at times be addressed in tandem. For example, a policy could exist that includes access restrictions (Authorization Integrity) and yet that the object's structural properties must not be changed (Intrinsic Conditional Integrity).

Our method of defining these integrity classifications includes only a limited number of scenarios. Based on scenario descriptions we repeated the analysis until we found the same integrity protection situation again. This way we observed important situations that contributed to a design of a first eSeal system, but cannot ensure completeness. Further on, the list is based solely on business scenarios analysis, whereas other policies may be found when analyzing other areas of life. As potential exploitation scenarios are within the business area, we do not consider this a significant system drawback. Subsequently, the remainder of this section continues within the business area.

2.2 Practical Considerations

The practical considerations and important requirements for the eSeal system design were derived from the inherent goals and properties of the business scenarios that were analyzed for the potential usage of eSeals. They essentially describe and confine the nature of the goods handled, the locations that they are transited between, and the interaction with humans.

- **Mobility.** The eSeal system should not introduce any handling constrains of objects, work without cabling, be small and unobtrusive.
- **Diversity.** Goods have different physical properties like size, shape, weight and experience different environmental conditions. Different values are in the interest to be sealed (e.g. time of transport, temperature). Therefore the eSeal system must provide a flexible platform to realize an electronic seal on a certain good.
- **Incomplete infrastructure coverage.** Physical goods can move through various situations and different locations or environments. Since the support through an electronic infrastructure (e.g. W-LAN, cameras) cannot be guaranteed in all cases, the eSeal system must be able to work offline and autonomous. It must have intensive contact to the object and experience its environment as genuine as possible through monitoring equipment and sensors.

3   eSeal System

Definition: *An eSeal is an electronic seal, which can be applied on physical goods in order to provide the guarantee of important aspects of the protection of those physical goods. The eSeal does not physically protect the sealed goods but can provide propositions and evidence of authenticity and integrity.*

The eSeal system, see Figure 2, consists of three conceptual layers: (1) the *Contractual*, (2) the *Logical*, and (3) the *Technical*.

Firstly, we regard an eSeal as a contract between an *Initiator* (a subject that applies the eSeal to a physical good) and a *Receiver* (a subject that assesses the evidence presented by the eSeal) stating the terms and conditions under which the authenticity and integrity of a physical good can be asserted. Secondly, the system provides logic for determining and presenting the "protection state" of the target goods to which it applies. Thirdly, the system is realized through particular technologies that meet the functional and quality requirements for its operational domain. In addition, we have also considered the actors that drive or benefit from the system's functionality. We have already mentioned the roles of the Initiator and Receiver, who are considered as the end-points of the activity chain and the key actors in the contractual aspects of the system. Supporting the Initiator and Receiver in monitoring the state of the eSeal system, and hence the contract, are *Checkpoints*. A Checkpoint is an intermediary actor that forwards system state to either of the contractual parties upon their query. A Checkpoint is considered the most proximate trusted source to the sealed goods at a particular time. Operational system state of the eSeal is either "valid" or "broken" and this is determined by the processing of delivery conditions, with which the system is initialized, and current conditions. A multi-sensory device, attached to the physical goods, which we simply refer to as the *"eSeal device"*, senses these current conditions.

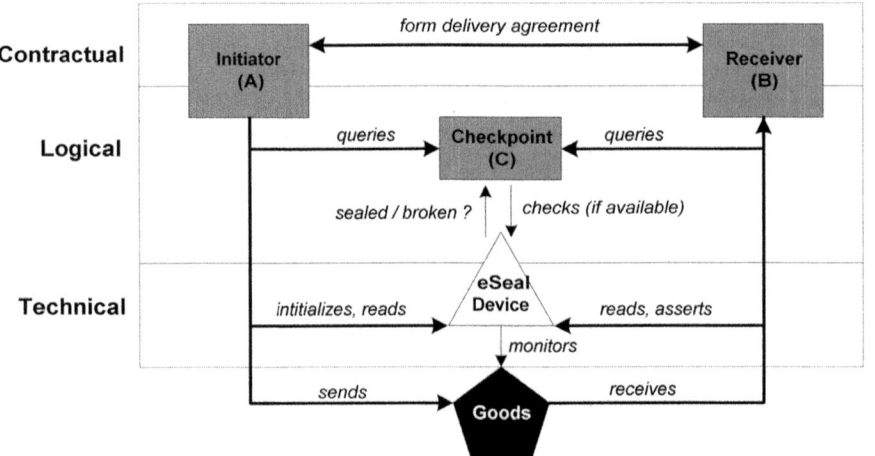

Fig. 2. eSeal System and Key Actors

3.1 eSeal Device

The eSeal device is a small embedded computer system directly attached to physical goods. The intention is to have an entity which can provide the Initiator or Receiver with a trustworthy statement about whether the operational system state also referred to as eSeal state is "valid" or "broken". An eSeal device implements three core functionalities: First, computation enables permanent updates of the eSeal's state using an algorithm derived from the contract between Initiator and Receiver. Second, a sensor system as part of the device supplies it with external information serving as input for the algorithm. Third, the device implements a communication functionality enabling the exchange of the eSeal's state and optionally additionally information with other parties in an authenticated manner. The eSeal device operates independently and without the support of a surrounded electronic infrastructure. Optionally, the eSeal device contains a timer to bind the eSeal's state to a timestamp and some storage capability to consecutively write the eSeal's state history and additional related information.

3.2   eSeal Activity Chain

The flow of the use of an eSeal, presented in Figure 3, involves at least three parts of the eSeal system: The Initiator, the Receiver, the eSeal device and optional Check-points.

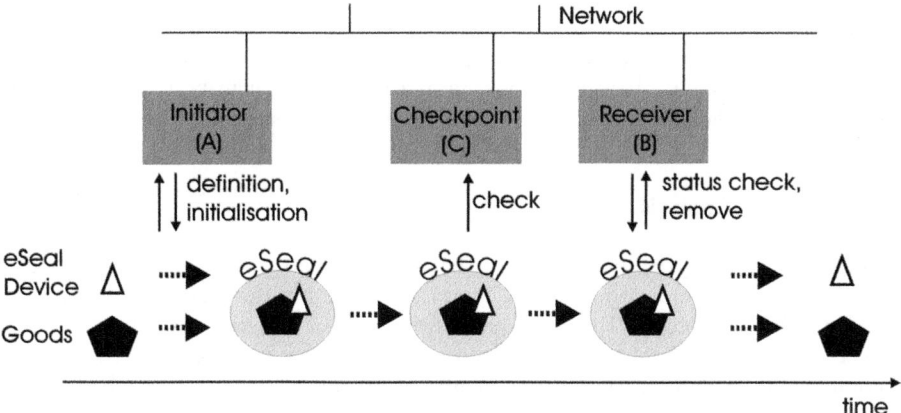

Fig. 3. eSeal Activity Chain

After the successful agreement and conclusion of a contract between Receiver and Initiator, the Receiver queries the Initiator to start the activity chain. The definition of the seal to be applied on the goods is derived from this contract.

The Initiator combines the eSeal device and the goods and initializes the eSeal device order to seal the goods which are subject to the contract and from this moment on, the eSeal device can be identified by the Receiver, Initiator and Checkpoints and carries the start state "valid". The initialization of the eSeal device includes the definition and download of all necessary information and algorithms that define the seal. The eSeal device can then continuously monitor the state of the seal.

If the Receiver or Initiator is interested in the actual state of the sealed goods, i.e. whether the seal is still valid or broken, they can query the state from Checkpoints or directly from the eSeal device using its communication interface. This enables both parties to prove whether there was an integrity breach or not.

The activity chain terminates when the sealed goods reach the Receiver. The Receiver and the Initiator query the state of the eSeal device for the states "broken" or "valid" and can decide whether the contract was held. If the eSeal device arrives "broken" at the Receiver, the contract partners can query the eSeal device for the reason of the breach of the seal. They can distinguish between breaches due to a contract breach or cases of attacks. In cases of severe attacks on the eSeal device which destroyed core functionalities of the eSeal device, this information might be lost. After the queries and contract examination the Receiver removes the eSeal device from the goods and thus deactivates the seal.

4   Operational Analysis and Challenges

When we discussed the eSeal approach in the second section of the paper, we mentioned two particular aspects of the problem domain that influenced the properties of the system architecture. These naturally have a significant bearing on the operational specification of the system, which is likewise separated. We therefore dedicated some resources to analyzing the security requirements and deriving a general functional protocol, and, secondly, analyzing the technical realization of the system based on the practical constraints. This also took into account the security requirements that emerged from the analysis, with respect to storage and processing.

4.1   Security Analysis

The Security Analysis considers the eSeal system actors, the nature of the goods to which it will apply, and the types of transactions and business scenarios that the system will be involved in. The eSeal protection goals of Integrity and Authenticity are once again revisited, but from a more detailed security perspective. There is large commonality with the concerns of authenticity and integrity in cryptographic analysis.

Authenticity: The receiver (B) must assert that a good or item (I) was really sent from a sender (A,) and is hence a genuine article, including that the electronic information also conforms to these properties. Threats include:
- A false initiator sends I by bearing A's identity (source masquerading)
- I or its electronic information is replaced in transit by a falsified item or data (replay attack)
- A false seal sends out item state information to A and B (seal masquerading)

Integrity: both the receiver (B) and initiator (A) must assert that item (I) (as well as its electronic information) is not tampered with while in transit, and that the correct handling policies are upheld. Threats include:
- I is tampered with (seal is broken) while left unattended, or by an authorised third party, therefore degrading quality of the product
- I is subjected to transit conditions that violate its handling policies

Other threats include the inevitable denial-of-service attacks through communications signal interference or continuous, unwarranted depletion of power resources. Additionally, in the case of highly sensitive information on the seal, confidentiality becomes another protection goal of eSeal. The communications protocol and power management features of the device address the denial-of-service attacks, while confidentiality is captured within the properties of the crypto protocols and physical handling policies enforced. These threats and their countermeasures, especially the asymmetric or public key protocol we use as our foundation, are well known in the field of security engineering [1]. However, it was a good opportunity to explore and assess the applicability of these standards within a domain where the physical and electronic protection goals are so tightly coupled.

4.2 The Detailed eSeal Communication Protocol

The protocol defined is based on the architecture depicted in fig. 1. It was specified in response to the security analysis, and details how the protection goals of the interaction between entity roles are captured. The protocol consists of 7 interaction phases, corresponding to the architecture depicted in fig. 1, but also of a set of security functions and elements defined below.

Security Functions and Elements

K_X: public key of an entity x
M: Query and status messages
n, q: Initial random sequence number, and sequence counter
P: Handling policy
$D_X\{\}$:   Decryption with private key of entity x
$E_x\{\}$:   Encryption with secret/ private key of an entity x
$H\{\}$:   Hash function
$S_X\{\}$:   Signing with private key of an entity x
$V_X\{\}$:   Verification of signature with public key of an entity x

1. **QUERY-ORDER**: receiver (B) sends an order request message (M_n), with which the initiator (A) can initialize an eSeal session. To avoid replay attacks at this stage, a signed hash of M_n, a random number n (used as a sequence number), and ipublic key (K_B) of the receiver (B) are also sent to the initiator (A). These are also encrypted with the public key of the initiator (A) - (i). Upon reception, the initiator (A) decrypts the packet using its private key – (ii), and then verifies the sender of the order, using the public key of B – (iii).

    $$B \rightarrow A: \quad E_A \{M_n, S_B\{H\{M_n\}\}, n, K_B\} \qquad (i)$$
    $$A: \quad D_A \{E_A \{M_n, S_B \{H\{M_n\}\}, n, K_B\} \qquad (ii)$$
    $$V_B \{S_B\{H\{M_n\}\}\} \qquad (iii)$$

2. **INIT-DEFINE**: initiator (A) starts the initialization process by defining a handling policy (P, which is a listing of context parameters), a statement of expected state-on-delivery (M_{n+1}), and by generating a key pair for the seal. The handling policy is encrypted with the private key of the seal to avoid electronic tampering. The seal is then electronically initialized with its private key (in protected memory), the handling policy, the public key of B (for communicating status updates to B with end-to-end authentication), and the expected state-on-delivery, which is hashed and signed by the private key of A.

    $$A \rightarrow Z: \quad \{P, n, S_A\{H\{M_{n+1}\}\}, K_B, M_{n+1}\} \qquad (iv)$$

    A then responds to B by sending a STATUS (see protocol operation 7), which includes sending the public key of the seal to B.

3. **SEAL**: Upon applying the seal to the item, this triggers the sensors to make the first check (see 6) in order to have an initial-sealed-state (M_{n+2}). The physical process of sealing also triggers an electronic process of encryption and signing of the initializing information and initial-sealed state respectively – (v). The seal can only be opened by parties that can respond to a challenge by the eSeal device,

such as the initiator (A) and receiver (B), as their public keys are known by the eSeal.

$$Z \; [I]: E_Z\{P, \; n, \; S_A\{H\{M_{n+1}\}\}\}, \; E_B\{M_{n+1}\}, \; S_Z\{H\{M_{n+2}\}\}, \; M_{n+2} \tag{v}$$

4. **QUERY**: This step in the protocol is equivalent to an ORDER. The only difference is that B may directly contact the seal, having received its public key, or it may need to go via a checkpoint E_A would therefore be replaced with E_Z , an operation on the eSeal itself, in (i), (ii) and (iii).

5. **CHECK**: Following an authorized party QUERY or internally scheduled query, the eSeal (Z) does a poll of its sensors and compares with the preferred context parameters specified in the handling policy (P). It then updates the last status (M_n) with current status (M_{n+q}), where q is equal to the sequence number of the query. There are three context states that the seal can be set to, and stated in M_{n+q}:

   - *VALID*: Current context match handling policy - seal remains intact
   - *DEGRADED*: Current context does not fully meet policy, but is within an acceptable bound – seal remains intact but records possible tampering attempt. For example, is currently in the hands of an unauthorized party.
   - *BROKEN*: Current context does not meet handling policy – seal is broken and relevant information is wiped from electronic storage

   The seal can also record the current handling party and labels them as *AUTHORIZED* or *UNAUTHORIZED* (unknown or black-marked). A higher-level notification is given off when the sealed item is being handled by an *UNAUTHORIZED* party.

6. **STATUS-RESPONSE**: There are two types of STATUS operations, which both transmit the CHECK to an authorized party. The first is a response to the authorized parties following a query. It is authenticated with a signature of the seal (S_Z). Additionally, depending on the policy, the status may be encrypted with the public key of the authorized party before forwarding. This is equivalent to forwarding the result of the crypto procedure in (v), where q = 2.

$$\begin{aligned} Z \rightarrow B: & \quad E_B\{S_Z\{H\{M_{n+q}\}\}, \; M_{n+q}\} & (vi) \\ B: & \quad V_C\{E_B\{S_Z\{H\{M_{n+q}\}\}, \; M_{n+q}\}\} & (vii) \\ & \quad D_B\{S_Z\{H\{M_{n+q}\}\}\} & (viii) \\ & \quad V_Z\{H\{M_{n+q}\}\} & (ix) \end{aligned}$$

7. **STATUS-DELIVER**: The second STATUS operation is when the item is physically delivered. The current handling party is set to AUTHORIZED, if B provides its public key K_B, i.e. responds to the eSeal's challenge. Furthermore, without K_B, procedure (viii) is not possible. If (viii) is not possible, then a notification is issued by the eSeal.

Important to note that in implementations where the microprocessor cannot support public key encryption, the eSeal challenge will have to be based on a symmetric approach. This would entail an earlier exchange of the eSeal secret key with the receiver and initiator, over a covert channel.

4.3 Technical Analysis and Realization

Reflecting back at the system architecture (Figure 2) the eSeal device is a central element of the eSeal system since this device is responsible for detecting integrity violation of the sealed goods. This section describes the technical details of the eSeal device and outlines requirements to prevent successful attacks which compromise the device. From section 3.1 the following functionalities are necessary in the eSeal device: computation, communication and sensing. Additionally, the device is supported by a power supply. The functionalities are implemented in different subsystems requiring separate appropriate protection against attacks. The figure below presents an overview about the components of an eSeal device.

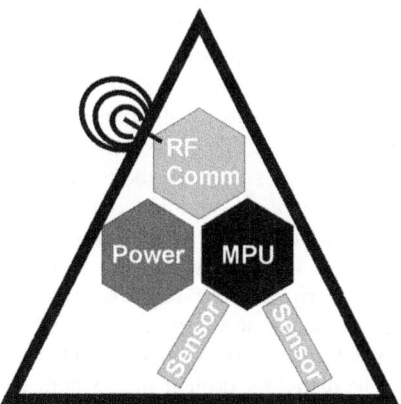

Fig. 4. Schema of an eSeal Device

All necessary computation functionality is implemented in the microprocessor unit (MPU). The MPU covers the tasks of cryptography (including de- and encryption, key management), permanent eSeal state determination and sensor value processing. Therefore the MPU contains the most sensitive data and present the most valuable attacking target. It is required that no invasive or non-invasive method will gain knowledge about the MPU's internal states. This complicated protection task is discussed in [9]. The authors describe there several ways to read out protected data from MPUs on SmartCards, but also effective countermeasures to those attacks Another requirement for the MPU is to hold a state within the MPU, which cannot be reproduced by any method once lost. This internal MPU state includes the eSeal device state and the integrity of the MPU itself. This state is wiped out of the MPU as soon as a seal breach or an attack is detected and will ensure that the seal cannot be reestablished. To our knowledge, there are currently two preferred MPUs on the market which fulfill these requirements. The first one is the DS5002FP [15] series from Dallas Semiconductor and the second one is the IBM 4758 architecture[14]. Both support countermeasures described in [9] to prevent non-invasive attacks. Additionally, both provide a protection of the MPU against invasive attacks using a physical shielding, e.g. a membrane to detect intrusion in order to avoid invasive attacks without notifications. However, publications like [2] and [5] point out that apart from hardware protection, the software layers in such processors need also be considered carefully. Otherwise, protocol attacks can make the hardware protection useless.

The communication hardware itself does not add further security vulnerability. The security tasks are part of the higher level protocols. A destruction of the communication or denial-of-service attack would prevent the receiver from reading the eSeal state. The receiver would consider the eSeal to be absent.

The sensors support the MPU to permanently update the eSeal state. Those sensors reside outside the tamper-proof shield of the MPU. The selection of the appropriate sensors generally depends on the target application. Sensors have to be selected from the requirements of the goods to be sealed and the required integrity situations. Outside the tamper-proof area of the MPU, sensors face attacks including manipulation of sensor values during the transport to the MPU or sensor cheating. In the latter case, the attacker tries to maintain the valid sensor conditions during his attack through creating the right environment in which the sensor is situated in. In order to accomplish attacks to the data transport from the sensor to the MPU, the transport has to be protected by either a physical protection such as shielding of cables and the sensor itself or by the use of crypto protocols for the data transmission. Latter will transform the sensor into another MPU based crypto system. One possibility how this can be realized is described in [11] for secure keyboard input in Next Generation Secure Computing Base (NGSCB) enabled computer. In order to approach the threat of sensor cheating, the MPU can regularly check the sensors' health state. This requires the sensor to record its operation conditions using further internal sensors of itself. This sensor-watches-sensor scenario can be replaced by a seal-watches-seal scenario, where an eSeal device can be supported by neighbored devices in order to verify its own reading. The physical arrangement of the goods to be sealed together with the eSeal device can also mechanically protect the sensors from an attacker. These considerations have to be made before initializing the eSeal since they depend heavily on the type of goods to seal and the expected attacks.

The eSeal device needs power source supplying all its components. In mobile scenarios, battery supply is appropriate to allow independent operation. Currently, the battery life-time determines the limits of the usage of an eSeal device since a power failure leads to the lost of the state of the MPU and therefore leads to the state "broken" in the eSeal device.

5   Applications and Implementations

We have recently started to implement the eSeal concept in various applications especially with the focus on the scenarios which were the basis of the integrity classification in section 2.1. One application for physical document integrity is considered in more detail.

5.1 General Implementation Details

Implementations of the eSeal devices are based on TecO's Smart-It Particle platform [3] providing the necessary functionality like sensing, computing and wireless communication of an eSeal device. Our eSeal prototype implementation adds more functionality where needed using the Particles hardware and software interfaces. The roles of initiator and receiver were taken over by regular personal computers. Connection between the eSeal devices and the Internet enabled personal computers are carried out via so-called XBridge devices which form a gateway between the wireless eSeal network and the Internet. Such Xbridge devices are installed at the site of initiators and receivers but also in certain checkpoints.

We developed a library and some hardware extensions for the Particle platform, namely to include special sensors needed for the eSeal applications. New software components focused on secure communication using the blowfish algorithm. Although the eSeal system design requires an asymmetric key algorithm, this will be available in the future implementation. The additional hardware we developed are capacitive sensors to supervise the integrity of compounds of goods. Using these kind of sensors we also investigated possibilities to detect invasive physical attacks to the eSeal device. Experiments are hereby still at the very beginning. Furthermore, eSeal devices have access to already built-in Particle functionality like Real-Time-Clock and the Cell-of-Origin location system[4].

5.2 A First eSeal Application for Document Integrity

In office environments documents are usually created in an electronic way. Nevertheless, for convenience or legal reasons they are also printed on paper. The DigiClip [6] is a digitally enhanced paper clip, which aims to bridge the gap between electronically created documents and their physical paper-based representation. It was developed to keep the state of an electronic document and its printed version consistent. Once clipped on a printed document (Figure 5) it is able to keep track of document locations on a room level granularity and to monitor various environmental and document specific contexts. Currently, it can detect contexts like "document put in a bag", or "page from/to document removed/inserted".

For the eSeal-based application the DigiClip device monitors the conditional integrity and environmental integrity of paper based document. We selected these two integrity situations because they represent two crucial document characteristics: the togetherness of all pages in a document and valid locations of a document. The device's capacitive sensor is able to detect the number of pages currently clipped and whether the clip is opened or not. Like the electronic file of the document keeps all pages within the document structure it is therewith possible to decide on the physical document whether all pages are still together or a page left the compound. The DigiClip's cell-of-origin location system enables the definition of areas where the printed document is allowed to stay. Like restrictions applied on the electronic document denying for instance move operations it is possible to apply such restrictions to the physical documents by limiting the handling to certain areas.

Fig. 5. DigiClip clipped on some Papers

Our scenario for using the DigiClip as an eSeal application was as follows: After an electronic document was printed it had to be transported from the initiator to receiver represented by personal computers in two different rooms. In between there were two checkpoints the clip had to pass and one other it was not allowed to pass. The initiator configured the DigiClip device to monitor the opening of the clip, the page count and the DigiClip device's locations along the path to the receiver. Therewith, the eSeal was established around the physical document. Its state was held in the memory of the Particle's micro controller. As long as the clip was not opened, the number of pages didn't change and the DigiClip device was on its way indicated by the checkpoints it has to pass, the structural and environmental integrity of the document was assured, i.e. the eSeal's state was valid. During the operation the device constantly monitored these integrity conditions. When the integrity was violated, meaning that the clip was opened or it was seen by the third checkpoint, the eSeal state was set to "broken" and reported back to the personal computers representing initiator and the receiver via Xbridge gateways in the checkpoints. When the clipped document reached its final destination the receiver personal computer queried the DigiClip device and could conclude the eSeal's state "valid" or "broken". All communication was encrypted using a symmetric blowfish algorithm because the micro controller on the Smart-Its particles is not powerful enough to practically implement advanced asymmetric algorithms like RSA. The shared secret therefore had to be exchanged over a covert channel, and out-of-band with respect to the device's communications. Using this first implementation we were able to detect both structural and environmental integrity breaches.

6   Related Work

There is other work which is related to our approach of an eSeal. Siegemund and Flörkemeier describe in [13] a scenario of smart product monitoring. Hereby, products are augmented with sensors to monitor exceptions like dropping of the product. This is then communicated to any mobile phone nearby without explicit pre-configuration. While the eSeal shares the use of sensors for detecting exceptions, it goes beyond this monitoring aspect. The eSeal state "valid" or "broken" is determined from conditions during initialization and current conditions. Sensor measurements are used to derive these current conditions. Further, the eSeal system design guarantees that only authenticated parties are able to query the eSeal's state and further that manipulations on the eSeal device are recognized.

The proliferation of electronic business processes has fostered the need to integrate physical goods into the electronic world. Especially in applications like supply chain management where goods are distributed among many different players, which might be spread around the world, it has become very important to electronically track such goods and to electronically assure their integrity and authenticity. As a consequence first electronic solutions like MacSema's ButtonMemory[10], Elogicity's eSeal [7], Hi-G-Tek's Active Hi-G-Seal[8] or Savi Technology's SmartSeal [12], which claim to seal physical goods, are available on the market. These solutions are based on various technologies like electrical contact in case of the ButtonMemory, RFID in case of Elogicity's eSeal, and GPS support in case of the Active Hi-G-Seal and the Smart-Seal. They mainly provide some tracking feature that makes it possible to monitor if your goods arrive at pre-defined checkpoints. Additionally, one can conclude whether someone access the device or the goods sealed by these solutions. Other integrity surveillance based on environmental conditions for instance is not achieved. Furthermore, except for the Active Hi-G-Seal, which uses 3DES, no other seal offers a secure communication. Our eSeal approach covers a wider scope towards other integrity conditions as well as a secure and authenticated communication to whom it is allowed to query the eSeal's state.

IBM's secure coprocessor, the IBM 4758, is guaranteed to work in a secure manner despite physical attacks [6]. In contrast to standard cryptographic accelerator chips this coprocessor puts cryptographic secrets and a tamper detecting and responding circuitry in a secure box. Any detected tamper event immediately results in loss of the cryptographic secrets. Hence this coprocessor unit can be considered as a sealed object, for which the integrity of condition is guaranteed. The scope of the seal is limited to the detection of intrusions into the secure box surrounding the coprocessor unit. Nevertheless the IBM 4758 can well serve as a hardware platform to built upon, for some forms of specialized eSeals as we introduced them in this paper.

7 Conclusion and Future Work

The background, approach, design, operational analysis and an applied example of the eSeal system have been presented in this paper. It has been shown that electronic counterparts may uphold the function of inert seals, in everyday applications. Furthermore, this primal functionality is extended by incorporating sensors, communications and micro processing, with the added capability of interaction with other information systems.

We foresee both economic and social impact if such an architecture were to be taken up by industry, and we are actively investigating such "take-up", by forming research and development projects and coalitions with industrial partners. Nevertheless, there is further work to do, as the extremities of reference implementations of this architecture have not been explored. There may be other application areas besides business and commerce. Sensors, microchips and communications capabilities will continue to evolve. Continuing experience commensurate with these developments will be disseminated throughout the research community.

References

1. Anderson, A. Security Engineering: A Guide to Building Dependable Distributed Systems. Published by John Wiley & Sons, 2001, ISBN 0-471-38922-6
2. Anderson, R., Kuhn, M.: Tamper Resistance - a Cautionary Note. The Second USENIX Workshop on Electronic Commerce Proceedings, November 18-21, 1996. Oakland, California. pp 1-11, ISBN 1-880446-83-9
3. Beigl, M., Zimmer, T., Krohn, A., Decker, C., Robinson, P.: Smart-Its - Communication and Sensing Technology for UbiComp Environments. Technical Report ISSN 1432-7864 2003/2
4. Beigl, M., Zimmer, T., Decker, C.: A Location Model for Communicating and Processing of Context. Personal and Ubiquitous Computing Vol. 6 Issue 5-6, pp. 341-357, ISSN 1617-4909, 2002
5. Bond, M.: Attacks on Cryptoprocessor Transaction Sets, Workshop on Cryptographic Hardware and Embedded Systems (CHES2001), 31st January 2001, Paris.
6. Decker, C., Beigl, M., Eames, A., Kubach, U. DigiClip: Applying electronic properties to physical documents. To appear in the Proceedings of the IWSAWC 2004, March 23rd 2004, Tokyo.
7. elogicity.com global track and trace solutions to all parties within the supply chain management process Available Online: http://www.elogicity.com/solutions.htm [Accessed: 07/11/2003]
8. Hi-G-Tek: Secured Cargo. Available Online: http://www.higtek.com/cargo2.htm [Accessed: 08/02/2004]
9. Koemmerling, O., Kuhn, M.: Design Principles for Tamper-Resistant Smartcard Processors. Proceedings of the USENIX Workshop on SmartCard Technology, 10-11 May 1999, Chicago, USA.
10. MacSema Inc.: MemoryButton Technology. Available Online: http://www.macsema.com/solutions.htm [Accessed: 08/02/2004]
11. Microsoft: Hardware Platform for the Next-Generation Secure Computing Base. Available Online: http://www.microsoft.com/resources/ngscb/documents/NGSCBhardware. Doc [Accessed: 08/02/2004]
12. Savi Technology: Securing the Smart Supply Chain. Available Online: http://www.savi.com [Accessed: 07/11/2003]
13. Siegemund, F., Flörkemeier, C.: Interaction in Pervasive Computing Settings using Bluetooth-enabled Active Tags and Passive RFID Technology together with Mobile Phones. In Proceedings of IEEE PerCom 2003 (IEEE International Conference on Pervasive Computing and Communications), March 2003, Fort Worth, USA.
14. Smith, S.W., Weingart, S.H.: Building a High-Performance, Programmable Secure Coprocessor. In Computer Networks, Special Issue on Computer Network Security, Vol. 31, pp. 831-860. April 1999.
15. Dallas Semiconductor: Datasheet to Secure Microprocessor DS5002FP. Available Online: http://pdfserv.maxim-ic.com/en/ds/DS5002FP.pdf [Accessed: 08/02/2004]

A Distributed Precision Based Localization Algorithm for Ad-Hoc Networks

Leon Evers[1], Stefan Dulman[1,2], and Paul Havinga1,2

[1] EEMCS Faculty, University of Twente, the Netherlands
[2] Ambient Systems, the Netherlands
{evers1,dulman,havinga}cs.utwente.nl

Abstract. In this paper we introduce a new distributed algorithm for location discovery. It can be used in wireless ad-hoc sensor networks that are equipped with means of measuring the distances between the nodes (like the intensity of the received signal strength). The algorithm takes the reliability of measurements into account during calculation of the nodes positions. Simulation results are presented, showing the algorithms performance in relation to its accuracy, communication and calculation costs. The simulation results of our approach yield 2 to 4 times better results in position accuracy than other systems described previously. This level of performance can be reached using only few broadcast messages with small and constant size, for each node in the network.

1 Introduction

Recent advances in digital radio communication technology, combined with the continuous increasing battery capacities and ongoing miniaturization and integration of digital circuitry, have opened up a set of new application areas for digital computing devices of various scale and purpose. The effects of cheap ubiquitous connectivity that GSM phones offer are already changing the world today. New applications are at the horizon, like the use of cheap, tiny, wireless connected, digital sensors that will be able to continuously monitor their environment, and communicate their findings in real time to whomever is interested. Large-scale use of these sensors, wireless connected in randomly deployed ah-hoc networks covering complete areas will have many uses, whether it be biologists monitoring live stock, farmers keeping track of their cattle, soldiers monitoring the battlefield, the employment of forest fire warning systems, and many others.

In this paper a distributed algorithm for location discovery is presented. It can be used in wireless ad-hoc sensor networks that are equipped with means of measuring the distances between the nodes (like the intensity of the received signal strength). The algorithm takes the reliability of measurements into account during calculation of the nodes positions. Simulation results are presented, showing the algorithms performance in relation to its accuracy, communication and calculation costs.

A. Ferscha and F. Mattern (Eds.): PERVASIVE 2004, LNCS 3001, pp. 269–286, 2004.

1.1 Related Work

These kinds of applications require a specific kind of devices, with a specific set
of requirements and capabilities, as well as limitations, as is described by Rentala
et al. in their survey on sensor networks [10]. Especially the sheer number, in
which deployment is desired, defines a set of limitations, which make the design
of these devices a particularly challenging one. Low cost and maintenance-free
use are among the most important properties, and restrict the technology to
make and operate them. So a strong focus to minimal need for hardware, energy
preservation, cooperation and autonomy are necessary in every step of the design
process.

One of the many research issues is the problem of location discovery. This
issue is an important one and has already had some attention in the past. In
[5] Hightower and Borriello present a comparative description on existing systems. Unfortunately, none of the described systems is designed for the specific
application targeted in this paper, and thus does not provide a good solution
for positioning in ad hoc sensor networks. The Radar system ([2]), specifically
designed for indoor use, depends on a central computer, performing the calculations, and requires a lot of manual calibration. The work described in [4], [15], [7]
and [13] as well depends on a centrally performed network-wide calculation, and
have strict requirements for the infrastructure set-up and/or node placement.

On the other hand, several systems have been designed with similar constraints on hardware and network configuration in mind. Three such systems
are described in [12], [6] and [11]. In [12], similar algorithms are described, but
an ultrasound ranging method is used, which is a more accurate distance measurement system, and doesn't deal with some of the problems that less accurate measurement systems pose, like RSSI used in this research. The systems
described in [6] and [11] on the other hand, are based on similar ranging techniques, and hardware requirements. These systems, however, because of their
rather low accuracy and coverage, might not be very useful for implementation
in many application areas. Their results, however, can serve as a comparison to
the solution proposed in this paper, which is done in Section 3.

The research described in this paper is a continuation of the work described
in [8]. The theoretical results of the calculations involved in the IQL algorithm,
as described in [8] have been tested inside a simulation during this research, and
some improvements to it have already been made.

1.2 Paper Outline

In this paper a solution for localization in ad hoc wireless networks is proposed.
We begin by describing the limitations and requirements of such a system in
Section 2. The general approach of the proposed system will be explained in
Section 3, and a more detailed description of the calculations involved is given
in Section 4.

With the described system, a simulation environment has been implemented,
to perform test on the performance. Section 5 gives a description of the details

involving the implementation, used to perform the tests, which are discussed in Section 6. Finally, a discussion on the performance in relation to other systems is presented in Section 7.

2   Preliminaries

The specific platform that our proposed system is intended for is that of a sensor network that can be deployed at random, consisting of a (possibly very large) number of similar sensor nodes, and only a very small amount of base stations or some other kind of (manually configured) fixed infrastructure. The sensor nodes will be small, cheap and battery operated, with short range radio frequency (RF) communication hardware, simple (slow) microprocessors, and additional sensing hardware.

This node is equipped with an 8 MHz 16 bits microcontroller (with 4 KB of data memory and 60 KB of program memory) and a radio transceiver with Received Signal Strength Indication (RSSI) capabilities.

Due to the nature of these nodes, some important design goals for the algorithms used to provide the localization data have been identified, much like the algorithms described and compared in Langendoen et al. [6]. Such algorithms should be truly distributed, as well as self-organizing, robust and energy-efficient. This means that the calculations should be made on each individual node, using as little as possible computation and especially communication, without relying on any fixed infrastructure or network topology, and being able to cope with changing conditions, such as node failures.

2.1   Environment

The kind of environment in which these algorithms have to work will be one where a large amount of randomly placed nodes forms a network using the short range RF transceiver to broadcast messages to only a small subset of the nodes in close enough vicinity to be able to receive those messages. Nodes that are able to directly communicate with each other will be called neighbors throughout the rest of this paper. The amount of neighbor nodes that each node can reach, called the connectivity from here on, can be variable by changing the transmit power of each node. A constant power level has been assumed in this research though.

A small subset of the nodes in the network, called anchor nodes, will initially be equipped with its own location, expressed by its coordinates relative to some network-wide coordinate system, either by manual configuration, or by using other location sensing techniques requiring extra hardware, like for example GPS. All other nodes will initially not know their location. Anchor nodes are assumed to have the same hardware capabilities, so factors like communication capabilities and energy consumption considerations will be equal to non-anchor nodes. Ideally, these nodes should be spatially distributed equal across the network, even though in certain application areas this might not be the case, or cannot be relied upon. Certain flexibility towards this property has to be assured by the localization algorithm. To minimize on installation and maintenance effort the fraction of anchor nodes in the network should be really small, and the location algorithm should be able to deal with this small amount of anchor nodes.

The results presented in this paper only focus on situations with fixed sensor locations, since this already proves to be enough of a challenge. Other environmental factors, like the positions of objects in between the nodes, might be changing, however, resulting in varying readings of RSSI measurements between pairs of nodes at different times. Systems where the nodes are mobile can use the described algorithms to update the nodes positions continuously. However, this has not been implemented yet and is the object of our future work. The presented results can be easily adapted and can be used as a guide when designing a mobile system though.

3 Solution Outline

As the basis of an algorithm to determine a node's location, some measurements have to be available. With the hardware and network structure described before, the received signal strength indication can be used to obtain an indication about the distance between a pair of neighbor nodes in the network. The specific calculations to translate the RSSI readings into the distance towards the sending node will not be addressed in this paper. It is assumed that such a calculation can be constructed and is available to the algorithm presented herein.

3.1 Precision of Measurements

If the distance to at least 3 neighbor nodes is known, as well as the locations of those nodes (for example because they are anchor nodes), the position in 2-dimensional space can be computed using triangulation calculations. Unfortunately, especially in indoors environments, the distances obtained from the measured RSSI will be quite imprecise, because of the fading and multi-path effects of the radio signals meeting with the various surfaces in the surroundings of the node. According to [3], this error can be as large as 50% of the measured distance. The error of the measurement does however conform to a Gaussian distributed random variable. As a descriptive value of the error distribution the standard deviation can be used. For maximum errors of 50% this means standard deviation of about 20% of the distance. This value will be referred to further on as η. Different environments with other error characteristics will result in different (possibly smaller) values of η.

By itself, with distance errors of this size, the computation through triangulation will contain a large error as well, especially because of the accumulation of the errors in subsequent calculations. This might render the result of the calculation practically useless. However, by using the connectivity of the network, which is usually more than 3 neighbors per node, the redundancy in the network can be exploited to improve on the results of estimating a nodes location. Other factors, such as the known properties of the error distributions, as well as obtaining multiple measurements between pairs of nodes instead of just one, can be taken into account as well to try to obtain a reasonable precision of the computed location.

3.2 Iterative Multilateration

For nodes with more than the minimally required 3 neighbors with known position, the "iterative multilateration" method, described by Savvides et al. [1] can be used to calculate the position with smallest error. Because of the known error distribution of the distance measurements, the error of the obtained location can be calculated as well. This can be modelled as a Gaussian distributed random variable, denoting the probability of the real location of the node to be within a certain range of the computed value, expressed with the standard deviation of the error distribution. In other words, the standard deviation serves as a measure of the precision of the location estimation.

This newly calculated position, combined with its precision, can now be used in subsequent calculations for other still undetermined nodes. This is done by combining the precision value of the newly located node with the error of the measured range between this node and its neighbor to obtain a total precision value on the distance between the node's real location and its neighbor. When a still undetermined node obtained at least three range measurements to already determined neighbor nodes, it can itself calculate its position. The precise calculations involved in this process will be described in detail in Section 4. This iterative multilateration can be used again to refine a node's position, when more neighbor nodes have their position calculated, or after the neighbor nodes have obtained refined position estimates.

This approach of refining a node's position based on the measured ranges to neighbor nodes is used as well in the refinement stage of the system described in [12], but because a precision value of a node's initial position is known as well in the scenario described above, initial position estimates of non-anchor nodes can be used to calculate a position estimate of undetermined nodes as well. When a possibly very imprecise (with large standard deviation value) location estimate is used in subsequent estimation calculations, the large error is accumulated in the results of the new calculation. Though, this result is provided together with an even larger standard deviation of the position error meaning that the lower accuracy is provided with the result of calculation.

3.3 Multi-hop Ranging

The problem that still arises is that at the start of the location discovery algorithm, only the anchor nodes will have a known location (with infinitely high precision). Because of the small fraction of anchor nodes, in many cases there will be no non-anchor node with at least three anchor nodes as its direct neighbors, allowing it to calculate its own position. To almost all non-anchor nodes the anchors are several hops away, and no direct range measurements can be obtained. In such situations no positioning can be achieved. It is possible though, to obtain a distance to the anchor nodes from all non-anchor nodes, by using the distances measured at the intermediate hops on the shortest path to each anchor. The multi-hop distance can be obtained by multiplying the sum of all single hop measurements by a precomputed bias factor. The standard deviation that belongs to the multi-hop distance is computed in the same way. In Section 4.2 the computation of these bias factors is explained.

4   Calculations

4.1   Iterative Weight Least Squares Estimation

The calculations of a nodes position from a set of range measurements between the node with unknown position and a set of nodes with known or estimated positions can be performed in a way similar to that of GPS [9] and the "iterative multilateration" described in [1]. The main difference with those systems and the one proposed in this paper is that the precision estimates of all data values, expressed as a standard deviation of the error distribution, are also taken into account in the whole calculation. This section deals with the description of the Iterative Weight Least Squares Estimation used in the proposed system.

Starting from an initial estimation, an improvement vector is calculated iteratively and added to the previous estimation until the improvement vector is smaller than a certain value. This vector is obtained through a weighted least squares estimation:

$$wA\mathbf{x} = w\mathbf{b} \tag{1}$$

where w is a weight factor, A is a matrix, and $\mathbf{x}$ and $\mathbf{b}$ are vectors.

To calculate the improvement vector we take the true position of the node to be calculated as $\mathbf{x} = (x, y)$ for a 2-dimensional system, or $\mathbf{x} = (x, y, z)$ in the 3-dimensional case. The initial estimation of the position is denoted as $\mathbf{x}^{est}$ and the positions of the n neighbor nodes as $\mathbf{x}_i$, for $i = 1..n$. The measured ranges to those nodes can be denoted as: $r_i = \|\mathbf{x}_i - \mathbf{x}\| + \epsilon_i$, $i = 1..n$, with ϵ_i denoting the measurement error, and the node's true position as: $\mathbf{x} = \mathbf{x}^{est} + \delta\mathbf{x}$.

In the same way, the distances to the estimated position are: $r^{est} = \|\mathbf{x}_i - \mathbf{x}^{est}\|$, $i = 1..n$ and

$$\delta r_i = r_i - r_i^{est} = \|\mathbf{x}_i - \mathbf{x}^{est} - \delta\mathbf{x}\| - \|\mathbf{x}_i - \mathbf{x}^{est}\| + \epsilon_i$$

$$\delta r_i \approx -\frac{(\mathbf{x}_i - \mathbf{x}^{est})}{\|\mathbf{x}_i - \mathbf{x}^{est}\|} \cdot \delta\mathbf{x} + \epsilon_i = \mathbf{1}_i \cdot \delta\mathbf{x} + \epsilon_i, \, i = 1..n \tag{2}$$

with $\mathbf{1}_i$ the direction vector of length 1 from the nodes estimated position to node i.

For each range measurement r_i a weight $w_i = 1/\sqrt{\sigma_i^{edge2} + \sigma_i^{node2}}$ is calculated, where σ_i^{edge2} is the variance of the range measurement to node i, and σ_i^{node2} is the variance of the position of the node i.

Matrix A consists of n rows, each filled with the direction vector $\mathbf{1}_i$, one for each anchor node involved in the calculation. Vector $\mathbf{b}$ is constructed as a column vector filled with the values δr_i, one for each anchor node. Note that the i'th row of the matrix A needs to correspond with the i'th row of vector $\mathbf{b}$ with respect to the anchor node the values are calculated off. The least squares solution of equation (1) will then look like this:

$$
\begin{bmatrix} \mathbf{1}_1 \cdot w_1 \\ \vdots \\ \mathbf{1}_n \cdot w_n \end{bmatrix} \cdot [\delta \mathbf{x}] = \begin{bmatrix} \delta r_1 \cdot w_1 \\ \vdots \\ \delta r_n \cdot w_n \end{bmatrix} \tag{3}
$$

This way the improvement δx is calculated. In general, the least squares solution of $\mathbf{x}$ for $A\mathbf{x} = \mathbf{x}$ is: $\mathbf{x} = A \cdot C \cdot \mathbf{b}^T$ where $C = (A^T \cdot A)^{-1}$.

Using covariance matrix C, a square matrix with the number of rows and columns equal to the dimensionality of the system, the node's standard deviation is calculated as:

$$
\sigma^{node} = \sqrt{\frac{1}{2} \sum_i \sum_j C_{i,j}} \tag{4}
$$

When the estimated position of the nodes is calculated like this, the optimal location will be calculated, based on the given ranges to and positions of the neighbor nodes.

4.2   Multi-hop Distances

Calculation of the distance between two nodes $\mathbf{A}$ and $\mathbf{B}$ that are connected only by a path of more than one hop can be performed indirectly. On each hop along the shortest path between the nodes, the range and corresponding standard deviation are measured. An estimation of the distance and precision between $\mathbf{A}$ and $\mathbf{B}$ can be made, however, by taking the sum of all these ranges. The distance calculated this way, called r_m will usually be larger than the true distance between the two nodes, and the standard deviation will be larger than the summed single-hop standard deviations σ_m, because of the error introduced by the less precise calculated distances. A better estimation of those values can be made, however, by statistically analyzing large sets of these summed measurements.

Every measured summed distance r_m can be seen as a sample of a random variable R_{r_m}. If this variable is of Gaussian distribution, it serves as a good bases to obtain estimation of the actual distance. The expected value of R_{r_m} is the mean of the distribution, $E(R_{r_m}) = \mu_R$. Normalizing R_{r_m} by the measured distance r_m results in a normalized random variable $Q = R_m/r_m$ for all values of

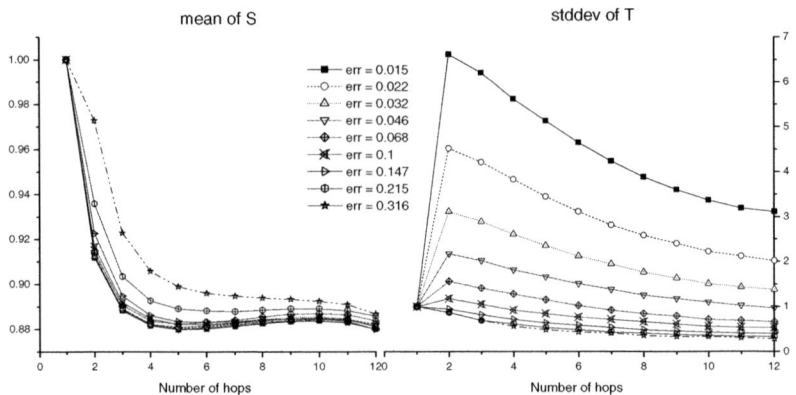

Fig. 1. Bias table values for μ_S and σ_T

r_m measured along paths with equal number of hops, from which the estimated value of the true distance r_t can be calculated as: $r^{est} = \mu_Q \cdot r_m$. The error between the measured distance and the estimated true distance $\epsilon_{r_m} = r_m - r^{est}$ can, in the same way, be seen as a sample of the random variable S_{r_m}. The standard deviation σ_S of S_{r_m} is the precision of the estimated distance r^{est}.

A normalized random variable T is defined as having samples ϵ_T, obtained by dividing all samples ϵ_{r_m} by the expected error η on the true distance r_t:

$$\epsilon_T = \frac{(r_m - r^{est})}{r_t \cdot \eta}.$$

The expected error η is an environment and hardware dependent value of the range measurement error (given as a percentage of the range) as mentioned earlier in Section 3.1. From the standard deviation σ_T of T the precision of the estimated true range r^{est} can now be calculated from σ_m and σ_T as: $\sigma^{est} = \sigma_m \cdot \sigma_T \cdot \mu_Q$.

The values for σ_T and μ_Q can be calculated offline, and stored in a table, containing one set of (σ_T, μ_Q) values for all possible hop lengths of the shortest hop paths between any pair of nodes. These values for single hop paths (for nodes that are each others direct neighbors) will of course be $(1,1)$. The value of σ_T does depend on the η, so for different values of η another table will have to be used. This value does however only depend on the hardware and environment used for the network, which will be known before deployment of the network, so only one table will have to be stored or in active use throughout the whole localization calculation. Figure 1 shows the values of σ_T and μ_Q for values of η and different hop counts.

5   Simulation

5.1   General Set-up

To test the performance of described system, a simulation environment was built using the OMNeT++ discrete event simulator [14]. The tests performed in the simulation environment are described is detail in Section 6. The network communication model used in the simulation makes use only of local broadcasts, where messages are delivered to all nodes within a certain and fixed range from the sending node. Colliding transmissions and message corruption are abstracted from (it is assumed that lower network layers will be in charge of providing a reliable broadcast service). The positions of all nodes are fixed during a simulation run, and are determined at initialization. The simulation environment uses a 2-dimensional coordinate system. Each node's coordinates are selected randomly and uniformly distributed within a square region of a size in units equal to the amount of nodes in the network.

During network initialization a number of nodes is selected as anchor nodes, in a way similar to the one used by Langendoen et al. [6]. The anchors are selected in a grid-like structure, where the nodes closest to the points of the grid will be chosen to be anchors. The grid is of size $s \times s$, $s \leq N$ with N the number of anchor nodes, and s the largest number that still satisfies the inequality. The rest of the anchors are selected randomly from the remaining nodes. This way of selecting the anchor nodes is chosen because it provides a more or less equal environment for all nodes, where the closest anchors will always be a certain maximum distance away. It is beneficent as well for the performance of the algorithm, because it proves difficult to obtain a correct estimation of a node's position if it is not surrounded by anchor nodes, as can be seen in the simulation results. Another reason to choose this strategy is because it allows easy comparison to the results described in [6].

The distance measurements between nodes are calculated from the signal strength of a transmission. This means that every received message also provides a range indication between itself and the sending node. In reality the calculation of this range infers a certain amount of error, as mentioned earlier. In the simulation the range measurement including error is modelled by drawing a random value from a Gaussian distribution with the true range as its mean, and a standard deviation of $\eta \cdot true\,range$. The implementation of the algorithm records all range measurements from each message it receives, and uses the average distance measured to a neighbor node in all further calculations. This way, as more messages are received, the measured distance will converge to an ever more constant value, which leads to better position estimates during refinement.

5.2   Algorithm Details

The protocol uses a two-phase approach, and relies on two corresponding kinds of messages being passed between the network nodes; start-up messages and refinement messages. During the start-up phase, a node attempts to calculate an initial position estimates, based on the distances towards the anchors. This initial position will then be improved to get a more accurate estimate during the refinement phase.

Start-up messages contain information about the distance and hop count, as well as the position of another node (with known, or at least estimated position) to the sending node. Refinement messages contain the position of the node that sent it. When the localization algorithm starts, usually when the node is switched on, or awakes from a sleep state, a non-anchor node starts by broadcasting an announcement message, indicating its presence and requesting for information from nearby nodes. On activation, an anchor node directly starts by broadcasting its own position. For its surrounding nodes, this is an indication of its presence as well. Upon receiving an announcement message, a node broadcasts all information currently known to that node. This way, the new node quickly learns the positions of all nearby nodes, and helps the newly connected node to catch up on the current state of the network. For all nodes in start-up state, all received messages with hop count lower than the maximum are stored, until enough distances (at least 3 in the 2-dimensional simulation) are known to be able to calculate its own position estimate. If an estimate can be calculated, the node will proceed to refinement state, and broadcast the position estimate just calculated.

Nodes in refinement state only collect refinement messages from direct neighbors. After receiving a position update from one of its neighbors, the node can re-calculate its own position, based on the new or more precise position just received, together with all other neighbor's positions. If this new position estimate is more precise than the previously known location (smaller σ), this new position estimate is broadcast in a refinement message. This strategy ensures finiteness of the algorithm. After each round of message passing, the improvement in calculated precision will be smaller. If a node cannot calculate a more precise position update, it will not rebroadcast its position. All other nodes will recalculate their positions based on fewer position updates from their neighbors, and the new estimate will be more likely not to be more precise, in which case the node will stop re-broadcasting as well. After a few rounds no node will be able to improve its position estimate, and the algorithm stops.

As the protocol progresses, nodes update their position estimates based on the improved position estimates of their neighbor nodes, and possibly a neighboring anchor node. These improved position estimates of the neighbor nodes in their turn, were calculated from the previous estimate of their neighbors. So, in effect, a node's new position is calculated indirectly from its own previous improved estimate. It is important that all nodes recalculate their positions on a similar rate, so ensure each node keeps a reliable calculated precision of its position estimate, relative to it's neighbors. This is ensured by allowing a node to re-broadcast only after a certain, network wide, fixed time after the previous position estimate was broadcast. This also keeps the number of broadcasts low.

Anchor nodes already know their position from the start. Because their position is obtained in a way external to this algorithm, they do not recalculate it.

As a result, for the majority of the time, they can be silent, and only react to announcement messages by rebroadcasting their fixed position. The precision of anchor nodes positions is obtained externally as well, and can be much greater than possible to achieve with the calculations of this localization algorithm. In the simulations, the anchor nodes precision is set to be infinitely high.

6 Results

With the simulator described above, a series of tests have been performed to obtain measurements on different performance factors of the system. The performance characteristics depend on several factors in the environment of the system, like the number of nodes, connectivity, error size, and number of anchors. These values can be set as input parameters for the simulator. This large set of inputs makes it impossible to thoroughly test the system through the complete range of all combinations of input parameters. Therefore, a set of parameter values is chosen, from which one value is changed every time to test the system's sensitivity to that particular parameter. These standard values are:

- Number of nodes: 225, placed on a square area of 15 units length;
- Radio range: 2.1 units, results in connectivity of about 12;
- Relative range error η: 10%;
- Number of anchors: 5% = 11 anchors;
- Multihop distance hop count: 4 hops.

These values are chosen to be equivalent with the parameter values of the standard scenario in [6], again for easy comparison of the respective results.

Because of random effects, produced by the use of randomly selected error values, node positions and possibly other factors, consistent results on all of the performed tests can only be obtained by averaging over large quantities of individual test runs. To ensure consistency, all tests performed have been repeated at least 50 times.

6.1 Distance Error

As a measure of the performance of the described algorithm, as well as an indication of the usefulness in specific application areas, the accuracy of the algorithm, as well as its precision are important indicators. The accuracy can be described as the average distance between a node's actual position and its position estimate. The precision describes to what extent the real position error is near this average distance error. This is calculated using the standard deviation of the error values. In order to test those values, an initial test has been performed using the parameter values of the standard scenario. Figure 2 shows the error distribution of the relative position error along with the average and standard deviation values, indicated with the dashed line and two dash-dotted lines at each side of it at $\mu - \sigma$ and $\mu + \sigma$. Note that, even though the majority of the error values stay below 10%, the average error value is around 17%. This is due

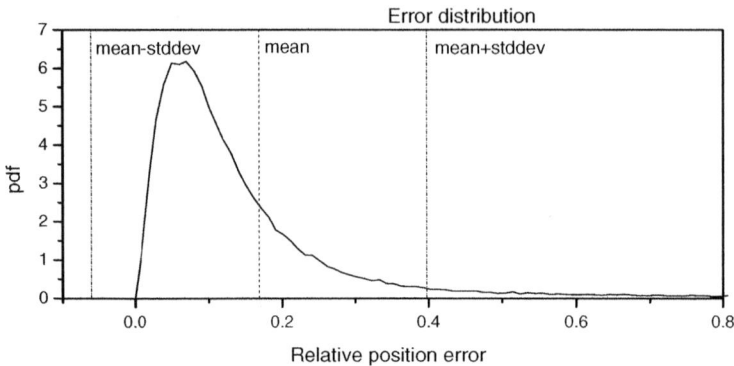

Fig. 2. Position error distribution for standard scenario

to the fact that a small number of nodes have relatively large errors, of up to and beyond the radio range. This fact is illustrated as well by the large standard deviation of about 0.23.

Even though, as the results show, each node's position cannot always be estimated very accurately, it does perform well on assigning a position estimate to all nodes. On average over the test runs executed in this first test a coverage factor of 98.9 % has been measured, meaning almost all of the nodes have obtained a position estimate, with the exception of the few nodes that do not connect to enough neighbors to actually start an estimation.

In the next series of simulations, the algorithm's sensitivity to variations in radio range, range error, and the fraction of anchors is tested. Figure 3 shows the results of these tests. In all three diagrams, both the average distance error and the standard deviation are shown, as well as the coverage factor.

The simulation results show that the system hardly is sensitive towards number of anchors, except for really small numbers. At an anchor fraction value of 1.8%, which is 4 anchors, no nodes are able to obtain a position estimate during the start-up phase. This is because the four nodes are placed at the corners of the area, and there is no node that has 3 anchors near enough to receive multi-hop distance information to them. This can be solved of course by increasing the multi-hop hop count, so the distance information from the anchors travels for a greater number of hops. From a value of 4% onwards the system shows little change in both relative error (average and standard deviation) and coverage.

As for the radio range, from a connectivity of about 8 to 10, the result is nearly the same, showing little change in relative range error. Note that the radio range and connectivity are related linearly, because of the uniform distribution of the nodes in the deployment area. In the results for both the anchor fraction and radio range variations, the average and standard deviation of the distance

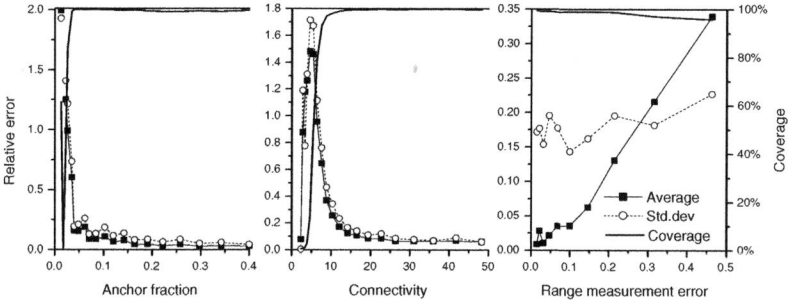

Fig. 3. Sensitivity towards anchor fraction, radio range and range error: connectivity (unmarked, right scale), average (square mark) and standard deviation (circle mark)

error have proportional values. This indicates that the distribution of the range error is similar to the one shown in Figure 2, scaled along the x axis.

With respect to the range error sensitivity, this is somewhat different. While the standard deviation keeps a more or less constant value around 0.27, the average error does change linearly with η. The size of the range error is directly reflected by the accuracy of the system, whereas the precision remains largely unchanged.

6.2   Network Communication Cost

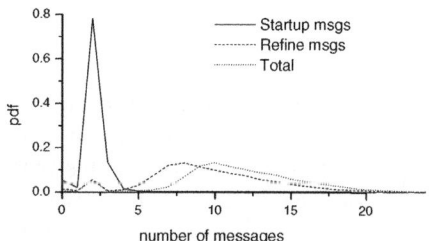

Fig. 4. Distribution of number of Startup and Refine messages for standard scenario

In a real-world implementation of an ad-hoc sensor network, energy preservation is of great importance. Because network communication is largely responsible for the energy consumption, this figure is of major interest as well. Some

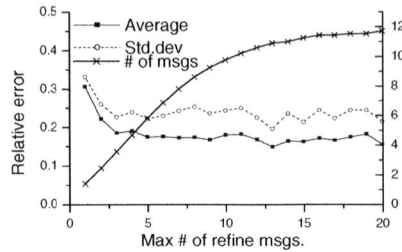

Fig. 5. Relative distance error and average number of Refine messages as a result of limiting the number of Refine messages

additional tests have been performed to quantify the network communication cost of this system.

Both the Startup and Refine messages have a small, fixed size. The network communication cost can thus be quantified by the number of messages broadcast per node. Figure 4 shows the probability distribution of both message types, and the total number of messages for the standard scenario. It is clear that the Refine messages take up the majority of the total number. Decreasing the number of Refine messages directly improves the total communication costs.

The algorithm can be adapted so that every node will only send a certain amount of Refine messages. The implications of this in relation to the distance error are shown in Figure 5. It clearly shows that after the first 3 or 4 Refine messages more communication hardly makes any improvement in accuracy and precision, while the average number of messages increases linearly. Limiting the number of Refine messages is an easy way to improve the total performance of the system.

6.3   Processing Requirements

Considering the small size and low cost of the targeted devices, only very limited processing resources will be available. The processing time to complete the localization process is an important aspect of this particular system.

As the central part of the system, the Iterative Weight Least Squares Estimation takes up the majority of processor cycles in the total calculation. This part of the calculation is therefore a good indicator of the total processing time used.

A number of factors are responsible for the total time each node spends calculating its (updated) position. At first, the size of the matrices used, and thus the spent calculation time, is proportional to the amount of neighbors involved in the calculation. Besides that, the number of iterations taken for the calculation to complete is a proportional factor to the total calculation time. At last, a third factor is the number of times the IWLS calculation is performed. In general, at least every time before a Refine message is sent, a node's position is re-calculated several times, until a more precise result is found.

The factors involved in calculation time, as mentioned above, have been examined, using the data sets generated for the earlier tests, described in Section 6.1. Figure 6 shows the number of times a calculation is performed per node, average number of IWLS iterations per calculation, average number of neighbors involved, and number of calculations per broadcast Refine message. The solid, unmarked line, on the right scale, shows the grand total of the number of calculations, number of iterations and number of neighbors multiplied. This is a number proportional to the total calculation time spent per node in the IWLS calculation for the whole localization algorithm.

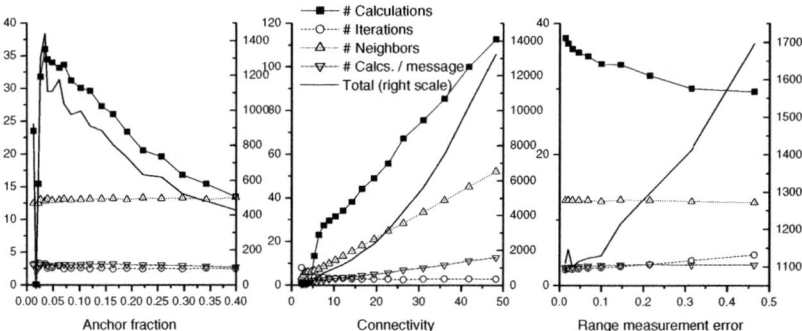

Fig. 6. Nr. of calculations for different parameter values. The number of calculations, iterations, neighbors and calculations per message are drawn on the left scale; the total calculation time per node is shown by the unmarked solid line, drawn on the right scale.

For the targeted devices, it is very important to only spend energy and processing time as long as it gives improvement to the positioning accuracy. Selecting the right set of parameters optimal to the deployment area to ensure this is should be considered carefully. Reducing the processing cost can be achieved by reducing either of the above-mentioned factors.

Comparison with Figure 3 shows that limiting the amount of neighbors in the IWLS calculation greatly reduces the processing cost, while this hardly has an influence on the range error performance of the system. From the graphs it is clear that the number of calculations per broadcast message is an almost constant factor.

7   Discussion

The previous chapters have covered the description of the algorithm used in obtaining the position estimate in a distributed network.

The implementation of the system presented in this paper is only a simple, straightforward one. As has been shown in the previous chapter, performance improvements can be made, by making small changes, based on close inspection of the results obtained by selectively changing certain parameters that are of importance in the whole calculation of the algorithm. As one of such improvements, multi-hop distances towards indirect neighbors could also be used in the refinements phase, instead of just to obtain initial position estimates. This could be especially useful for networks with a low connectivity. It does promise to pay off in terms of performance increase, to spend some time fine-tuning the algorithm, based on a particular application area.

As noted before, similar algorithms have been designed, and can serve as good comparison material. Langendoen ([6]) provides a comparative test for combinations of different algorithms. Even though the general approach is somewhat different, the refinement steps are almost identical to the one used in our system. The data after refinement are comparable to the results presented in this paper.

The multi-hop distances and IWLS calculation are quite similar to the sum dist and lateration methods described in [6] and others, with the exception that use is made in our system of the precision indications available with the messages.

From the results presented in [6], it is clear that the refinement algorithm only is not a very useful addition to the first two phases except in very specific conditions. The use of refinement dramatically reduces the coverage of the whole system to levels below 50% in all cases except when the distance measurement errors are very small ($\eta < 0.05$). Under these conditions, the position error can drop to 25% or even 20%, although coverage of just 60% can be reached. Without the refinement phase, the various combinations of stage 1 and 2 algorithms can only reach a position error of 42%, and less for higher connectivity, but with 100% coverage. Having full coverage is very important, of course, since acquiring a position estimate is the intended goal of these algorithms.

Our results, on the other hand, show better position estimates, while practically full coverage can be reached, in almost all situations. In all but the most extreme environments, a position error of 25% or less can be achieved, while keeping 95% coverage. Large distant measurement errors or low connectivity do make the numbers less feasible, showing the application limits for this algorithm. But in more optimal conditions, position errors as low as 10% can be reached, with near 100% coverage. Table 1 summarizes the results for the algorithms mentioned above.

The improvement in position error with our proposed solution is most likely caused by the availability of more measurement data, namely the precision indication, even though not a lot can be said about this in general, since the obtained location precision is a result of many interrelated factors.

In terms of communication overhead, our solution proves to be competitive as well. In [6] it is stated, that the initial flooding (including calibration) of the

Table 1. Comparison of the presented algorithm with the ones described in [6], on relative position and connectivity

	Condition	Pos.err.	Coverage
Traditional alg. + refine	$\eta < 0.05$	20-25%	50-60%
Traditional alg. phase 1+2	$conn. > 12$	35-42%	100%
Precision based algorithm	$\eta < 0.25$	15-25%	95-100%
Precision based algorithm	$\eta < 0.1, conn > 12$	10-15%	98-100%

network costs about 3.5 to 4.8 messages per node, depending on the algorithm used. This is comparable to the average number of startup messages, which is 2.1 in our solution. The amount of messages needed during refinement is largely dependent on the limit imposed on the system, and in both algorithms, the resulting position error is constant from about 5 messages onwards. All systems use messages of a small, constant size.

It is hard to compare the calculation requirements of all algorithms, especially because only simulations are available up to this pint in time of both the system described in this paper and the ones in [6]. Besides that, Langendoen et al. do not give any metrics about calculation time or complexity in their paper. At this point, no conclusions can be made about how the different algorithms compare in terms of calculation time.

It is shown that, using RSSI as a distance measurement system, position errors of as low as 10% can be reached. Even though this could be not sufficiently accurate for use in all kinds of situations, it does provide good enough results to at least be able to know about the networks topology, and coverage area of the individual nodes. Altogether, it can be concluded that the goal of obtaining position information of nodes in a distributed ad-hoc network, making use of only rather inaccurate measurement techniques, is within reach.

8   Conclusions and Future Work

In this paper we presented a distributed localization algorithm for ad hoc wireless networks, which takes into account the precision of measurements. It is aimed at networks where the nodes are static. The algorithm takes a two-step approach: first an initial location estimation is made, based on distance measurements obtained from RSSI readings. Subsequently, refinements are calculated. Both steps, including the calculations involved, are described in detail, after which the performance is discussed. In a 225-node network, of which 5% are anchors, with 10% range error, a relative distance error of about 16% can be achieved, with a nearly 100% coverage. In general, the results of this approach yield 2 to 4 times better results in position accuracy than other systems described previously. This level of performance can be reached with just 10 or fewer messages broadcast per node in the network, which are of small, constant size. Details about the calculation cost are discussed as well, and some suggestions are given on how to optimize the performance of the algorithm for real world implementations.

The results provided in this paper are all obtained from simulations. Further research will have to be done to find how the algorithm performs in a real world situation, where certain parameters could be different from the assumptions made in this paper. A hardware platform is being designed that can be used to further test the system.

The described system is applicable only in networks where the nodes are static, but many applications include mobile nodes in the network. By itself, that poses many new problems. The nature of our system is designed in such a way that with only little adjustments, it can also be made useful in networks with moving nodes. In the future, more research can be done to make this possible.

References

1. M.Srivastava A.Savvides, C.-C.Han. Dynamic fine-grained localization in ad-hoc networks of sensors. In *7th ACM Intl. Conf on Mobile Computing and Networking (MOBICOM), p.166-179*, 2001.
2. Paramvir Bahl and Venkata N. Padmanabhan. RADAR: An in-building RF-based user location and tracking system. In *INFOCOM vol.2*, pages 775–784, 2000.
3. Jan Beutel. Geolocation in a picoradio environment. In *MS Thesis, ETH Zurich, Electronics Lab*, 1999.
4. N. Bulusu, J. Heidemann, and D. Estrin. Gps-less low cost outdoor localization for very small devices. In *IEEE personal communications*, pages 28–34, 2000.
5. Jeffrey Hightower and Gaetano Borriella. Location systems for ubiquitous computing. *IEEE Computer*, 34(8):57–66, 2001.
6. Koen Langendoen and Niels Reijers. Distributed localization in wireless sensor networks: A quantitative comparison. In *Computer Networks (Elsevier), special issue on Wireless Sensor Networks*, 2003.
7. L.Doherty, K.Pister, and L.El Ghaoui. Convex position estimation in wireless sensor networks. In *IEEE INFOCOM, Anchorage, AK*, 2001.
8. L.Evers, W.Bach, D.Dam, M.Jonker, H.Scholten, and P.Havinga. An iterative quality based localization algorithm for adhoc networks. In *Department of Computer Science, University of Twente*, 2002.
9. P.Misra, B.P.Burke, and M.M.Pratt. Gps performance in navigation. In *Proceedings of IEEE, vol.87, nr.1, pp.65-85*, 1999.
10. P.Rentala, R.Musunuri, S. Gandham, and U. Saxena. Survey on sensor networks. Department of Computer Science, University of Texas.
11. Chris Savarese, Jan Rabaey, and Koen Langendoen. Robust positioning algorithms for distributed ad-hoc wireless sensor networks. In *USENIX technical annual conference, pp.317-328, Monterey, CA*, 2002.
12. Andreas Savvides, Heemin Park, and Mani B. Srivastava. The bits and flops of the n-hop multilateration primitive for node localization problems. In *WSNA, Atlanta, GA, pp.112-121*, 2002.
13. S. Simic and S. Sastry. Distributed localization in wireless ad hoc networks. Tech. report, UC Berkeley, 2002, Memorandum No. UCB/ERL M02/26., 2002.
14. Andras Varga. The omnet++ discrete event simulation system. In *Proceedings of the European Simulation Multiconference (ESM'2001)*, 2001.
15. Roy Want, Andy Hopper, Veronica Falcão, and Jonathan Gibbons. The active badge location system. In *ACM Transactions on Information Systems*, pages 91–102, 1992.

Adaptive On-Device Location Recognition

Kari Laasonen, Mika Raento, and Hannu Toivonen

Basic Research Unit, Helsinki Institute for Information Technology
Department of Computer Science, University of Helsinki
{Kari.Laasonen, Mika.Raento, Hannu.Toivonen}@cs.Helsinki.FI

Abstract. Location-awareness is useful for mobile and pervasive computing. We present a novel adaptive framework for recognizing personally important locations in cellular networks, implementable on a mobile device and usable, e.g., in a presence service. In comparison, most previous work has used service infrastructure for location recognition and the few adaptive frameworks presented have used coordinate-based data. We construct a conceptual framework for the tasks of learning important locations and predicting the next location. We give algorithms for efficient approximation of the ideal concepts, and evaluate them experimentally with real data.

1 Introduction

Location-awareness has several applications in ubiquitous computing. Location-triggered reminders are a simple example (e.g., [1]), adaptive systems that use location as a part of the input for adaptation are more ambitious. Location is also an important clue about a person's communication context, and giving out this information in a form of presence service can enable more efficient communication (see, e.g., [2]).

We present an adaptive framework for building location-awareness from cell-based location data, especially GSM-network cell information. The goal is to be able to learn, within the mobile device, personally important places and routes between them without knowledge of the physical topology of the network.

Most previous work on location-awareness is based on pre-defined location infrastructure and rules about the use of this infrastructure, although the applications do not always strictly demand it. Instead, we are interested in building a learning, adaptive framework for individual location recognition. There has not been very much work in this area, maybe the best-known examples being Marmasse and Schmandt's comMotion [3] and Ashbrook and Starner's work [4]. Both of these approaches use coordinate-based location data, provided by GPS.

Our contributions are twofold. First, we create a conceptual framework for identifying important locations and routes from cellular network data (Sect. 2). The important locations that we can automatically recognize from a user's location data are called *bases*. If the user is moving, we try to recognize the route and aim at predicting the base the user is heading to. Since routes sometimes

A. Ferscha and F. Mattern (Eds.): PERVASIVE 2004, LNCS 3001, pp. 287–304, 2004.

fork and may lead to several bases, we group bases into areas and use these as route targets if a single base cannot be determined with high enough confidence.

Second, we give efficient methods for analysing the observed cellular data on a mobile device, e.g., a cellular phone (Sect. 3). Implementing the adaptive location-awareness system on the mobile device is challenging, mainly because of limited resources. A major advantage is that this enables the building of location-aware systems without additional service infrastructure. Several privacy issues are also avoided, since the location data is held within a device owned by the user.

An experimental evaluation of the concepts and algorithms is given in Sect. 4, using real GSM data from three users, covering a period of six months. Section 5 contains conclusions and outlines future research issues.

2   Locations, Bases, Areas, and Routes

Our goal is to automatically recognize locations that are somehow significant to the user and routes between those locations. As an example application, consider a presence service. We want to describe the current whereabouts of the user. If the user is at a well-known, important place, this is a useful description. If he is not, knowing where he is going to (and coming from) is an informative description of his context. For example, assume that Bob is waiting for Alice at a restaurant. Checking Alice's presence information he could see either that Alice is still at work, has left work and is heading towards the restaurant (or an area including the restaurant) or is heading somewhere else. Based on this information Bob has a fairly good idea of whether he should remind Alice of the meeting or not.

2.1   Locations and Bases

Recognizing locations from GSM cell data is challenging, for a number of reasons:

- Cells can be very large, up to some kilometers in diameter, especially in sparsely populated areas.
- Areas covered by base stations overlap, so that several cells may be seen in a single location.
- Overlap of cells and radio signal shadows can cause cells to be non-contiguous areas.
- There is no one-to-one correspondence between a physical location and the cell used by a phone, e.g., due to changing radio interference.

Other types of cellular networks are likely to have similar properties. An advantage of GSM cellular data is that it is available almost everywhere.

The location data available for a single device consists of a time-stamped sequence of transitions between cells. A correponding cell graph serves as an abstract representation of the cell topology, without reference to any physical locations.

Definition 1 (Cell graph). The cell transitions of a given user define a directed unweighted cell graph $G_c = (V, E)$, where the set V of vertices consists of all observed GSM cells, and there is an edge $(c_i, c_j) \in E$ if and only if a transition between c_i and c_j has been observed.

Several overlapping cells can be frequently seen at a single location. In areas with a dense GSM network, it is typical that even if the user stays in one room, the serving cell may oscillate between two or three alternatives. On the other hand, physical back and forth movement most often also indicates a single semantic location, for example when moving around in an office. When such oscillation is observed, we therefore cluster several cells together.

Definition 2 (Cell cluster). Given a cell graph $G_c = (V, E)$, a set $C \subseteq V$ of cells is a cell cluster, if and only if

1. The cells form a subgraph of diameter at most 2 in the cell graph G_c.
2. The average length of a visit to the cluster is $t\_avg_C > |C| \max_{c \in C} t\_avg_c$.
3. Any proper subset of C does not satisfy condition 2.

The first condition simply requires that all cells in a cluster are near each other. The second condition tests oscillation: the average time spent visiting a cluster is larger than the sum of the individual times only when the user moves back and forth between the cells in the cluster ($|C| \max_{c \in C} t\_avg_c$ is an upper bound for the sum of individidual averages, and makes the condition relative to the most important cells). Without the last minimality condition some extra cells could possibly be included in cell clusters.

If the user is at a cell that belongs to multiple clusters it is unclear which of the clusters he really is at. There are several alternative ways of dealing with this; for simplicity, we recursively combine all the clusters that have shared cells. This leads to a partitioning of the set of cells to distinct locations. The term "location" is used in this formal meaning in the definitions and algorithms that follow.

Definition 3 (Location). Given a cell c and a cell graph $G_c = (V, E)$, cell c is at location

$$\mathrm{loc}(c) = \begin{cases} \{c\}, & \text{if } c \text{ does not belong to any cluster;} \\ \mathrm{cl}(C), & \text{if } c \text{ belongs to cluster } C, \end{cases}$$

where $\mathrm{cl}(C)$ is the transitive closure of overlapping clusters:

$$\mathrm{cl}(C) = C \cup \bigcup_{C': C \cap C' \neq \emptyset} \mathrm{cl}(C').$$

The set of locations is

$$\mathcal{L} = \{\mathrm{loc}(c) \mid c \in V\}.$$

The goal is that locations, as defined above, are the smallest reliably distuingishable units in a GSM cell data. Bases, or important locations, can now be defined simply as locations where the user spends a large portion of his time. However, we want to give more weight for locations visited more recently, in order to adapt to changes in users' movement patterns.

Definition 4 (Bases). The (weighted) time spent in location L is

$$
\text{time}(L) = \int_{t_0}^{t_{\text{now}}} \text{at}_L(t) r^{t_{\text{now}} - t} \, dt,
$$

where t_0 is the time of the rst observation, t_{now} is the current time, $\text{at}_L(t)$ is an indicator function which has value 1 if the user is in location L at time t and 0 otherwise, and $0 < r \leq 1$ determines the rate of aging. We assume time is measured in days.

The set B of bases consists of a minimal set of locations that cover fraction p of all (weighted) time:

$$
B = \underset{B' \subseteq \mathcal{L}}{\text{argmin}} \left\{ |B'| : \sum_{L \in B'} \text{time}(L) \geq p \int_{t_0}^{t_{\text{now}}} r^{t_{\text{now}} - t} \, dt \right\},
$$

where $0 < p \leq 1$ de nes how large a proportion of the total (weighted) time the bases must have.

In other words, bases are the minimal set of locations in which the user spends proportion p of his total time, taking into account the weighting.

Giving priority to recent events could be accomplished in many ways. Our exponential weight function can be approximated efficiently and incrementally, and its smoothness means that there are no radical changes in the results with advance of time (like there would be with a window with sharp edges).

The aging rate r is determined heuristically. Basically the aging should allow regular events to show their regularity, and not assign overly great weights to daily events. Probably most regular visits happen at least once a week, so a rate that would allow two week old events to have reasonably high weights would be appropriate. Reasonably high weights can be construed as $r^{t_{\text{now}} - t}$ being of order 0.25 at $t = t_{\text{now}} - 14(\text{days})$. This gives an estimate $r = 0.9$, which is used throughout this paper.

The proportion p of total time has to be selected as well. We do not present any analytical means here either, but will present some estimates based on our test data in Sect. 4. Too high a proportion will allow purely transitional cells in the set of bases. If too low a proportion is used, not all real significant locations fit in. In our application model, where the user confirms bases by naming them, the former case puts a higher burden on the user but is likely to lead to a more useful set of bases.

2.2   Routes and Areas

The user is not always at a base. Quite often when they are not, they are on their way from one base to another. If we can determine where they are going, this is useful as presence information. For adaptive (user-modelling) applications the from-to base pair can be used as a state to characterize the current context.

The use of cell-based location data again presents some issues:

- We do not know the physical topology of the cells, instead we only know about the times of cell changes.
- Cell changes do not only result from changes in locations, as interference may bump the phone from one cell to another.
- We do not always observe the cell change in the same actual location when moving in different directions. To minimize oscillation (and so network traffic) there is some lag in changing cells to favor the current one.

The first characteristic defines to a large extent what we can infer from the data, and prevents some analytical approaches. The two others make the data fairly stochastic in nature. Our route learning and prediction algorithms are based on sequences of cell transitions.

Routes taken by people may fork or pass by bases (this is typical of public transport as well as main car routes). When travelling along a certain stretch of a route it can be very hard for anyone to predict just from the movement where the person will stop, or which fork they will take. To address a realistic and useful problem, we will group bases into areas of nearby locations, and use these to indicate the approximate direction of movement.

For an area to be a good indication of direction, the bases belonging to it have to be physically close to each other. We do not know the physical (geographical) topology, but we can approximate it with the travel times of the user. These times are used as distance measures between bases, and the bases are then clustered into areas with a density-based clustering algorithm similar to DBSCAN [5]. The base graph represents the base topology.

Definition 5 (Base graph). The bases define a weighted, undirected graph $G_b = (B, E, w)$ where B is the set of bases, and where observed transitions between bases define the edges E, i.e., if the user leaves base $b_i \in B$ and without visiting any other bases arrives at $b_j \neq b_i$, then there is an edge (b_i, b_j) in E.

The weight $w(b_i, b_j)$ is given by the median of observed travel times between b_i and b_j in the graph. The travel time starts when the user leaves the last cell belonging to b_i (b_j) and ends when the user enters the first cell belonging to b_j (b_i). The distance is calculated symmetrically using travels in both directions.

Median is favored because of its robustness in the face of outliers. Especially large travel times that are not representative of the true distance are observed, since the user does not always move continuously from one base to another, but may stop on the way. A minimum would be too vulnerable to a single abnormal transition. If the number of observations grows too large, the median can be calculated approximately [6].

We next define an area of density (travel time) t as a set of bases where any base can be reached from any other base in the area by recursively following edges of weight at most t.

Definition 6 (Area). Given a base graph $G_b = (B, E, w)$ and a density (travel time) t, the areas (with respect to t) are the connected components of the undirected, unweighted graph $G = (B, E')$ where $E' = \{e \in E \mid w(e) \leq t\}$.

We use a number of densities to obtain a hierarchy of areas. In practice we use a three-level hierarchy, with travel times of $t_1 = 3, t_2 = 10$ and $t_3 = 60$ minutes. These densities are the only parameters to the area clustering. They have been chosen to represent reasonable human perceptions of movement.

Ashbrook and Starner [4] utilize a similar method to group locations, but with only a single layer. These are directly used in their route learning instead of the individual locations. They have a physical topology underlying their location data and have chosen a half mile radius as the threshold, which is (if walking) slightly higher than our first level of $t_1 = 3$ minutes.

We can now formally define the route prediction problem using the concept of areas.

Definition 7 (Route prediction problem). The input consists of the full history of cell transitions $\{(c_0, t_0), \ldots, (c_n, t_{now})\}$, and a sequence $\{A_{t_0}, \ldots, A_{t_m}\}$ of partitionings of bases to areas with time densities t_i where $A_{t_0} = B$ ($t_0 = -1$) and $t_i \leq t_{i+1}$. Let the previous base visited by the user be b. The task is to output the next base or area the user will visit: the prediction is an area $a \in A_i$ with maximal i such that $b \notin a$, and an estimated probability u for that prediction.

In other words, the task is to predict the roughest possible area that is different from the previous one. The rationale for this is that making detailed predictions for remote locations is not feasible, but identifying the direction is. In practice, of course, the next base is often in the same area and will be predicted as such. Table 1 illustrates how different kinds of predictions can be interpreted, and how a presence service could describe the user's location to others in various circumstances.

Ashbrook and Starner [4] propose a route recognition model that builds a Markov model of movement between important locations. This is not directly applicable in our work, since we have much fewer bases than they have important locations. The idea of using Markov chains, however, is relevant. The problem of predicting short-term movement in cellular networks has been of interest in systems that use paging. Bhattacharya and Das [7] propose a path learning algorithm for cellular networks based on variable-order Markov chains. We use similar ideas, but do not restrict ourselves to predicting just the next cell but instead try to find the next base.

3   Algorithms

We next consider algorithms to solve the base and route learning problems in a mobile device with limited resources. We assume the availability of 1–2

Table 1. Description of current location in different cases

Analysis	Example description
At base	At Work
Not at base, prediction within area with high probability	Left Work 15 minutes ago, heading towards Itäkeskus
Not at base, prediction outside current area with high probability	Left Home 30 minutes ago, heading towards Parents/Jyväskylä center/Summer cottage
Not at base, prediction with low probability	Left Work 10 minutes ago, possibly heading towards Viikki sport center
Not at base, no prediction	Left home 30 minutes ago

megabytes of memory and a peak processing power of 50–100 MIPS. This means that we must store fairly compact statistics of the data and do quite straightforward online processing, and that we are willing to compromise some accuracy in favor of computational efficiency. We also cannot use the available processing power for long periods to avoid draining the battery of the mobile device.

These practical considerations and the requirement of using online algorithms mean that some of the definitions in Sect. 2 cannot be used as is. In particular, algorithms for detecting bases and building locations differ from the corresponding definitions, because the algorithms have to make decisions online, without having access to the full data.

3.1 Bases

The input arrives in cell transition events that contain the new cell identifier c and a timestamp t. The goal is to be able to tell if a cell forms a base or a location, or is part of an existing location. To do this online, we have the following per-location state: $time(L)$ is the total time spent in location L, and $count(L)$ is the number of visits to L. Cell transitions within a single location do not affect the count. The average stay time is defined as $avg\_stay(L) = time(L)/count(L)$.

Let B be the set of bases. To update B when cell transition events occur we run algorithm BASEEVENT (Algorithm 1). This routine performs two tasks. First, it updates location statistics (lines 2–4) to determine locations that are visited often or where the user stays for long periods of time. These two factors are combined (line 7) with a weight function that has the form $weight(count, time) = time \cdot count^2/(count^2 + 1)$. The dependence on $count$ is significant only when the count is small. This helps us ignore cells seen only a few times, and has proved necessary in the implementation. Aging according to Definition 4 is approximated outside this algorithm by multiplying each $time(L)$ by the scaling factor r once per day. Finally, the set B is rebuilt on lines 7

BASEEVENT(c, t)

Input. A cell identifier c and a timestamp t.

State. Set $\mathcal{L}$ of locations, associated statistics, set $B \subset \mathcal{L}$ of bases, previous event $\{c', t', L'\}$, base weight threshold p.

1: $L \leftarrow$ the location containing c
2: $time(L') \leftarrow time(L') + (t - t')$
3: **if** $L \neq L'$ **then**
4:    $count(L') \leftarrow count(L') + 1$
5:    MERGEEVENT($L', t - t'$)   ▷ Build locations
6:    $c' \leftarrow c; t' \leftarrow t; L' \leftarrow L$
7: $total \leftarrow \sum_{x \in \mathcal{L}} weight(count(x), time(x))$
8: $v \leftarrow$ array of nodes $x \in \mathcal{L}$ sorted into descending order by $weight(x)$
9: $B \leftarrow$ first k entries of v such that $\sum_{i=1}^{k} weight(v[i]) \geq p \cdot total$

Algorithm 1. Detecting bases from cell transition events.

to 9. Here the number of bases is determined by taking the proportion p of total weight (cf. Definition 4).

The second task of BASEEVENT is to merge neighboring locations to create new, larger locations, according to Definition 2. This is done using a greedy approximation algorithm MERGEEVENT (Algorithm 2). It follows the most recently seen cells and already formed locations. The history m has to have room for enough items to support merging between μ locations. It turns out that in practice a large majority of mergings occurs between two existing locations, and a few occur between three locations. Accordingly, the maximum size μ of the history was set to four. However, this does not mean that the algorithm cannot produce larger locations. Instead of forming them in a single step, most locations are built from a series of pairwise merges.

In most cases, once a location is formed, it is treated as if it were just a large cell. The algorithm keeps several pieces of information about recent locations: time(L) is the total time spent in location L while we have tracked it, max_stay is the longest single stay in the location, avg_stay is defined above, and count is the number of times this location has been seen during merge tracking. Whenever we see a new location (lines 7–16), we consider the previous locations for merging. To be merged, we choose the smallest subset of recent locations that fulfills all the required conditions (line 11).

Definitions 2 and 3 give conditions for clustering based on the average and maximum times spent over all data. In the online algorithm, we make decisions based on a single data point for the candidate cluster, since we cannot store statistics for all possible groupings. That a single observation fulfills these conditions is a required but not a sufficient condition for the full data. The additional conditions on time spent and number of visits try to ensure that the data point is not significantly unrepresentative of the full (unknown) distribution.

MERGEEVENT(L, t)

Input. A location identifier L and a time t stayed in L.

State. List m of recent locations, with associated statistics; maximum size μ of m.

```
 1: if L = m₁ then      ▷ Same as the previous location
 2:     Update time(m₁) and max_stay(m₁)
 3: else if L = mᵢ for some i > 1 then      ▷ Recently seen location
 4:     Update time(mᵢ), count(mᵢ) and max_stay(mᵢ)
 5:     Move mᵢ to the front of the list m
 6: else   ▷ Outside the set of tracked locations
 7:     k ← 2; merged ← false
 8:     while k ≤ |m| and not merged do
 9:         s ← {m₁, ..., mₖ}
10:         τ ← Σᵏᵢ₌₁ time(sᵢ)
11:         if τ ≥ |s| max avg_stay(sᵢ) and τ ≥ |s| max max_stay(sᵢ)
                and τ ≥ 10 min and min count(sᵢ) ≥ 2
                and some item in s is already a base
                and graph formed by cells in s has diameter D ≤ 2 then
12:             Merge cells in s into a location
13:             merged ← true
14:         k ← k + 1
15:     Remove the last entry of m, if |m| = μ
16:     Add a new entry with {time = max_stay = t, count = 1} to m
```

Algorithm 2. Building locations by merging existing locations.

3.2 Areas

We use a simple form of density clustering to build areas. Recall that only bases are considered for area clustering. When we enter a base, we update the base graph (Definition 5). The vertices of this graph are the bases, and the (undirected) edges are the observed transitions between bases. Each edge carries with it the average transition time, used as the simplest possible approximation for the median.

Area clustering is performed once per day. The basic clustering step takes a weighted graph $G_b = (B, E, w)$ and a density threshold t. We start the search at an arbitrary node $b \in B$. Then we recursively follow all edges $e \in E$ where the transition time $w(e) \leq t$. All the visited nodes are placed in the same area cluster. If any nodes remain, we choose another node and start the search again and build a new cluster. Nodes whose distance to any other node is larger than t become singleton clusters.

3.3 Routes

Route prediction depends only on cell transitions. When transitions occur, we store them in an event history H. This history contains pairs $h_i = (t_i, c_i)$ of time t_i and cell c_i.

When the user arrives at a base b, the history is used to construct new entries in the route prediction database as follows. We take sequences with a window size k from the history. The first sequence would be $s_1 = h_1 \ldots h_k$, then comes $s_2 = h_2 \ldots h_{k+1}$, and so on. We then store the associations $s_i \to b$ in the database. Later we can retrieve all the associations that correspond to a cell identifier sequence $c_1 \ldots c_k$. Associations are stored repeatedly for decreasing values of k until we reach $k = 1$. When searching for a match, we similarly use progressively shorter sequences until a match is found. Typical initial values of k are about 2 or 3. If $k = 1$, there is no history to tell about the direction of movement; if $k > 4$, most of the sequences we observe will be unique, making it difficult to predict future actions.

We will also experiment with the possibility of using time of day to make more specific predictions. For this purpose, we want to utilize the time distributions of the last transition in the stored sequence s_i. To be able to later reconstruct the distribution of times, in addition to base b we store triplets (n, s, q), where n is the number of occurrences of base b, and s is the sum and q the square sum of the event times.

Prediction of the next base is performed by Algorithm 3. The idea is to take a sequence of recent events as a key and find all the bases stored in the database. If a certain event sequence has led to more than one base, we need to choose the base with the largest probability.

PREDICTBASE(H)

Input. A history $H = (h_1, \ldots, h_n)$ of cell transition events.
Output. A pair (b, u) with a base b and its probability u, window size k.

1: $t \leftarrow$ current time
2: **for** $i \leftarrow k$ **downto** 1 **do**
3: 　　$r \leftarrow i$ most recent events in H
4: 　　$A \leftarrow \{$associations $x \to (n, s, q, b)$ where $x = r\}$
　　　　　▷ s is the sum and q the square sum of n time values.
5: 　　**if** $A \neq \emptyset$ **then**
6: 　　　　$sum \leftarrow w \leftarrow 0$　　▷ Sum and the best weight w seen yet
7: 　　　　**for all** $(n, s, q, b) \in A$ **do**
8: 　　　　　　$\mu \leftarrow s/n; \sigma \leftarrow \sqrt{q/n - \mu^2}$
9: 　　　　　　Assuming $T \sim N(\mu, \sigma)$, let $w \leftarrow n \Pr(t - \alpha \leq T \leq t + \alpha)$
10: 　　　　　　**if** $w > w$ **then**
11: 　　　　　　　　$(b, w) \leftarrow (b, w)$
12: 　　　　　　$sum \leftarrow sum + w$
13: 　　　　**return** $(b, w/sum)$

Algorithm 3. Predicting the next base.

There are two factors that influence the probability: the number of times a base has been seen, and the time distribution. A certain base may be more probable in the morning than in the evening, or the user may have different routes in the weekend than during the work week. These aspects are handled by assuming that the event times follow a normal distribution. This is the simplest possible assumption, as we only need to store the triplet (n, s, q) mentioned above to recreate the distribution. However, experimental evaluation shows that the effect of the time distribution is not as large as it might appear at first.

On line 9 we compute the probability of an association, given the current time of day. Using n and the sums s and q we reconstruct the distribution parameters μ and σ. Then we find the probability of the range $[t-\alpha, t+\alpha]$, where $\alpha = 30$ min. The day of the week can be handled similarly (omitted here). We obtain two probabilities which are combined with the number of occurrences n. If the time distribution is not used, line 9 can be simplified by setting the weight w equal to n; it is also unnecessary to maintain the sums s and q.

As an alternative to the presented method we could also predict the next cell instead of the next base. This corresponds to finding the stationary distribution of a kth order Markov chain. This model is more expensive to evaluate and has problems with inevitable loops in the transition graph.

Algorithm 3 yields an estimated probability u for the most likely next base. We can distinguish unsure predictions, with small value of u, from more confident ones. Given a threshold value u', we say that the prediction is confident or has high probability if $u \geq u'$. The final output of the prediction, a base or an area, will be determined as in Definition 7 (see Table 1 for examples).

3.4 Limiting Memory Use

We attach a timestamp to each stored data item, be it a location with accumulated time or a cell sequence. This timestamp is set to the current date and time each time the information is updated. If we run out of resources (or reach a predefined memory limit) we start to remove information starting from the oldest items and continue until the desired level of memory usage is achieved. We assume that there is enough memory to keep at least the "normal" day-to-day schedule of the user in memory.

3.5 Possible Improvements

Bhattacharya and Das present an information-theoretical model for selecting the order of Markov chains to use for each path [7]. This way they try to use an optimal amount of information: if there is a long unique chain of cells in a path, that chain is identified. This could be used in our algorithms as well. We do not want to use too long sequences, though, since we are not really interested in matching the total path travelled, only enough of the recent history to give the direction of travel.

Cell transition sequences are fairly stochastic. For example we have observed that a train trip has only sporadically exactly same sequences when the trip is repeated. Since the sequences do not necessarily match exactly, our route matching is not always able to give good results. We should try to find a suitable metric to define best matches between sequences of cells.

4   Experimental Evaluation

Evaluation of the location model is not straightforward. While the concepts of bases and routes seem intuitively attractive and the clustering of cells and bases is justified by the problem setting, the quantitative quality of the resulting bases and areas is difficult to judge. The route learning has at least one possible objective measure: the accuracy of the predictions based on route recognition.

4.1   Data Gathering

We have gathered cell transition data from three volunteers. The data has been collected for six months with software that runs continuously on the mobile phones the persons use normally (both at work and at leisure). Two of the persons (1 and 2) have fairly simple movement patterns that mainly consist of moving between home and work during the week and some weekend trips. The third user moves somewhat more during the week and visits a larger number of locations.

4.2   Locations and Bases

There are two aspects in judging the quality of the clustering of cells into locations: how well the definitions in Sect. 2.1 work and how well the online algorithm approximates them. Both are somewhat subjective since we do not know the "correct" clustering. Because in the data gathering cells have been named according to the perceived locations, some idea of the quality is given by checking whether the cells in a locations have the same name or not. By looking at the data and the resulting locations, we can say that the original definitions give quite good results. The online approximative version (Algorithm 1) is not too far off. Mostly the problem with the online algorithm is with slightly too large locations: very frequently visited locations tend to assimilate some neighbouring cells as well. Less frequently visited locations are cleaner.

Table 2 shows the bases discovered for person 3. The proportion p (of total time spent in the bases) used in these experiments was 0.8, yielding 37 bases. The aging parameter r is fixed at 0.9 for all tests. The quality of the bases found seems fairly good. Both the stable, recurring locations of everyday life (like work, home, leisure, friends and family) are found, as well as more transient locations on trips (like accomodation in different places). Only few of the bases found are unclear. When p is raised to 0.85 the number of bases for this person grows to 86 and contains quite a few more unclear items.

Figure 1.A shows the number of bases found with different values of p. Larger values potentially allow us to recognize the current location of the user more often. If we assume that the user has to name the bases found, the larger the

Table 2. Description of bases found for person 3

Home	Work
Friends' home	Girlfriend's home
Parents	Girlfriend's parents
Shopping center	Vammala town center
Girlfriend's summer cottage	Student association house
Viitasaari accommodation	Family firm office
Summer cottage	Helsinki center west
Vammala town center 2	Helsinki center east
Vammala church (friends' wedding)	Viitasaari town center
Restaurant	Accomodation in Porvoo
HIIT/ARU office	Porvoo town center
Tvärminne conference center	Sister's home
Friend's grandparents	Friend's home
Accomodation in Tartto, Estonia	A student association
Restaurant	Area near Helsinki center (unclear)
Accomodation in Uppsala, Sweden	Restaurant in Uppsala
Restaurant in Stockholm, Sweden	Unclear
Restaurant in Stockholm	Previous work place
Unclear	

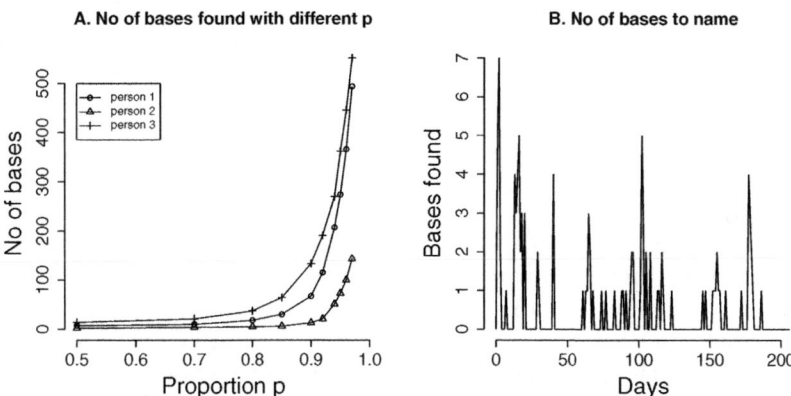

Fig. 1. A: Effect of p on number of bases. B: Number of bases found per day for person 3 with $p = 0.85$

number of bases, the larger the cognitive load on the user. So the chosen p has to balance these facts. Values between 0.8 and 0.9 seem to be reasonable. From the quality of bases found 0.8 would seem best for these users, but 0.85 is not unacceptable, although the data is by no means exhaustive. Adaptively finding a suitable value would be an interesting research topic. When looking at the actual lists of bases, it seems that raising p over 0.85 starts to introduce a large number of locations that do not have a clear meaning to the user in question.

Although the total number of bases the user has to name is important, so is also the peak naming load. Figure 1.B shows the number of bases a user has to name in a day within the data gathering period, with proportion $p = 0.85$. The figure is for person 3, who has the highest number of bases. The naming is distributed fairly evenly over the observation period, although there are some peaks that would probably be annoying to the user.

4.3   Areas

The area clustering seems to work quite well. The first level of the area hierarchy with density $t_1 = 3$ min picks up areas within towns or cities, the second level ($t_2 = 10$ min) picks individual cities and the third level ($t_3 = 60$ min) regions of countries. For example with the bases of person 3, the first level areas have grouped four districts of Helsinki into their own areas, as well as the town center of Jyväskylä. The second level has grouped the bases in each town in Finland into areas and the third level regions in Finland like Central Finland and the area around Kuopio.

With areas the online algorithms directly implement Definition 6, with the exception that areas are recalculated once a day. The quality of the area clustering depends only on how the approximation of cell clustering affects the selection of bases. The results with online and offline algorithms are almost identical for $p = 0.8$.

The next section shows that the use of areas improves the route recognition results significantly.

4.4   Routes

We evaluate the route prediction by calculating the prediction after each cell transition (unless the user arrives at a base) and by comparing the prediction to the actual base that was reached next. The route learning is done online, so that only data seen up to this point is used in the prediction. We test different variations of the basic algorithms given in section 3 and different parameter settings. The parameters r (aging) and t_1, t_2, t_3 (area densities) have been fixed in these experiments to 0.9 and $(3, 10, 60)$ minutes respectively. In the experiments where p (proportion of total time for bases) or window size are not being varied, values 0.85 and 2 have been used, respectively.

We leave out two cases from the evaluation: when the user is not moving in a well-defined direction and when the next reached base has not been seen before. We say that the user is not moving in a defined direction, if the most recent history of n cells only contains $k < n$ unique cells. For the evaluation we have selected after some experimentation $n = 6$ and $k = 3$. There is no clear evaluation criteria for paths (bases) that have not been seen before, so these are left out.

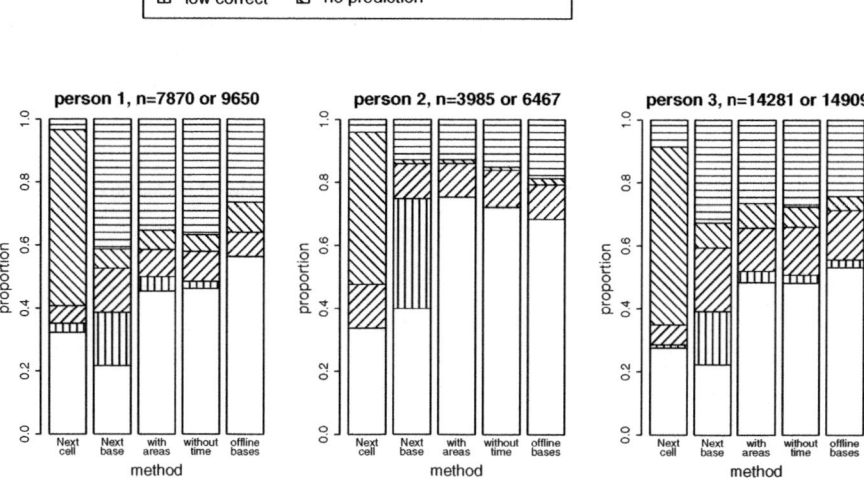

Fig. 2. Route recognition accuracy with different variants of the method. The number of predictions made is n.

Figure 2 shows how well different variations work. The following counts are used in the graphs: M atch is correctly predicted next base or area with high confidence (the probability u estimated by Algorithm 3 is over 0.3), bw correct is correct prediction with low confidence. Low fail is incorrect prediction with low confidence, no prediction means that there was no match for the current sequence of cells, and fail is the count of wrong predictions. The methods are: W ithout tim e is the version that produces the best results overall. It uses Algorithm 3 without the time distribution, and the area clustering for improving accuracy of predictions. The other methods show the effects of varying the algorithm. The first two methods, N ext cell and N ext base do not use areas. N ext cell calculates the probabilites of the next cell and repeats until a base is encountered (using both sequence frequencies and the time distribution) and N ext base directly calculates the probability of the next base. The remaining methods all use "Next base". W ith areas uses the area clustering (as do the rest). W ithout tim e is the same except that it uses only the sequence frequencies, and finally o ine bases presents the results if we calculate both clustering of cells and bases with all the data offline, before running the online route recognition. The impractical offline version helps to evaluate the effect of of the approximations of the online algorithms. The number of predictions n is the same for all the online versions, but lower for the offline variant because the cell clustering is different.

The route prediction results justify the use of the area clustering; it raises the accuracy significantly. It is interesting that using time-of-day affects the results minimally, and not necessarily to the better. This is probably due to assuming a single-mode Gaussian distribution, which is not always true for the data. The last method shown that uses precalculated cell clusters and bases shows that the online approximation is not quite optimal, but it is not too far off either. Calculating next cells instead of bases is not justified, because in addition to being much more expensive computationally, it gives worse prediction results.

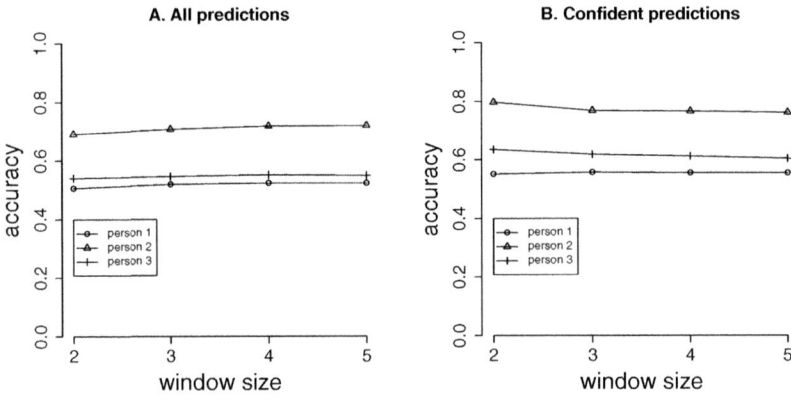

Fig. 3. Route recognition accuracy with varying window size

4.5   Selecting Parameters

Figure 3 shows the effect of the window size k on the route recognition accuracy. The method used is "Without time". Figure 3.A shows the ratio of correct predictions made (both low and high confidence) against all predictions, Fig 3.B the ratio for predictions made with high confidence (estimated probability over 0.3). Lenghtening the window increases the average accuracy of the predictions slightly, but the fact that the longest matches are used results in higher confidences for false predictions as well. Elsewhere in this evaluation we have used window size $k = 2$ as it seems to give the best overall results.

That the prediction accuracy decreases when the window size is increased (especially the for the predictions with high probability) may seem contradictory. The effect is due to overfitting: Algorithm 3 finds the longest possible matching sequence, and assigns probabilities to only sequences of that length. Using all possible matches, but assigning lower weights to shorter matches, could be a useful compromise.

The choice of the proportion p affects both the quality of the bases and the accuracy of the route recognition. Figure 4 shows the effect on prediction accuracy. It seems that to reach prediction accuracy above 0.5, which can be

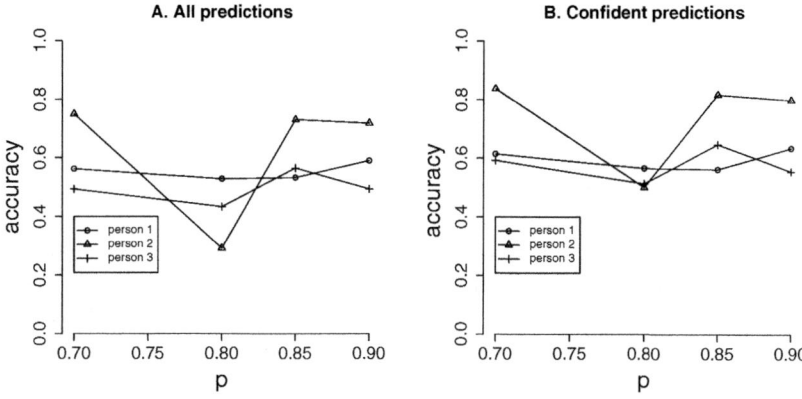

Fig. 4. Route recognition accuracy with varying p

seen as a very rudimentary baseline (that we are correct more often than not), p should be set to 0.85 for our three test persons. However, the accuracy seems to behave in a rather non-monotonic way as a function of p.

5   Conclusions and Future Work

We have presented methods for learning significant locations—bases, areas of nearby bases, and routes between bases from cell transition data. The methods work on a mobile device, reducing privacy issues in the analysis of the data. No new infrastructure or expensive sensors are needed for this kind of location analysis. The results indicate that with some user interaction we can provide interesting presence information from cell-based location data.

The accuracy of the inferred location is limited both by the data and our methods. The impact of this depends on the application. For adaptive applications our framework can be seen as a feature-extraction layer, whose accuracy could be improved using other available data, e.g., from sensors. In presence services, humans seem to be able to augment the location information with other presence data and background knowledge. For example, if the location given by the presence system for Bob is "At home", Alice may infer with fairly high confidence that Bob is on free time at the moment, and either at home or near his home.

The route prediction algorithms presented give adequate results for some applications, but could be improved upon. The algorithms presented use exact substring matching. To improve the recognition accurary we plan to look into string matching techniques that are more suitable for this kind of stochastic data.

Some of the accuracy problems mentioned above have to do with the nature of the GSM cells. Since cells are large, a single cell may contain several significant locations, and often extend to the area surrounding a location. Identifying the individual locations within a single cell cannot be done from the GSM cell data alone. Using the same methods on networks with smaller cell-size, like Wi-Fi and UMTS, would enable us to delineate bases more accurately and add detail to routes. As a preliminary step in this direction we plan to gather data which includes neighbouring GSM cells, so reducing the size of cells to intersections of cells.

Probably the most natural alternative to cell-based locationing would be GPS, but it is not without problems either. GPS is not widely available in mobile devices. Further, GPS signal disappears in large parts of urban areas due to the buildings shadowing the signal. Finally, coordinate-based data has the same problem of identifying locations and routes that are useful in personally meaningful location-awareness.

References

1. Dey, A.K., Abowd, G.D.: CybreMinder: A context-aware system for supporting reminders. In: Handheld and Ubiquitous Computing: Second International Symposium, HUC 2000, Bristol, UK, September 2000. Proceedings. Volume 1927 of Lecture Notes in Computer Science., Springer (2000) 172–186
2. Want, R., Hopper, A., Falcão, V., Gibbons, J.: The active badge location system. ACM Transactions on Information Systems (TOIS) 10 (1992) 91–102
3. Marmasse, N., Schmandt, C.: A user-centered location model. Personal and Ubiquitous Computing 6 (2002) 318–321
4. Ashbrook, D., Starner, T.: Learning significant locations and predicting user movement with GPS. In: International Symposium on Wearable Computing, Seattle, WA (2002)
5. Ester, M., Kriegel, H.P., Sander, J., Xu, X.: A density-based algorithm for discovering clusters in large spatial databases with noise. In: Proceedings of the 2nd international Conference on Knowledge Discovery and Data Mining (KDD 96), AAAI Press (1996)
6. Manku, G.S., Rajagopalan, S., Lindsay, B.G.: Random sampling techniques for space efficient online computation of order statistics of large datasets. In: Proceedings of the 1999 ACM SIGMOD international conference on Management of data, ACM Press (1999) 251–262
7. Bhattacharya, A., Das, S.K.: LeZi-update: an information-theoretic approach to track mobile users in pcs networks. In: Proceedings of the fifth annual ACM/IEEE international conference on Mobile computing and networking, ACM Press (1999) 1–12

Accommodating Transient Connectivity in Ad Hoc and Mobile Settings

Radu Handorean, Christopher Gill, and Gruia-Catalin Roman

Department of Computer Science and Engineering
Washington University in St. Louis
Campus Box 1045, One Brookings Drive
St. Louis, MO 63130-4899, USA
{radu.handorean,cdgill,roman}@wustl.edu

Abstract. Much of the work on networking and communications is based on the premise that components interact in one of two ways: either they are connected via a stable wired or wireless network, or they make use of persistent storage repositories accessible to the communicating parties. A new generation of networks raises serious questions about the validity of these fundamental assumptions. In mobile ad hoc wireless networks connections are transient and availability of persistent storage is rare. This paper is concerned with achieving communication among mobile devices that may never find themselves in direct or indirect contact with each other at any point in time. A unique feature of our contribution is the idea of exploiting information associated with the motion and availability profiles of the devices making up the ad hoc network. This is the starting point for an investigation into a range of possible solutions whose essential features are controlled by the manner in which motion profiles are acquired and the extent to which such knowledge is available across an ad hoc network.

1 Introduction

Conventional wired networks are increasingly being extended to include wireless links. It is important to distinguish between wireless links that are essentially stable in time and space, and those that are not. With wireless technology claiming an ever increasing share of the market, the interaction patterns between hosts need to be re-examined to address additional issues raised by this new environment. Former assumptions that made things easy or even possible in the wired setting may simply cease to hold in a wireless world. Wireless technology allows devices to become mobile. The topology of the network can change while applications are running and interactions between such mobile hosts are in progress.

1.1 Conventional Wired Networks

Modern wired computer networks provide a high degree of reliability and stability. The interaction between two different hosts in a wired network is expected

A. Ferscha and F. Mattern (Eds.): PERVASIVE 2004, LNCS 3001, pp. 305–322, 2004.

to succeed and it is not believed that the normal operation of a host can make it, even temporarily, unavailable to the rest of the network. Security and other considerations may cause an *application* residing on a host to become unavailable during its normal functioning but even such intervals tend to be reasonably well bounded in duration and frequency. Furthermore, the interaction between hosts is stable: the party a host interacts with is always found at the same IP address or under the same name, even when the name is resolved to different addresses at different times (*e.g.*, obtain its address from a DHCP server).

The stability of such an environment allows for efficient configuration of the traffic in the network. The total set of *possible* routes is mostly static, "hard-coded" in configuration files and scripts that do not change very often. While on-the-fly adjustment of information about the set of *available* routes is possible, such adjustments generally have a reactive flavor, *e.g.*, in response to network congestion. Fundamentally, modern routing protocols in wired networks depend at least indirectly on a stable set of physically reliable routes.

1.2   Mobile Wireless Networks

Nomadic networks, like the cellular telephony infrastructure or the pervasive wireless networks at universities, ensure communication between mobile devices such as phones, laptops, and PDAs. Nomadic networks do this by maintaining connectivity between each device and some part of a fixed infrastructure that acts as a communication liaison between such devices. Wireless devices connect to the infrastructure via access points connected to other access points connected to other wireless devices as they travel through physical space. Connectivity among devices is maintained by routing algorithms that deliver messages between mobile devices over the fixed infrastructure. The physical position of each device accessing the fixed infrastructure is important for routing packets to the closest access point to the device. The devices are always connected, but move in space.

In ad hoc networks the infrastructure for communication is made entirely of mobile hosts that interact in a peer-to-peer manner and route traffic for each other. There is no fixed infrastructure on which the mobile hosts can rely for message delivery. As long as two devices are connected (in direct contact or via multiple hops), they can exchange information and interact as if they were wired. In ad hoc networks, however, a stable multi-hop connection has severe temporal limitations, as well as spatial limitations due to the ad hoc routing algorithms that have to be carried out by hosts themselves.

1.3   Transition from Wired to Wireless Networks

Communication in ad hoc networks is significantly more challenging than in wired settings. Issues like multicast and routing, solved in wired networks, become difficult in ad hoc settings. Much effort has been dedicated to addressing these issues, to achieve a level of functionality comparable to the wired networks.

Multicast: In [1] multicast communication is approached from the perspective of ad hoc environments. The shared-tree mechanism [2], [3], [4] is analyzed and upgraded to exhibit an adaptive behavior, better suited for the changing environment. Flooding is considered in [5] as a possible approach to implementing multicast for highly dynamic environments. Multicast session ids can be assigned dynamically to hosts [6] and their ordering used to direct the multicast flow, organizing hosts in a tree structure rooted at the source of the multicast.

Routing: The loop-free, distributed, and adaptive routing algorithm presented in [7] is designed to operate in highly dynamic environments. This source-initiated, temporally ordered, routing protocol focuses its route maintenance messages to only the few nodes next to the occurrence of a topological change. Destination-sequenced distance-vector routing [8] and its extension ad-hoc on-demand distance vector routing [9] are improvements to the Bellman-Ford algorithm to avoid loops in routing tables. Sequence numbers help mobile nodes distinguish obsolete routes from the new ones. The wireless routing protocol (WRP) [10] addresses asymmetric links and uses piggybacking for route information updates.

This paper presents an approach to increasing communication capabilities in mobile ad hoc networks. It makes three main contributions to the state of the art in ad hoc networks. First, it formalizes the problem of message delivery in the presence of disconnections between hosts, and illustrates the formalism with examples. Second, it gives an algorithm for reliable message passing in the face of temporally discontinuous connectivity, exploiting host motion and availability profiles. Third, it offers an analysis of and refinements to that algorithm, based on the amount of information needed about the temporal patterns of connectivity.

The rest of this paper is organized as follows. Section 2 discusses issues of transient connectivity in mobile ad hoc networks, which motivate this work. Section 3 formalizes the problem of exploiting knowledge of host motion and availability for message delivery in ad hoc networks. Section 4 describes our solution and Section 5 analyzes it in terms of message, storage and computation complexity. Section 6 discusses the implications of our work. Section 7 presents related work, and Section 8 offers concluding remarks.

2 Transient Connectivity

In mobile ad hoc networks, communication paths are created, changed and destroyed at a much greater rate than in conventional or nomadic networks. The message routing infrastructure is composed of mobile participants, and therefore is itself subject to disconnections. A new approach is needed to describe the connectivity and communication paths between mobile hosts in ad hoc networks. In ad hoc networks, the mobility of hosts introduces the question of whether two hosts can be connected at all. The attention shifts from maintaining stable *connectivity* paths (*i.e.*, paths that route the traffic between two hosts that are connected to the network at the same time), to whether two hosts can *communicate* in an environment with transient connectivity (*i.e.*, a message can be delivered from one host to another over a set of hosts which may or may not form a *continuous* path from source do destination at all times during the delivery).

In conventional networks, two hosts can exchange messages if and only if a stable path exists between them in the network topology. The ability to exchange messages in that setting is reflexive, symmetric and transitive, *at each hop and along the entire path*. However, in mobile ad hoc networks paths are not stable over time, as the connectivity between any two hosts can vary due to the motion of those hosts or due to their availability (*e.g.*, due to power-saving modes).

When two hosts are in direct contact, messages can flow between them symmetrically, *i.e.*, as peer-to-peer communication. Ad hoc routing extends this symmetry across multiple hops, but requires all hosts along the path to be connected throughout some continuous interval of time.

However, there are reliable message delivery paths in ad hoc networks that are not symmetric, and thus cannot be exploited by either peer-to-peer or ad hoc routing approaches. In particular, messages that traverse a link before it goes down may be delivered on that path, while messages that reach that link after the disconnection will not. Therefore disconnection introduces non-symmetric message delivery semantics for any path longer than one hop.

We leverage characteristic profiles of the motion and availability of the hosts as functions of time to accomplish disconnected message delivery. Thus, we do not need to rely on the availability of a reliable end-to-end communication route during the process of communication. The only thing we can rely on is pair wise connectivity of hosts, which can be inferred from the characteristic profiles.

2.1   Connectivity Intervals and Message Paths

Figure 1 shows an example of temporally discontinuous intervals of direct connectivity between hosts a, b, and c. At time t_0, hosts a, b, and c are disconnected. From t_1 to t_2, hosts a and b are connected and can exchange messages. From time t_2 to t_3 all hosts are disconnected, etc.

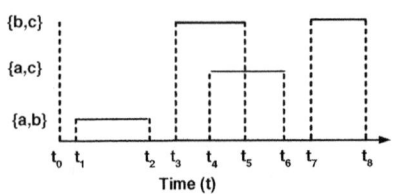

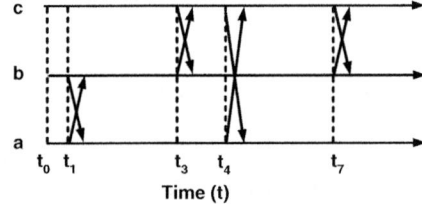

Fig. 1. Connectivity Intervals   **Fig. 2.** Possible Message Paths

We define a *connectivity interval* for any two hosts to be a continuous time interval they can use for effective communication. In practice, a connectivity interval is likely to be shorter than the total time the two hosts can communicate directly, since the beginning and the end of the direct connectivity period may incur overhead spent to establish contact and to disconnect without loss of data or inconsistency of state. However, because such overhead is a function of the connection and disconnection protocols used, for the sake of generality in this paper we assume the connectivity interval for two hosts is identical to the interval during which they can communicate. It is relatively simple to extend the formalism in Section 3 and the algorithm in Section 4 to consider such sources of overhead in specific applications, as we describe in Section 6.

Figure 2 shows the possible message exchanges that can take place between hosts a, b, and c in intervals in which the connectivity is available as shown in Figure 1. When two hosts are connected, a message can hop from one to the other, and the sequence of connectivity intervals over time can be transformed into a directed graph.

We define a *message path* between any two hosts to be any sequence of hops that a message can follow over time from the source host to the destination host. This should be understood differently from the ad hoc routing definition of a path. In ad hoc routing the entire path has to be defined from start to end at a given moment in time. Our approach allows ad hoc routing to occur when hosts are directly connected but also allows message delivery, even when (potentially all) hosts are disconnected during an interval between when a host receives a message and when it transmits it to the next host.

2.2   Discontinuities and Non-symmetric Paths

Two hosts that are directly connected have symmetric message passing capabilities. For example, in Figure 1 hosts a and b can exchange messages during interval $[t_1, t_2]$, as b and c can during $[t_3, t_5]$, and so on. However, if no direct connection exists between two nodes then messages may be able to pass between them in one direction but not the other.

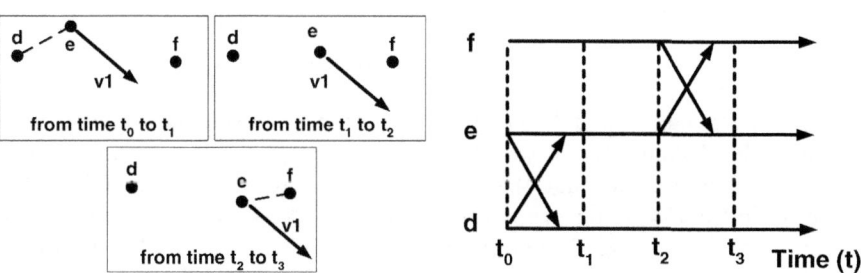

Fig. 3. Disjoint Mobile Connectivity   **Fig. 4.** Non-symmetric Message Paths

Figure 3 illustrates two disjoint connectivity intervals between three hosts d, e, and f, where d and e are connected over $[t_0, t_1]$, and e and f are connected over $[t_2, t_3]$. Figure 4 shows the non-symmetric end-to-end message passing capability resulting from the connectivity intervals shown in Figure 3, *i.e.*, messages can be delivered from d to f but not from f to d.

3   Problem Statement and Formalization

The problem addressed by this paper is to determine whether given

- a set of hosts N,
- a source host $p \in N$,
- a destination host $q \in N$,
- an initial moment t_0, and
- *characteristic profiles* (here, of mobility and availability) over all hosts in N,

we can construct a path so as to ensure message delivery from p to q. In Section 4 we provide an algorithm based on this formalization, to decide whether for two given elements of the set, a message can travel from the first to the second under the restrictions imposed by the characteristic profiles of all nodes in the set. We also seek to build the set of paths a message could follow from source to destination, if any.

We express the *mobility* of a host in terms of a function that maps time $t \geq t_0$ to its physical position: $m_p(t)$ characterizes the mobility of host p. We formalize the *availability* of a host (*e.g.*, due to power saving modes) as a boolean function over time: $\alpha_p(t)$ is true if and only if host p is willing to communicate. Finally, we formalize a *characteristic profile* $\pi_p(t)$ for host p as a function from time to the pair consisting of p's mobility and its availability: $\pi_p(t) = < m_p(t), \alpha_p(t) >$.

We formalize the direct connectivity of any two hosts p and q in N as a time-dependent boolean relation $\Gamma_{pq}(t)$ abstracted from their characteristic profiles $\pi_p(t)$ and $\pi_q(t)$. $\Gamma_{pq}(t)$ is true if and only if at time t both hosts are available and their Euclidian distance is less than R_c, the communication range of a host (assumed to be known, constant, and the same for all hosts), and false otherwise. We express this as:

$$\Gamma_{pq}(t) = |m_p(t) - m_q(t)| < R_c \wedge \alpha_p(t) \wedge \alpha_q(t)$$

We now formalize an end-to-end path relation κ between hosts p and r over the intervals of connectivity *generated* by the relation $\Gamma(t)$. In doing so, we remove the requirement that the entire communication path be defined from source to destination uninterruptedly. We say that hosts p and r are in relation κ if and only if a message leaving from p can be delivered to r (directly or over a path accounting for disconnections). The formal expression of this definition is:

$p \; \kappa \; r \Leftrightarrow \exists \; t_0...t_k$ monotonically non-decreasing moments in time, with t_0 representing the current moment, and $\exists \; n_1...n_k$ hosts chosen [1] from the set of hosts N, such that $n_1 = p$, $n_k = r$ and $\Gamma_{n_1,n_2}(t_1) \wedge ... \wedge \Gamma_{n_i,n_{i+1}}(t_i) \wedge ... \wedge \Gamma_{n_{k-1},n_k}(t_{k-1})$.

---

[1] A given host in N may be chosen repeatedly in the sequence.

At any given instant t, the direct connectivity relation $\Gamma(t)$ is reflexive and symmetric, in that (1) a host is always connected to itself and retains messages across intervals of unavailability, and (2) if two hosts are connected they can communicate bidirectionally. As Figure 4 illustrates, the end-to-end path relation κ is reflexive but is neither symmetric nor transitive.

4 Solution

Figure 5 shows a more complex scenario involving eight hosts, a, b, c, d, e, f, g, and h. The discussion begins with the moment t_0 when host a wants to send a message to host h. The top half of Figure 5 depicts the connectivity intervals graphically, while the bottom half shows the pairs of hosts in $\Gamma(t)$ during each interval (for readability we show only one of each two symmetric pairs). The timeline in Figure 5 is divided into discrete intervals according to the times when at least one pair of hosts connects or disconnects. We assume sufficient time during each interval for ad hoc routing to occur, $i.e.$, messages can be exchanged among all pairs of hosts in the transitive closure of Γ during that interval - the widths of the intervals appearing in any of the figures in this section are not proportional to the duration of the interval.

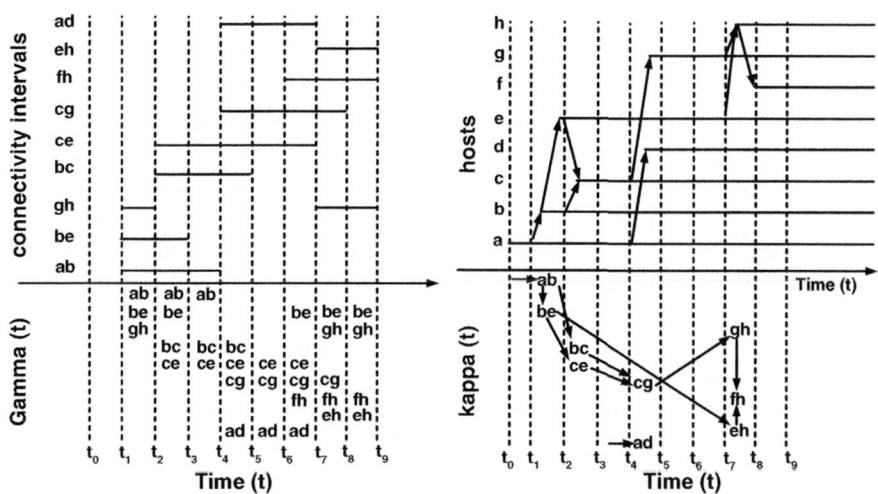

Fig. 5. Example Connectivity   **Fig. 6.** Building the Paths Tree

4.1 Global Oracular Algorithm

The first problem we solve is to show that if a source host a has full knowledge of the characteristic profiles for all hosts, it can compute a message delivery path from a to some destination host h if such a path exists. Our first algorithm for message passing in ad hoc networks depends on knowledge that is global with respect to space and time, *i.e.*, the motion and availability profiles for all hosts.

This strong requirement restricts the applicability of this first algorithm. Nonetheless, there are many situations in which the assumption of global knowledge holds. For example, we can easily imagine the connectivity segments illustrated in Figure 5 being generated by the motion profiles of a set of robots in an industrial factory, whose paths bring them temporarily (and possibly repeatedly) into wireless radio contact. Each robot might need to send messages to other robots telling them to slow down or speed up steps of a common task, to communicate failures, or to pass along sensor information about their environment. While in some settings it would be possible to connect the robots with a wired network, in others the range of motion or the cost of deploying and maintaining a wired network over numerous hosts may make a wired solution less attractive. Other applications with potentially well defined motion and availability profiles in which wired networks are impossible or costly to integrate include satellites, light rail systems, and ground-controlled unmanned aerial vehicles.

Using the connectivity intervals depicted in Figure 5, the following algorithm describes how to build a tree of message delivery paths from source a to destination h. In the path tree, the root is the source host for the message and any path from the root to a node in the tree represents a path a message *could* follow until it is delivered to the destination or gets stuck without any chance for progress.

We start building the tree from the root. Each node contains the ordered sequence of hosts the message has visited. We add children to each node for each previously unvisited host the message can reach next from that node until there are no more children to add to any node. Using the previously defined formalization, we express this as follows. For example, assume a node contains the marking $< a, b, e >$, because nodes a, b, and e, have already received the message. Because during the interval $[t_2, t_3]$ Γ_{bc} holds, we add a child node to the tree, attached to the $< a, b, e >$ node: $< a, b, e, c >$. We avoid adding a previously seen host (which could result in undesirable looping of messages or unbounded recursion of the algorithm itself) by recording the sequence of hosts seen so far. We also label each edge in the tree with the earliest point in time when that transition can be taken, again based on the ordering of connectivity intervals in relation κ.

Once the tree is built, a depth-first search ending in a leaf of the tree containing the destination node will reveal the path the message needs to follow from source to destination. Further analysis on the entire set of paths the tree may reveal can allow for optimizations in terms of the number of hops or the set of hosts the message will visit (*e.g.*, the latter can be important for security reasons). If we label each edge of the tree with the earliest moment when the extension can take place (when the child node is added to the tree), then the end-to-end delivery time or the time spent on each host can also be considered as constraints in the path selection decision.

Figure 6 illustrates how our algorithm runs on the example input data, which consists of an initial time t_0, a set of hosts a, b, c, d, e, f, g, h, a source host a, a destination host h, and the characteristic profiles of the hosts, $\pi_a(t)$, $\pi_b(t)$, $\pi_c(t)$, $\pi_d(t)$, $\pi_e(t)$, $\pi_f(t)$, $\pi_g(t)$, and $\pi_h(t)$. The top half of Figure 6 shows which hosts store and forward the message, and when they do so. The bottom half of Figure 6 shows how the connectivity intervals in Γ are concatenated to form the end-to-end path relation κ.

During the interval $[t_1, t_2]$ hosts a, b, and e form a temporary ad hoc routing network, and the message can be passed from a to b and from b to e. During the interval $[t_2, t_3]$ the message can be passed from either b or e to c. During the interval $[t_4, t_5]$ the message can be passed from a to d and from c to g. During the interval $[t_7, t_8]$ the message can be passed from either e or g to h and then from h to f.

The resulting tree of all message delivery paths from source host a is shown in Figure 7. If we specify a particular destination host such as h, the subtree of interest is one in which all leaves are labelled with a sequence of host names ending in h, illustrated by the dashed line in Figure 7.

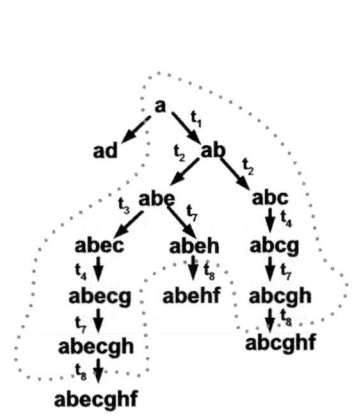

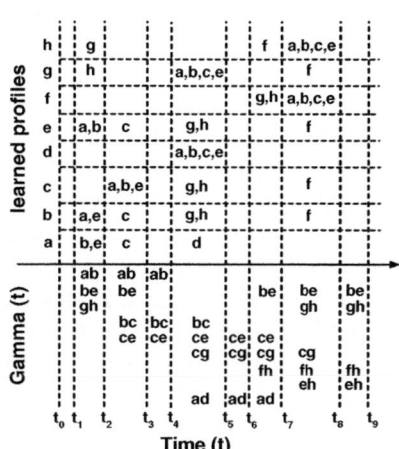

Fig. 7. Computed Paths Tree   **Fig. 8.** Learning Profile Information

4.2   Learning the Future

In this section, we relax the assumption that any given node starts with global knowledge, and consider how well a node can do if nodes accumulate and exchange profiles opportunistically each time they meet. In particular, we describe how partial information held by each node can still allow nodes to construct message paths, and show that any path so constructed is valid.

The approach presented in the previous subsection assumes global and accurate knowledge of the information used to compute the end-to-end paths. While there are situations where the assumption of global knowledge is reasonable, in this section we relax that assumption while still assuming perfect information in the characteristic profiles that are known. This allows our algorithm to be used in applications where hosts are highly mobile and global information may be difficult to obtain.

The message delivery paths that *precede* the moment when a host wants to send a message are especially important, as they represent prior opportunities for hosts to have exchanged information about their future behavior. At the point of wanting to send a message, the source host can then exploit the information available up to that point and plan its message delivery path accordingly, if possible. Several optimizations can be done by a host learning information that is not known to other hosts, including the ability to shortcut a path planned by another host when a better plan is discovered – we call this particular optimization a *path update*, and describe its use in greater detail in Section 4.3.

A key feature of this version of our algorithm is that a host can send not only information about itself, but information that it has learned from other hosts. We extend the algorithm in Section 4.1 so that when two hosts meet, in addition to exchanging messages whose message delivery paths go from one to the other, each host also sends profiles learned since their last encounter. If two hosts have not met previously, each sends the other all of the profiles it knows. At any point in time, a host can compute message delivery paths over the set of hosts whose characteristic profiles it knows at that point. The algorithm to construct those paths is thus the same as in Section 4.1, although the domain of hosts over which the algorithm operates is restricted to the subset of hosts whose profiles are known locally.

Assuming the same intervals of connectivity depicted in Figure 5, Figure 8 illustrates how the hosts learn each others' profiles. For the sake of this discussion, we start at time t_0 with each host knowing only its own profile, and show how characteristic profiles can be learned as time passes and interactions between hosts occur. Profiles are passed between hosts during all intervals from t_1 to t_8 except $[t_3, t_4]$ and $[t_5, t_6]$. It is also interesting to observe how the non-symmetric message delivery paths in κ result in non-symmetric profile information across the hosts. For example, f, g, and h all learned the profile for a, but a did not learn their profiles.

It is particularly interesting to note that in Figure 8 even if a message can start from a at time t_0 and be delivered to h as early as t_7, a is not able to discover that information. Thus, it is important to consider not only which message delivery paths exist between the hosts, but which hosts learn those paths and when. Only paths that are learned prior to when a message needs to be sent can be exploited by this version of our algorithm. The partial information held by the source node when it needs to send a message determines which paths the source node can compute. If a source node cannot compute a valid path due to incompleteness of its information, even if that path should exist, then the only alternative is for that source node to flood messages speculatively.

4.3   Recording the Past for Future Efficiency

With only local information, a source host may fail to compute any message delivery path to a particular destination. In that case, the source host may resort to Epidemic Routing [11], passing the message labelled only with the desired destination host (and not a planned path to that destination host) to each new host it encounters.

For this reason, in addition to maintaining and exchanging sets of characteristic profiles, it can be useful to record the history of hosts each message has visited. In this section we extend the algorithm presented in Section 4.2 as follows. We store the sequence of hosts a message has transited within the message itself. When a host forwards a message that does not have a planned message delivery path, it adds its identity to the set of hosts recorded in the message. When the message is delivered, the receiver also learns its history. All hosts share a common decision function to rank such paths. A host that knows of a better path than the one in a message they receive (we assume a planned path is better than not having one) will always perform a path update, recording the new plan in the message and sending it to the next host in the plan when it meets it.

Three kinds of information can be provided by recording a message's history. First, if we record the set of hosts a message has transited, a host that receives the message can learn of other hosts having the message even though it knows nothing of their characteristic profiles. Second, from the same information, a host can infer at least partial information about which other hosts have been in contact previously, *i.e.*, if combined with hosts exchanging profiles the message history can give each host that receives it at least a partial representation of the information each previous host knows about the others. Third, if we also record the time at which messages traverse each host, and similarly stamp other information such as when host profiles were exchanged, then information about the past can be correlated directly with information about the future. Recording the past history of messages can thus increase the scope of information at each host, and offer future performance improvements, particularly in message complexity.

As Figure 8 illustrates, hosts may know different subsets of the global set of characteristic profiles. Using additional information which it has but which the sending host does not have, an intervening host may be able to route a message along a better path it computes, instead of the path originally given to it by the sender. Hosts using only local information can therefore select paths that are closer to optimal if, as Section 4.2 describes, they are allowed to update path plans for messages sent from other hosts.

For example, assume the governing path selection heuristic is "fewest hops" and source host a computes its best path to destination host e through the set of hosts whose profiles it knows, $\{b, c, d, e\}$, as $abcde$. Suppose that just after receiving the message from host a and then disconnecting from host a, host b encounters a new host, f, which knows it can deliver the message directly to e. In that case, host b could update the planned path to be $abfe$ and route the message through host f. Note that an intervening host can only learn of a better path by meeting a host whose profile was unknown to the sender: otherwise the sender would have already computed that better path.

Recording the past can further improve future efficiency by eliminating double delivery if the two message delivery paths intersect. The two paths will intersect at the destination but earlier intersection points (if any) can help cancel one of the redundant delivery paths before the message reaches the destination thus limiting unnecessary use of resources.

While dropping duplicates can potentially improve efficiency, it can also be a pitfall leading to deadlock in the message delivery protocol. For example, consider the scenario in Figure 9. Host a is the source of a message that leaves once through b and once through c (e.g., through Epidemic Routing), on two message delivery paths towards a common destination h. The two path intersect at d and f. It is possible that the message coming on the route $a-b-d-e-f-h$ reaches e when the message coming on the route $a-c-f-g-d-h$ is at g. In this situation, d will drop the message from g because d has already seen it. Similarly, f will drop the message from e. This leads to deadlock in the message delivery procedure.

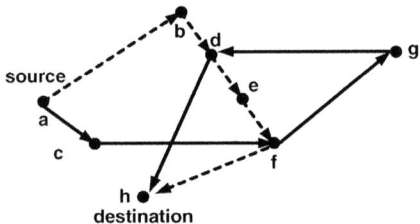

Fig. 9. Potential Deadlock in Message Delivery

It is possible to avoid the deadlock exemplified in Figure 9 by partially ordering all message delivery paths from a given source to a given destination, and dropping messages only on paths where another path from the source to the destination appears *earlier* in the partial order. The greatest potential for savings occurs when the paths are totally ordered, so that the message is delivered only along a single path, and the message coming along the highest priority route passes first through an intersection point.

It is possible to achieve a partial ordering by simply assigning a non-negative integer to each host, concatenating the host integers along each message delivery path to form an $|N|$-ary numeric value for each path with earlier hosts in the path representing higher-order digits, and having a host only drop the message if it knows a lower-valued path can deliver it reliably. If we make the host numbers unique, then the paths become totally ordered.

Unfortunately using this scheme alone unfairly causes traffic to be routed preferentially through lower-numbered hosts which could lead to overloading particular hosts or sub-networks. Such an arbitrary host numbering should therefore only be used if needed to break ties between message delivery paths in other more suitable partial orders. Two such partial orders are reasonably evident, one based on host bandwidth and one based on times at which messages are transmitted.

First, hosts could be numbered according to decreasing bandwidth and then arbitrarily among hosts with equivalent bandwidth. The host numbers at each hop would be then concatenated as before to form the path numbers. Even though the message routing is still unfair in this case, it is distributed more proportionally according to the capabilities of the hosts. As with the arbitrary host numbering scheme, shorter paths are preferred to longer ones, and higher-bandwidth hosts are preferred to lower-bandwidth ones at each hop.

Second, path numbers could be generated by concatenating the times of the message transmissions, but with the *later* transmission times representing higher-order digits in the concatenation and the assembled digits placed to the right of the radix as a floating point fraction, rather than as an integer. Although unlikely, it is possible two message delivery paths could be temporally equivalent, in which case they can be arbitrarily distinguished to form a total order by concatenating a unique lowest order digit to the end of each. This construction does not prefer any hosts in particular, nor does it prefer paths according to the number of hops they take. Rather, this construction prefers paths where the delivery of the message to the destination is earliest, then paths where the delivery time to the penultimate host in the path is earliest, and so forth.

5   Complexity

In this section we provide an analysis of the algorithms presented in Section 4. We examine three kinds of complexity: *message* complexity, which characterizes the number of messages that are sent and the number of hops they traverse, *storage* complexity, which characterizes the amount of storage that hosts in the network must maintain, and *computational* complexity, which characterizes the cost of computing information at each node in the path tree and overall.

5.1   Message Complexity and Path Optimality

Because edges in a global path tree (such as the one shown in Figure 7) map one-to-one to message transmissions by the global algorithm, the message complexity along any distinct path in our global algorithm is bounded by one less than the height of the tree, *i.e.*, $|N| - 1$. The number of transmissions needed to propagate a message from a source host to *all* other reachable hosts is also bounded by $|N| - 1$ because in the presence of global information, transmissions are only made to hosts that do not have the message, and after a transmission the number of hosts that do not have the message is reduced by one.

With local information, the same worst case bounds also hold. However, best-case and average-case message complexity may be worse with local information since a source host might not find the path with the fewest possible hops due to incomplete information. This same argument applies in general to arbitrary heuristics for selecting the best path. The profile information about other hosts that a host knows defines the subset of the global domain over which it can compute Γ and κ. Optimality of the paths it selects is thus subject to the constraints on its local information, *i.e.*, paths that a host cannot compute with its local information are not selected.

As we have seen in Section 2.1, the connectivity interval is shorter than the interval of time two hosts can communicate. In practice there is an overhead associated with the communication protocol, which we did not address in this paper. Messages carrying characteristic profiles are part of this overhead and therefore do not count (or do not influence significantly) the message complexity results derived from the analysis above.

5.2   Storage Complexity

The principal issue for storage complexity is that of maintaining characteristic profiles on the hosts. The best case occurs when a closed form can be found for equations describing the characteristic profiles. For example, the characteristic profiles of a set of small solar-powered monitoring satellites in orbit around a space station might be described as a function of their orbits and the phases of those orbits when they are in sunlight. In such a case, the storage complexity is effectively linear in the number of profiles stored.

A similar argument can be made about the complexity of profiles that repeat over time, but for which a closed form may not exist or for which storage needed for the expression exceeds the storage needed to represent the profiles explicitly. In these cases, the profiles can be stored as tables of values with entries for each relevant segment of time in the repeated period. Although the storage requirements for these profiles are again effectively linear in the number of profiles, the overhead per profile is expected to be higher than for closed form equations.

The worst case storage complexity occurs when no temporal structure of the characteristic profiles can be used to represent them efficiently. In that case, it may not be possible to represent the entire future of the hosts, because the timeline is potentially unbounded. In some cases, *e.g.*, for factory robots that are powered down at the end of a product assembly run, it may be possible though expensive to store the profiles over a meaningful segment of time, though the storage required is linear in the number of physical movements and availability transitions. Similarly, if characteristic profiles evolve over time due to environmental influences or independent decisions by hosts, *e.g.*, for coordinated mission re-planning among mobile infantry squads, it may be appropriate for each host to store profiles for each of the other hosts only up to the time of some pre-defined future meeting.

5.3   Computational Complexity

Computing even a single reliable path between source and destination hosts a and h in our global algorithm can be factorial in the number of hosts $|N|$. Specifically, because paths may not be symmetric, the number of nodes in the path tree that must be computed is a function of the permutations of hosts in the subset of N with a and h removed, $N - \{a, h\}$, plus one for the path tree node containing only a.

If it is necessary to compute the *best* path between each pair of hosts, an upper bound on the computational complexity to compute all paths in our global algorithm is given by the formula $(|N| - 2)! + 2$. Fortunately, the global nature of the information means that if it is in fact possible to obtain global and accurate information about the hosts characteristic profiles, then it is also possible to compute the path tree *off-line*, choose a path for each message, and simply give the paths to the appropriate source hosts.

In contrast, the Epidemic Routing protocol does not need to do any additional computation besides sending the message, so its computational complexity is constant. Thus, our global algorithm and the Epidemic protocol represent different points in the space of possible trade-offs between message and computational complexity for hosts in mobile ad hoc networks.

6   Discussion

Section 2 mentioned that in practice communication intervals between any two hosts will necessarily be shorter than the intervals during which they are connected. Extending the formalism described in Section 3 and the algorithm described in Section 4 to consider such overhead is mainly a matter of re-defining the relation $\Gamma(t)$ so that it holds over the segment of the communication interval not consumed by overhead. However, the overheads themselves may be non-trivial functions of the protocols and factors such as transmission latency may also need to be considered for each application. Sources of overhead must be identified and such factors must be incorporated in our message passing model. Clock synchronization is an important issue, especially for the algorithm with local information. The fact that host b tells host a that host c will be in a certain place at some moment t has real value as long as for hosts a and c moment t represents the same time. In this paper we assume clock synchronization across all hosts. As future work we will examine the implications of transient connectivity and of incomplete local information for clock synchronization protocols.

Another key area of future work is to examine the performance of the techniques described in this paper in the context of specific mobile ad hoc networked systems. We plan to conduct simulation studies based on motion profiles from a range of applications such those mentioned in Section 4 (*e.g.*, robotics, rail systems, unmanned aerial vehicles, satellites) to quantify the costs and benefits of our approach in real-world settings. We expect these studies to yield insights into further optimizations and trade-offs for the techniques described in this paper.

One important extension to the current research is to consider *message sizes.* The main question addressed in this paper was a simple decision problem: can we deliver a minimum-size message from a host to another? While this is an important issue, many real life applications handle a variety of information, and the sizes of messages have an important effect on transmission timing. If the message can be partitioned, and a single sufficient path cannot be found, parts of the message can be sent on several paths and reassembled at the destination. Constraints related to the order in which the message parts are received pose additional challenges. If the processing of the data at the destination has a prefix property (*i.e.*, its parameters are dynamically updated such that the processing of a segment of information depends on what has happened before), ensuring in-order delivery is a must.

In practice, the quality of information, especially for motion profiles learned from other hosts, may also depend on its freshness. We assess the performance of a host in predicting the correct future paths in terms of the *completeness* of the information (for which hosts it knows motion profiles - a subset of the total graph) and the *accuracy* of each profile it has. As future work, we will examine the effects of limiting the quantity of information we can learn about the future, as well as considering a deterioration factor associated with the closeness of the upcoming events, *e.g.*, information about the near future may be more accurate than information about the far future. Repeated characteristic profile information exchanges can help update the information about a known future profile, in case the real evolution differs from the initially declared one.

7   Related Work

Many other approaches to message routing in mobile ad hoc networks have been proposed. Most of the early work concentrates on using complete route information from source to destination at a certain moment in time. Such algorithms account for mobility by adjusting the route the messages will follow. However, the asymmetry and the disconnections frequently encountered in mobile ad hoc settings are serious challenges to the applicability of many such approaches. For example, DSDV [8], its extension CGSR [12], and WRP [10], are the most prominent members of the table-driven family of routing protocols. They all maintain tables with routing information for each pair of source-destination hosts. When a host needs to send a message it will have to consult the routing table dynamically maintained and choose a complete end-to-end route for the message. The drawback of this approach is the overhead of maintaining the tables and the assumption that the routes remain stable once the transmission begins.

Another important family of protocols is the source-initiated on-demand routing protocols. Members of this family are AODV [9], DSR [13], and ABR [14]. All of them initiate route discovery on-demand, when the source of a message needs to send that message to some remote destination. The assumption is that a potential route doesn't change and doesn't break during the delivery, remaining entirely defined, end-to-end, throughout the entire process.

Trying to relax the requirement that the route stays stable for the entire delivery period and that the communication is symmetric (e.g., the source sends discovery messages and the destination has to answer before the source sends the data), a new family of protocols is addressing disconnected communication. The main characteristic of this family of protocols is that they do not require the entire route to be defined at a certain moment in time and to remain defined for as long as the delivery takes place. The technique used in most approaches is to store-and-forward messages from one host to the other when connectivity allows it. Our approach does the same thing, allowing for next-hop or multi-hop message delivery if connectivity allows it over a given interval.

Generally, protocols addressing disconnected communication can be separated in two groups: proactive and reactive. Reactive protocols adjust their delivery strategy to the environment. Proactive protocols take steps towards exploiting communication opportunities when and where possible, if they were not already available.

Among the reactive protocols, we have mentioned Epidemic Routing [11], one of the first efforts in this area, in which hosts broadcast a message to all nearby hosts, hoping for eventual message delivery. Hop counts and buffer size limitations may restrict the excessive use of resources. In [15], the authors add the notion of a message lifetime and also develop four strategies for dropping messages in the case of buffer overflow in the same Epidemic Routing algorithm. Partial and total ordering schemes are used in [16] to ensure message convergence towards a destination when alternate routes can be chosen on the spot. Nodes are labelled such that the destination has the lowest id and messages are sent from a node to any other node with a lower id until it reaches the destination.

From the proactive family of protocols we mention two papers. In the first one [17], the authors modify the hosts' trajectories to ensure a sequence of 1-hop communication segments from source to destination. In [18], designated message *ferries* take the messages from source and deliver them to destination.

8   Conclusions

In this paper we have shown that reliable delivery paths can be established between hosts in an ad hoc networking environment, even in the presence of disconnection. The quantity and the quality of the information known *a priori* influences the selection of message paths. Host motion and availability profiles as functions of time are two of the most important parameters that influence message delivery in mobile and ad hoc networks. These profiles can be known *a priori*, discovered at run time, or updated on the fly. The algorithms for message delivery in the face of transient connectivity that we have presented in this paper can help deliver messages when other routing algorithms fail.

Acknowledgements. This research was supported by the Office of Naval Research under MURI research contract N00014-02-1-0715. Any opinions, findings, and conclusions or recommendations expressed in this paper are those of the authors and do not necessarily represent the views of the research sponsors.

References

1. Chiang, C., Gerla, M., Zhang, L.: Adaptive shared tree multicast in mobile wireless networks. In: Proceedings of GLOBECOM '98. (1998) 1817–1822
2. Carlberg, K., Crowcroft, J.: Building shared trees using a one-to-many joining mechanism. ACM Computer Communication Review (1997) 5–11
3. Ballardie, T., Francis, P., Crowcroft, J.: Core based tree (cbt) an architecture for scalable inter-domain multicast routing. In: Proceedings of the ACM SIGCOMM. (1993) 85–95
4. Deering, S., Estrin, D.L., Farinacci, D., Jacobson, V., Liu, C.G., Wei, L.: The PIM architecture for wide-area multicast routing. IEEE/ACM Transactions on Networking 4 (1996) 153–162
5. Ho, C., Obraczka, K., Tsudik, G., Viswanath, K.: Flooding for reliable multicast in multi-hop ad hoc networks. In: Proceedings of the 3rd International Workshop on Discrete Algorithms and Methods for Mobile Computing and Communications, Seattle, WA (1999) 64–71
6. Wu, C., Tay, Y.: Amris: A multicast protocol for ad hoc wireless networks. In: Proceedings of IEEE MILCOM'99, Atlantic City, NJ (1999)
7. Park, V.D., Corson, M.S.: A highly adaptive distributed routing algorithm for mobile wireless networks. In: Proceedings of INFOCOM'97. (1997) 1405–1413
8. Perkins, C., Bhagwat, P.: Highly dynamic destination-sequenced distance-vector routing (DSDV) for mobile computers. In: ACM SIGCOMM'94 Conference on Communications Architectures, Protocols and Applications. (1994)
9. Perkins, C.: Ad-hoc on-demand distance vector routing. In: MILCOM '97 panel on Ad Hoc Networks. (1997)
10. Murthy, S., Garcia-Luna-Aceves: An efficient routing protocol for wireless networks. Mobile Networks and Applications 1 (1996) 183–197
11. Vahdat, A., Becker, D.: Epidemic routing for partially connected ad hoc networks. Technical Report CS-200006, Duke University (2000)
12. Chiang, C., Wu, H., Liu, W., Gerla, M.: Routing in clustered multihop, mobile wireless networks. In: IEEE Singapore International Conference on Networks. (1997) 197–211
13. Johnson, D.B., Maltz, D.A.: Dynamic source routing in ad hoc wireless networks. Mobile Computing 353 (1996)
14. Toh, C.K.: A novel distributed routing protocol to support ad-hoc mobile computing. In: Fifteenth Annual International Phoenix Conference on Computers and Communications. (1996) 480–486
15. Davis, J.A., Fagg, A.H., Levine, B.N.: Wearable computers as packet transport mechanisms in highly-paritioned ad-hoc networks. In: International Symposium on Wearable Computing. (2001)
16. Roman, G.C., Payton, J.: A termination detection protocol for use in mobile ad hoc networks. (to appear in Special Issue of Automated Software Engineering on Distributed and Mobile Software Engineering)
17. Li, Q., Rus, D.: Communication in disconnected ad hoc networks using message relay. Parallel and Distributed Computing 63 (2003) 75–86
18. Zhao, W., Amma, M.H.: Message ferrying: Proactive routing in highly-partitioned wireless ad hoc networks. In: Ninth IEEE Workshop on Future Trends of Distributed Computing Systems. (2003)

Microbiology Tray and Pipette Tracking as a Proactive Tangible User Interface

Harlan Hile[1], Jiwon Kim[1,2], and Gaetano Borriello[1,2]

[1] University of Washington
[2] Intel Research Seattle

Abstract. Many work environments can benefit from integrated computing devices to provide information to users, record users' actions, and prompt users about the next steps to take in a procedure. We focus on the cell biology laboratory, where previous work on the Labscape project has provided a framework to organize experiment plans and store data. Currently developed sensor systems allow amount and type of materials used in experiments to be recorded. This paper focuses on providing the last piece: determining where the materials are deposited. Using a camera and projector setup over a lab bench, vision techniques allow a specially marked well tray and pipette to be located in real time with enough precision to determine which well the pipette tip is over. Using the projector, the tray can be augmented with relevant information, such as the next operation to be performed, or the contents of the tray. Without changing the biologist's work practice, it is possible to record the physical interactions and provide easily available status and advice to the user. Preliminary user feedback suggests this system would indeed be a useful addition to the laboratory environment.

1 Introduction

Many work environments can benefit from integrating computing devices so that they can provide users with important information where and when they need it. In addition, keeping track of what users are doing can enable proactive applications that automatically record actions for later analysis or prompt users about the next steps to take in a procedure. Of course, automatically discerning the many actions users may perform can be a daunting task for an automated system. Fortunately, the structure of most work environments helps to make this a tractable problem.

Our work focuses on augmenting the cell biology laboratory [1]. Biologists perform their tasks at a collection of specialized work areas. Each work area has specialized equipment and material and is usually dedicated to one particular task. Our experience is with the laboratories of University of Washington's Cell Systems Initiative with whom we collaborate on this work [2]. There, we have identified 8 different tasks and their corresponding work areas, including centrifuging, incubation, titration and dispensing, thermocycling, etc. In this paper, we highlight our work related to the tasks that involve the dispensing of

A. Ferscha and F. Mattern (Eds.): PERVASIVE 2004, LNCS 3001, pp. 323–339, 2004.

liquid into well trays. Well trays are a convenient array of small containers, each holding a different combination of reagents and cellular material. Some steps in the experiments act on an entire array at once (e.g., incubation) to generate data for different values of reagent parameters.

Figure 1 shows an outline of an experiment plan that highlights the tasks that involve mixing liquids in the well trays. The tool being used is TeraLab, a product of TeraNode [3], a commercialization of the Labscape tools we previously presented at this conference [4]. TeraLab captures an experiment plan (the icons on the left represent the eight different types of steps) and associated parameters (type of material, amounts, links to images, etc.). The experiment plan of Figure 1 shows the cross-products of materials that are dispensed and mixed into the containers of the well trays. Note the vertical and horizontal arrangements to generate all possible combinations of materials. An expanded view of some of this detail is shown in Figure 2.

However, there are many more details missing from this experiment plan. When a biologist is at a workbench dispensing materials into the well trays (see figure 3), he is likely to be working on only one of many experiments. Therefore, we need a way of identifying the tray so that it can be connected with its corresponding experiment. We also need a way to know which materials are being dispensed and how much material is being pipetted into a particular well in the tray. The trays have a coordinate system that helps biologists keep track of what material they put where, but they must jot this information down on a piece of paper or on the tray itself. This is a common source of errors or ambiguities when the experiment is written up later as many shortcuts and abbreviations leave room for interpretation. In addition, the process of recording this information on paper distracts the biologist from the flow of the procedure. Interpretation is further complicated because the notes tend to be incomplete. In some cases it is not even possible to take notes since paper is not permitted in certain parts of the laboratories because it can be a contamination agent.

We seek to make this recording process automatic through the use of a variety of sensors that can capture the important aspects of the biologist's activities. To this end, we have developed a method for attaching RFID tags to reagent bottles and their lids. An antenna embedded in the workbench can be used to sense which bottle has been lifted, whether its cap has been removed and when it has been replaced, and when the bottle is returned to its original location [5]. We have also instrumented a pipette to wirelessly transmit its aspiration and dispensation events and report the quantity of liquid drawn or dropped [6]. This paper reports on the missing piece of automating this workbench, namely, determining into which well the liquid was dropped. In addition, we help the user keep track of what was placed in each well.

We present a video tracking system that tracks the well tray and pipette. It can precisely determine over which well the pipette tip is placed by the user. Correlating this with pipette events and movement of reagent bottles allows us to keep track of how much and what type of liquid was dispensed into a particular well in a particular tray. This information is used to annotate the

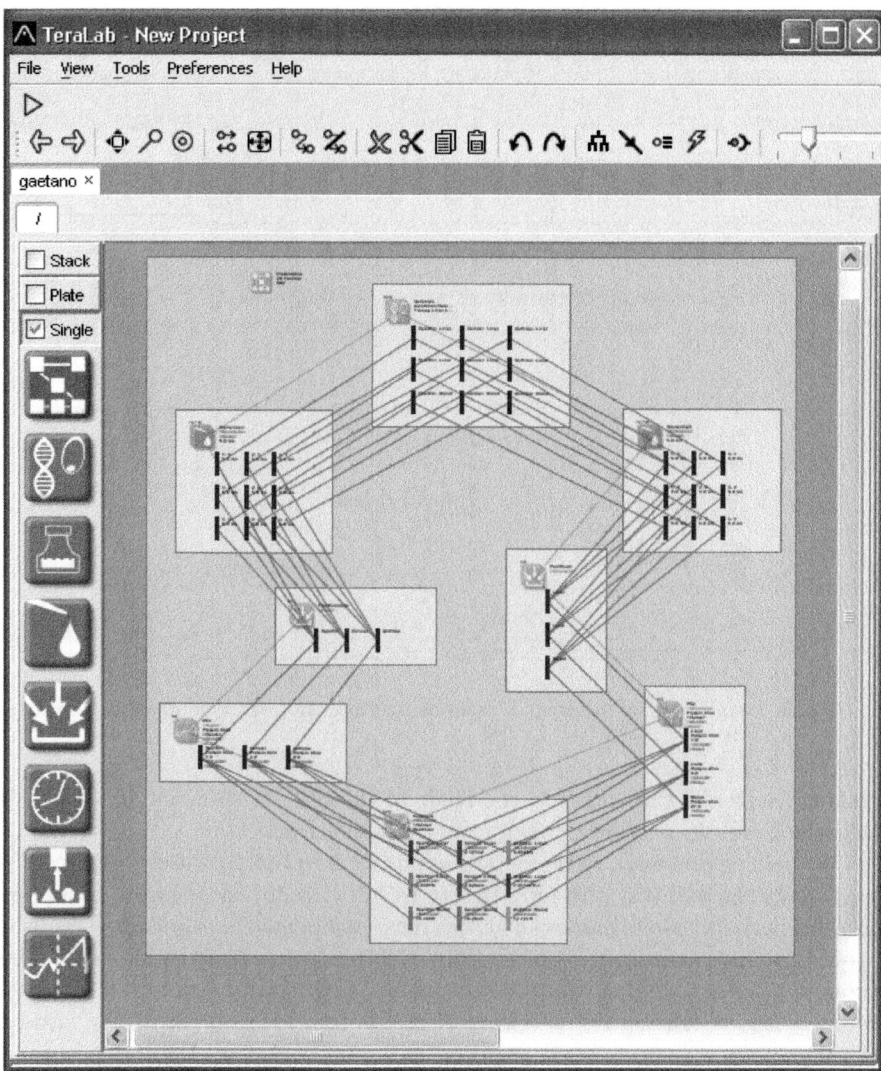

Fig. 1. Overview of a portion of a biology experiment expressed in TeraLab (Teranode).

TeraLab experiment plan and allow the user interface to continuously display to the biologist which steps have been completed and which are left to be done. Furthermore, we can project the contents of the well directly onto the workbench (immediately above the well tray) by using a projector set up in parallel with the video camera. The projector can also provide information about which well to dispense into next and generally aid the biologist in keeping track of their work even through interruptions. To make the tracking algorithm robust, we color-

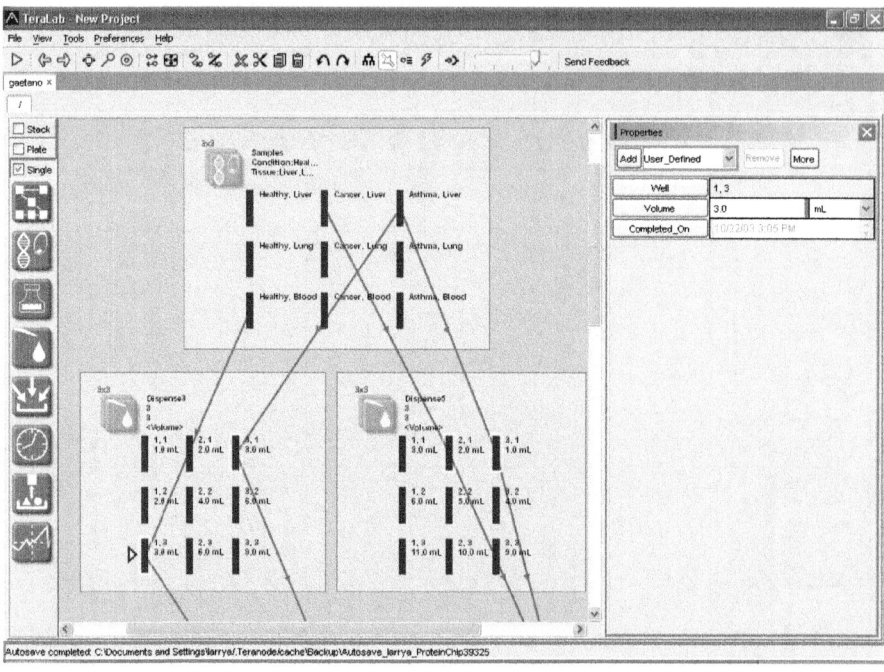

Fig. 2. Detail view of a TeraLab dispensing step. (not all dependencies shown)

code the edges of the tray and the shaft of the pipette (but not its disposable tip, which we can precisely position by extrapolation).

We believe this work to be a novel application of tangible user interfaces. We have made the well tray and pipette into I/O devices for the computers supporting this activity. Most importantly, we have augmented an already established procedure and require little or no adaptation on the part of the biologist. Their processes are not altered and the computing is invisibly inserted into the laboratory without causing undue distraction. Information is presented in situ and the objects of the work itself are used as input (well tray ID and position of pipette tip) and output devices (well tray location and contents projected onto the workbench surface).

In the following section, we will review related work. In section 3 we will discuss design and implementation of both the vision tracking system and the augmenting display. We will then present some analysis of our system, including feedback from our target user group, and wrap up with future work and conclusions.

2   Related Work

There is a variety of previous work that share some similar aspects with our project. These break down into two categories that follow the two main components of our system: augmented reality, where extra information is displayed onto existing surfaces, and physical interactions, where physical objects are used to interact with a computer.

Everywhere Displays [7,8] are steerable projectors that allow interfaces to be projected onto surfaces all around a room. A paired camera allows the system to notice occlusions and move the display and to look for user interactions with the display. Although some uses of the Everywhere Displays involve displaying environment relevant information, the main focus is to allow the computer display to be moved from its fixed location on the desktop and to allow the user to interact with the virtual objects in the display using fingertip tracking. There are also several other systems that use a projector to display an interface and a video camera to track interactions, often using hands or fingers. For example, FingerPaint [9] tracks fingertip location to be used in a drawing program, and EnhancedMovie [10] is a video editing system that allows the use of both hands to manipulate video clips. These systems all allow interaction with virtual objects using hands as pointing devices without involving other physical objects. Augmented Surfaces [11] use visually unique tags on physical objects to give them presence in a shared virtual display, extending interactions beyond the computer screen. Although the setup and techniques of this system are similar to ours, the goal is quite different, providing extended display or allowing access to virtual objects by physical association. In contrast, our system does not try to make virtual objects available for interaction. Instead it tracks interactions

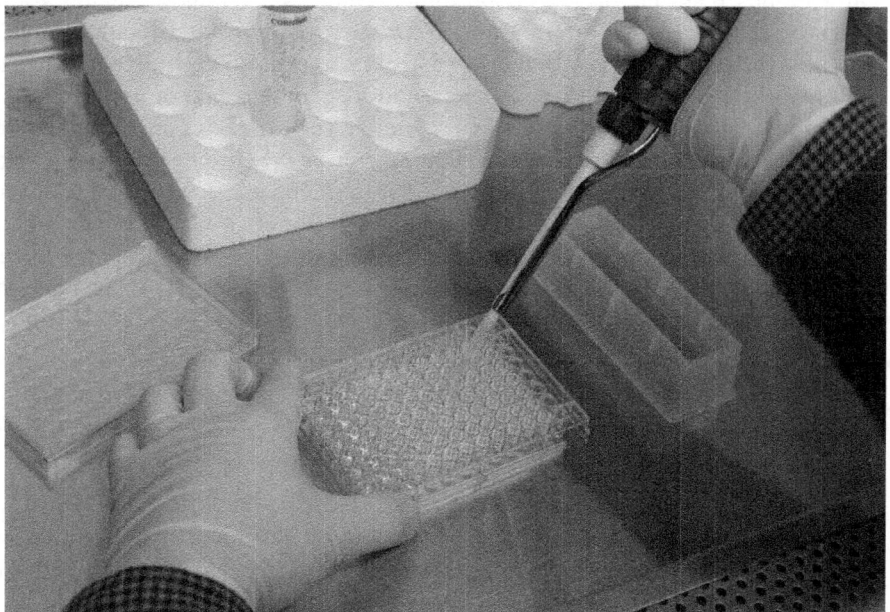

Fig. 3. A laboratory environment where a microbiology experiment is being conducted.

between physical objects used in an existing task and augments the environment with information about those physical objects.

There are also systems that focus on using physical interactions with objects as direct input to the system. These Tangible User Interfaces often try to attach virtual devices or abstract concepts to physical objects. Tangible Bits [12] and the metaDESK [13] take the standard desktop environment and convert some familiar components into physical versions, which can be directly interacted with. BUILD-IT [14] is an example of merging virtual and physical interactions in a so-called Natural User Interface, where users collaborate on a design by interacting with physical blocks. SiteView [15] allows users to create rules about their environment by manipulating physical representations of the rule components. All these systems try to give a physical representation to interfaces that are currently virtual. We take the opposite approach, and attach virtual associations to an existing physical environment.

Work that has been done on merging augmenting displays with physical interaction tracking is closest in spirit to our own. One well known example of this is the DigitalDesk [16] which blurs the line between physical and electronic documents. Using a camera and projector system similar to our own, it tracks a user's interaction with paper documents and can provide augmented services on top of the traditional interactions. In an idea similar to the Double-DigitalDesk [16], Tele-Graffiti [17] tracks paper interactions to allow remote collaboration through a sketching interface. Total Recall [18] captures whiteboard annotations which can later be rewound to be viewed in-place or on a separate display. The Designers' Outpost [19] captures existing site design practice in post-it note and whiteboard sketch form to support enhancements like remote collaboration, versioning, and extended information. Work on the Augmented Reality Toolkit [20] has created a variety of augmented reality applications using special tags to locate objects in relation to a camera and display information on them [21]. These projects and our work all share a similar goal of taking an existing task and enhancing it by augmenting the environment based on the actions that take place. However, they are all in very different domains and have different forms of interaction. In particular, a distinguishing feature of the laboratory domain is that an experiment plan (see figure 1) is often known before the experiment is performed. This important characteristic allows our system to proactively suggest actions and flag errors, something that is not available in the above mentioned systems.

3   Design and Implementation

3.1   Overview

Our system is composed of several components to handle display of information and sensing of objects using familiar computer vision techniques. Our physical setup (see figure 4) consists of a projector and camera mounted above a workspace, similar to many other systems. A semi-automated calibration routine computes a homography between the camera and projector image spaces, giving the system the ability to locate an object in the camera and project information about it onto the matching physical benchtop location. Our implementation currently looks for 96-well trays (8 by 12 array of wells) and a 100 microliter pipette. In order to identify and locate these, we have marked them with patterns of bright colors (see figure 5). Processing the colors and patterns gives us location and orientation of the objects, which can then be augmented with the projector display. The next subsections will detail the normal processing path of our system.

3.2   Color Segmentation

The use of brightly colored tags allows us to separate the markings from the rest of the image, making it easier to process the patterns by first performing color segmentation (see figure 6B). We are primarily interested in the hue of the color rather than the luminance of it, as the latter is easily influenced by lighting conditions. Following previous work [22], we convert from the camera's RGB color space to an HVC color space. This allows us to classify a color based on its hue. Furthermore, we take into account the color's saturation level (chroma), and filter out extreme luminances (value) because very bright or very dark regions often have poor color information. In order to run this as quickly as possible, we create a lookup table to directly map RGB values to their color segment, as has been done for robot vision [23]. We spent considerable effort choosing colors that the camera could pick up well, and we were able to hand-tune the results of semi-automatic color calibration in order to get more robust segmentation. These calibration steps should only be needed once per installation and do not need to be done by the users.

3.3   Grouping Pixels

Once each pixel is classified into a color segment, the next step is to group pixels into larger units. Connected components are found by looking in the neighborhood of a pixel for pixels of the same color and grouping them into one component. Each marking on an object should be found as one component. Although it is common to blur the image to smooth out connected regions, we found our results sufficient even without blurring. Regions are also filtered for size and shape, so that very small and very large regions that do not match expected ranges for our markers are ignored. These steps help filter out clutter that may be caused by other items in the workspace. For example, even if the container in figure 3 matches a marker color, it will not cause confusion because its size and pattern do not match known profiles. Once these connected regions are found, they are organized into linear groups. At first, we used a Hough transform to find lines at different angles through the image and locate what components were along this line. This was relatively slow, so we switched to a direct method that fits a line to a group of components. More specifically, line segments are fit to the three closest neighboring components, and approximately collinear segments are

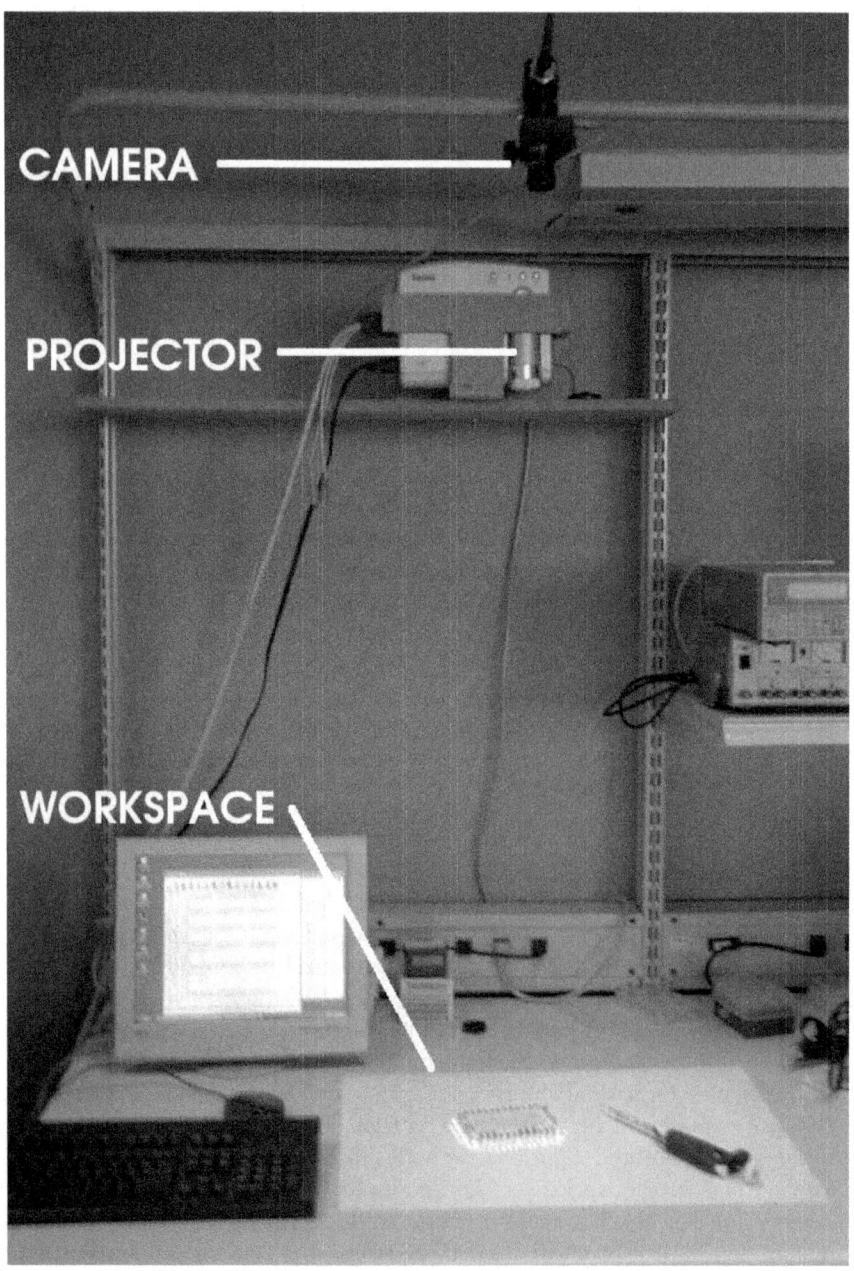

Fig. 4. Our setup with a camera and projector above the workbench. The camera tracks the tray and pipette, and the projector allows us to augment the tray with information.

clustered into groups. Then a single line is fit in a least-squares sense through all components in the group.

3.4 Recognizing Tray Patterns

Once lines are detected from the previous step, we first create a list of tray candidates by looking for groups of 4 lines that form a rectangle of an aspect ratio close to that of the tray (see figure 6C). To handle partial occlusions, we also consider groups of 2 or 3 lines that may form part of such a rectangle. Allowing a little margin for the angles between the sides and the aspect ratio enables us to detect the tray even when it is tilted out of the plane.

We then validate each tray candidate by matching the colors of components along each side with the actual patterns on the tray. The patterns are designed in a way that enables us to uniquely identify individual marks within the tray coordinate system, given sufficient number of visible marks. The marks are laid out with a regular spacing in alternating colors to indicate orientation and the origin of the tray coordinates (see figure 5). Also note that our marking system does not interfere with the standard form factors or usage models of the tray.

To see if a tray candidate matches the patterns, we first calculate an estimate of the inter-mark spacing for each side using the side's length, the expected number of marks, and the median of the spacings between neighboring components. If the entire side is missing, we borrow the spacing estimate from the other sides. Using the estimated spacing, we then assign each component to its corresponding mark on the patterns in all possible orientations of the tray. Finally, we compare the average color of each component with that of the assigned mark. All tray candidates that match the patterns are considered a valid tray.

The tray can be located even in the presence of considerable occlusion. We are able to detect the tray with at least parts of two sides visible (with sufficient number of marks to distinguish the sides and locate the marks within each side), and within a reasonable range of tilt that covers typical usage.

3.5 Locating Pipette Tip

The disposable tips used on the pipettes are transparent because laboratory workers need to be able to see liquid pulled into the pipette. Because of this, we could not put opaque markings directly on the tip of the pipette. Instead, we use a simple technique to determine the location of the tip in relation to the body of the pipette. A set of regularly spaced markings allow us to locate the body, and one marking at a different spacing allows us to determine the orientation (the handle end versus the tip end). Once the locations of the points and direction of the pipette is found, the tip can be located by using the spacing between the markings. In our setup (see figure 5), the tip is two spacing-widths away from the closest mark; this ratio holds even as the pipette is tilted out of the plane. Since there are fewer marks than on the tray, there is less redundancy and locating the pipette is not as robust to missing markers; the pipette tip can be located with one missing marker, as long as it is not the marker that determines orientation

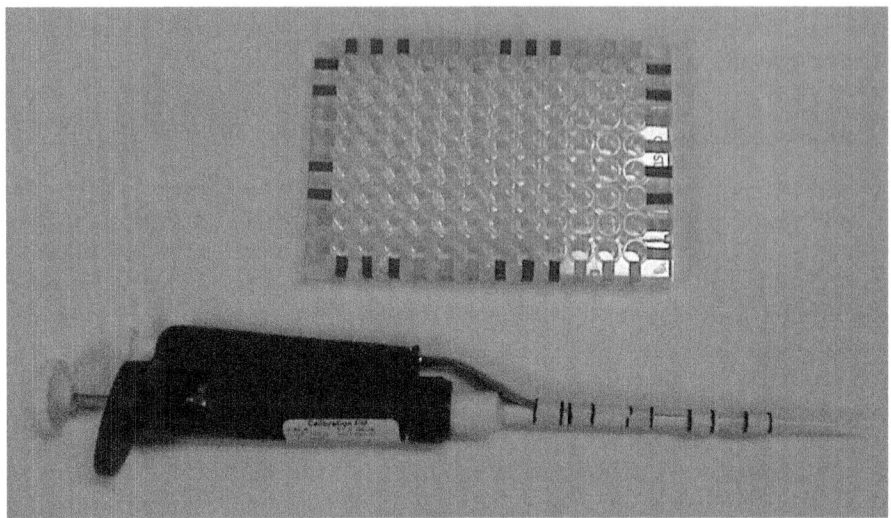

Fig. 5. Red, green, and yellow markings used on tray and pipette.

(see figure 6F). The pipette tip is located within a few pixels of the true location. This precision is on the same scale as the radius of the wells on the tray.

3.6   Interface Display

Once the tray and pipette are located in the camera image, this information is displayed back to the user directly on the work surface. The camera space is transformed into the projector space using a previously computed homography, so that projected elements land in the correct location. A simple homography is suitable because the tray and pipette are only used very near the surface of the table. A box is displayed around the tray, with the rows and columns labeled, to give acknowledgement that the tray has been located, and a red circle is drawn at the pipette tip. When the location of the pipette tip corresponds with the location of a well in the tray, the row and column labels are highlighted and the well is marked with crosshairs. This allows the user to easily and peripherally determine that an interaction is recognized correctly. If the user hovers the tip over the well for about a second, an information bubble about the well pops up, displaying its history. This currently only includes a list of the times at which material was added to the well, but integration with other systems should let us record what the material was and how much of it was added, as well as display the future planned contents of the well.

4   Analysis

Our image processing and user interface is implemented in Java, with input from a firewire camera running at 640 by 480 resolution. This setup can run at

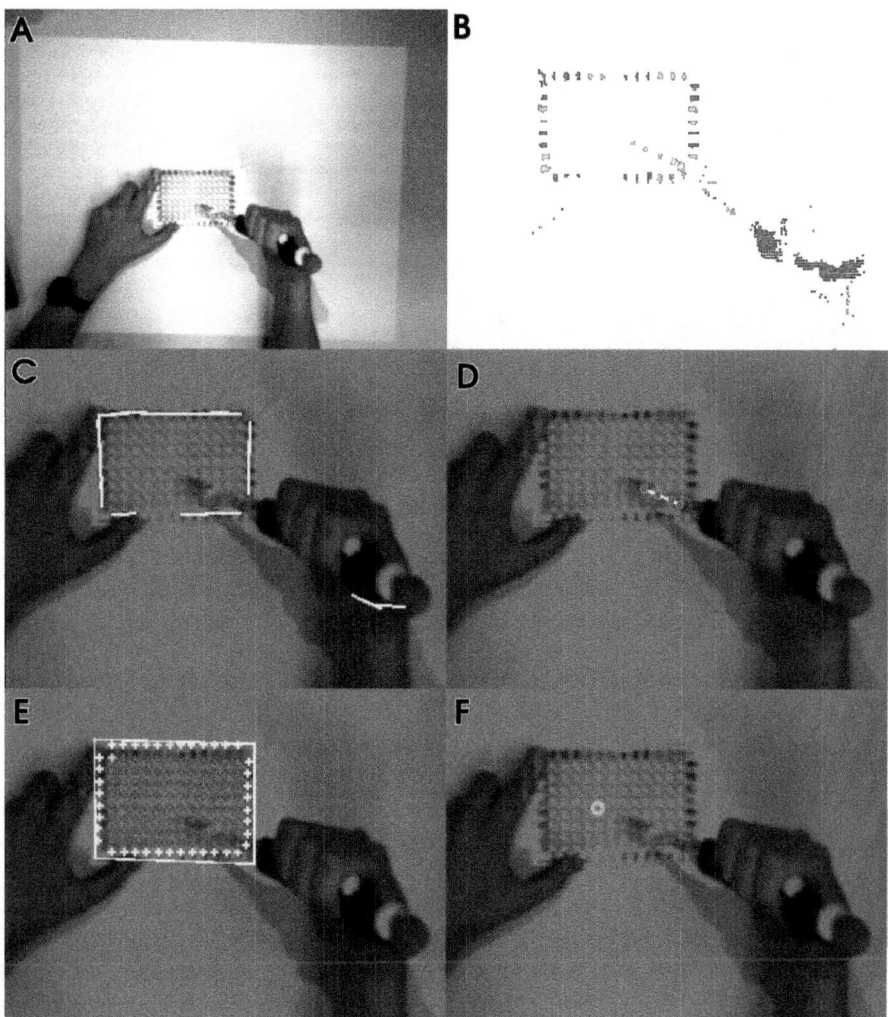

Fig. 6. The steps involved in locating a tray and pipette. A: the full view as seen from the overhead camera. B: the color segmentation of the relevant portion of the image. C: candidate lines for a tray location. D: candidate lines and marker points for a pipette. E: a located tray, showing well location and origin. F: a located pipette tip.

about 4 frames per second, but a surprising amount of time is spent displaying the interface. With the interface turned off, just running the image processing components, the system runs at 11 frames per second. Using an optimized Java virtual machine, or switching languages, would likely boost the system speed. With unobstructed views of the tray and pipette, the tray is found about 99% of the time, and the pipette about 95% of the time. The source of errors in these cases seems to be color variations in images retrieved from the camera.

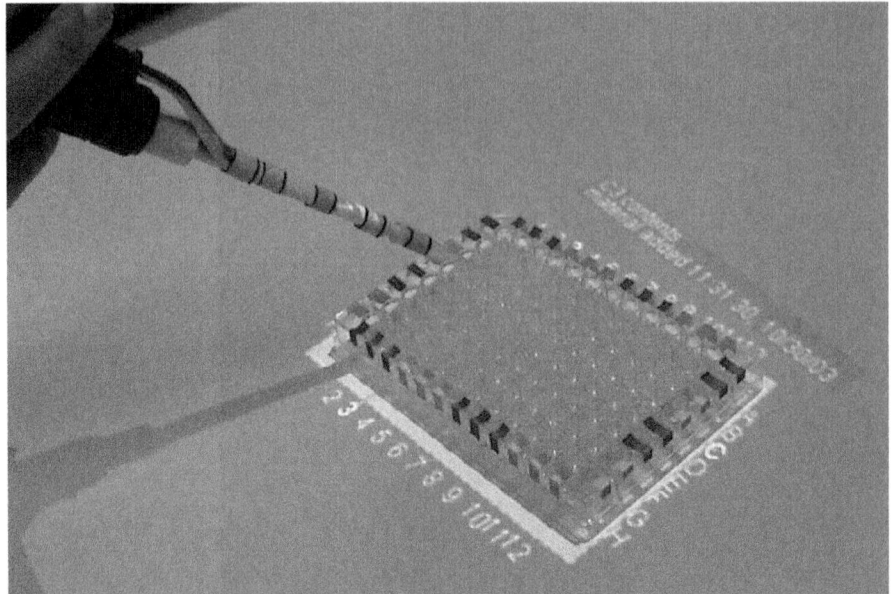

Fig. 7. Display reacting to pipette and tray. Note how the coordinate system of the tray is recognized and displayed, and how the pipette location is highlighted. Information about the selected well is displayed above the tray.

Our location routines are strict enough that false positives are very infrequent, and false negatives could be smoothed out to present a more stable interface to the user.

We asked three of our collaborators at University of Washington's Cell Systems Initiative [2] to test out the system. They are all experienced laboratory workers (not computer scientists), and are the target audience for our application. However, as members of our team they have an interest in seeing the application in daily lab use and are not the harshest critics. We invited them to get familiar with the system by performing tasks similar to those used in their own experiments, and to give us their opinions of the system and how it would fit with their lab practices. Each session lasted approximately one hour. This preliminary evaluation generated useful feedback, which will be discussed here.

During development and testing, we were using the pipette as a new pointing device, rather than a tool to move liquid. Because of this, we were holding the pipette in significantly different ways than would be done in the lab. Our participants told us it is necessary to make sure the liquid makes it into the well, and does not stay clinging to the pipette because of surface tension; this means the pipette tip is often brought into contact with the bottom or the side of a well. In order to touch the bottom of the well, the pipette must be held at a steep angle, often causing the camera's view of markings on the pipette to be obstructed. We originally chose to put the camera directly overhead to minimize distortion, but it is now clear that this will not work given the standard laboratory practice. Since most of the operations are in the plane of the benchtop, it should not be problematic to change the camera angle to give a clearer view of the pipette. Having the pipette tip in contact with the edge of a well occasionally confused the selected well with its neighbor, since the tip was now close to the center of both. This could be helped by the location algorithm suggesting a point slightly in from the tip, instead of the very tip of the pipette. Some simple experiments could tell us what part of the pipette was in the center of the well for the average use case.

Despite the problems caused by obscuring the camera view of the pipette, all of our participants were able to quickly adapt their technique to one that allowed the system to locate the pipette. Their updated technique was performed at approximately the same speed as their original. The visual feedback from the system also gave users awareness that their action had been recognized, and our users reported that the display was not distracting with regard to carrying out the experiment. To confirm the usefulness of feedback, we had our participants transfer liquid to the tray with and without feedback. Without the feedback, there was no indication that the location of the pipette had been correctly recorded for a dispensing event. With the feedback enabled, the users got a view into the state of the system so they could tell when a location was recognized and would be correctly recorded. This knowledge allowed them to adjust accordingly, usually in subtle ways or simply by waiting for acknowledgement of recognition. We found that feedback made a drastic difference; in one case, when our subject was not provided with feedback, 14 of 16 locations for dispensing events were not found, compared to only one of 16 not found with the feedback enabled. Our system was also able to keep up with our participants' motion, even though it was only running at 4 frames per second.

All of our participants felt that this system would be useful in their lab environment. Even in its current state, they felt that it would help them keep track of what operations they had done, and help them recover from interruptions. They indicated that a secondary display that could show them the status of all wells in a single view (see figure 8), would make it easier for them to find where they left off, rather than having to hover over each well to query its contents. They also liked the idea of using proactive prompting to show them which operation and which well should come next, or flag them with warnings if they were poised to deposit liquid into the wrong well. They also confirmed that they could save time if their actions were documented automatically as they did them, instead of having to manually record their process on paper, and then later transfer it back to a computer.

5   Future Work

We feel that there are several areas of future work: integration with TeraLab, incorporating more sensors, improving speed and robustness, and in-depth user

Fig. 8. Another potential view of information about the well tray. The display on the right of the tray shows an abstract status of all wells, with detailed information still available by selecting a single well.

studies. Some of these possibilities are discussed in more detail in the following paragraphs.

Integrating the experiment plans of TeraLab (see figure 1) with our system will provide much more information for the user interface. The display will then be able to indicate the next step to a user, or alert the user if their actions do not match with the expected steps. TeraLab's recording facilities will allow transparent storage of all the actions performed on each tray directly alongside the experiment procedure: for later recall and tracking, for use in formal write-ups, or for sharing.

Use of additional sensors will allow us to record more information and associations with the experimental data. RFID tags on the trays would allow us to uniquely identify trays and pull up their stored history for display. Tagging reagent containers would also provide information about what was dispensed. Conversations with the lab workers helped us to identify some difficulties in extending the system in these ways. They told us they often bring many reagents to the table at once, and they often put them in intermediate containers before they deposit them into the wells. They also sometimes hold the container they are dispensing from in their hand, rather than leaving it at a fixed location on the table. This means a simple RFID system to tell which reagent is at the bench is insufficient, and the camera may even be unable to find a consistent location for the source. However, biologists are trained to never open more than one bottle at a time to minimize contamination risks (a better structured environment than medication use, where similar RFID systems have been proposed [5]). We believed that determining the last opened reagent in a structured workflow such as this can be quite accurate in determining the reagent in use. In more complicated cases where reagents are mixed or diluted, a known experiment plan can be very helpful in telling the system what to expect, and minimal user prompting can be used where necessary.

We are also aware that some experiments involve transferring material from one tray to another, and while our vision system can find separate trays, it cannot distinguish between them. An RFID reader would be able to report the presence of multiple trays, but not location. However, using multiple RFID antennas could provide a finer location resolution to help associate identification with a visual location of a tray. The merging of these two systems could provide accurate identity and location information, similar to systems used for people tracking [24]. We might also consider a visually based unique tagging system like CyberCode [25] to identify and locate trays. If occlusions of the pipette markings continue to be a problem, a system like Mimio [26] with an added direction sensor may be a possibility for resolving pipette location. We also plan to investigate using multiple cameras to improve accuracy; multiple views would help with occlusion and glare problems. A small camera mounted on the pipette could give a more detailed view of the tray, and thus a more accurate location for the pipette tip in relation to the tray. Additionally, there are different types of pipettes (of different volumes and with different sized tips or multiple tips) that should be trackable using the techniques discussed, but they will need to have unique features so they can be distinguished from each other. Different tray sizes can easily be accomodated using RFID tags to include size, aspect ratio, and well diameter information.

Although our participants were able to use our system, we feel that faster and more robust tracking would improve the experience. We expect the speed of our system can be improved considerably by switching computing platforms and through further optimization. As mentioned, the use of multiple cameras may give better accuracy and robustness. We may also explore different markings besides color, perhaps infrared, or rely more on pattern than on color. More temporal information could also be included in our system to help improve reliability. It may be beneficial to incorporate uncertainty into our framework too, so the accuracy of the system can be taken into account when reasoning about events. For example, a threshhold can be set for when prompting is necessary, or a known experiment plan and user habits can provide prior knowledge of where to expect a pipette tip.

We would like to conduct further user studies in a real laboratory environment with users that are not part of our group. By asking users to perform common experiments in a realistic setting, we could study how our system influences work patterns. The use of different information displays could be studied to determine what was best suited for a given task. Creating a configurable information display would allow different displays to automatically be shown for different tasks or user preference. It would also be valuable to find what level of user prompting was acceptable and sufficient.

6   Conclusion

We have built a vision-based system for tracking the interactions between microbiology well trays and pipettes in a laboratory environment. The preliminary evaluation and feedback from our target users indicate that we have built the beginnings of a useful system. Using the projector to provide feedback allowed the users to readily adapt to the limitations of our system. We believe that improving the frame rate and adapting the camera setup for typical lab use will make our system more robust. We also think a more developed user interface will improve the usefulness of the system. There is still considerable integration necessary before the system can actually be deployed in a lab. Most importantly, we need to connect to the Teralab experiment plan and data recording facilities to allow the actions performed in the experiment to be automaticalled recorded and saved. A scheme to determine what reagents are being used is also necessary, and will likely be based on a combination of sensors and interactively prompting the user.

Acknowledgements. Raymond Kong, Qinghong Zhou, and Michael Look have our appreciation for their time and patience in helping us evaluate our system. Bob Franza generously provided access to his lab and staff at the UW's Cell System Initiative. Larry Arnstein and TeraNode provided access to the TeraLab tools. Intel Research Seattle provided lab space and hardware for this project as well as a summer internship for Jiwon Kim.

References

1. Arnstein, L., Hung, C.Y., Franza, R., Zhou, Q.H., Borriello, G., Consolvo, S., Su, J.: Labscape: A smart environment for the cell biology laboratory. Pervasive Computing Magazine **1** (2002) 13–21
2. http://csi.washington.edu: University of Washington CSI website (2003)
3. http://www.teranode.com: TeraNode website (2003)
4. Arnstein, L., Grimm, R., Hung, C.Y., Kang, J.H., LaMarca, A., Look, G., Sigurdsson, S.B., Su, J., Borriello, G.: Systems support for ubiquitous computing: A case study of two implementations of labscape. In: Pervasive Computing Conference. (2002)
5. Fishkin, K., Wang, M.: A flexible, low-overhead ubiquitous system for medication monitoring. In: Intel Research Technical Report 03-011. (2003)
6. Arnstein, L., Sigdursson, S., Franza, B.: Ubiquitous computing in the biology laboratory. Journal of Lab Automation **6** (2001)
7. Kjeldsen, R., Levas, A., Pinhanez, C.: Dynamically reconfigurable vision-based user interfaces. In: 3rd International Conference on Computer Vision Systems. (2003) 323–332

8. Pinhanez, C.: The everywhere displays projector: A device to create ubiquitous graphical interfaces. In: Ubiquitous Computing (UbiComp03). (2003)
9. Crowley, J.L., Coutaz, J.: Vision for man machine interaction. In: EHCI. (1995) 28–45
10. Ishii, Y., Nakanishi, Y., Koike, H., Oka, K., Sato, Y.: Enhancedmovie: Movie editing on an augmented desk. In: Ubiquitous Computing (UbiComp03). (2003)
11. Rekimoto, J., Saitoh, M.: Augmented surfaces: a spatially continuous work space for hybrid computing environments. In: Proceedings of the SIGCHI conference on Human factors in computing systems, ACM Press (1999) 378–385
12. Ishii, H., Ullmer, B.: Tangible bits: Towards seamless interfaces between people, bits and atoms. In: CHI. (1997) 234–241
13. Ullmer, B., Ishii, H.: The metadesk: Models and prototypes for tangible user interfaces. In: ACM Symposium on User Interface Software and Technology. (1997) 223–232
14. Fjeld, M., Bichsel, M., Rauterberg, M.: BUILD-IT: An intuitive design tool based on direct object manipulation. Lecture Notes in Computer Science **1371** (1998)
15. Beckmann, C., Dey, A.K.: Siteview: Tangibly programming active environments with predictive visualization. In: Intel Research Tech Report 03-019. (2003)
16. Wellner, P.: Interacting with paper on the DigitalDesk. Communications of the ACM **36** (1993) 86–97
17. Takao, N., Shi, J., Baker, S.: Tele-graffiti: A camera-projector based remote sketching system with hand-based user interface and automatic session summarization. In: International Journal of Computer Vision. Volume 53. (2003)
18. Sanneblad, J., Holmquist, L.E.: Total recall: In-place viewing of captured whiteboard annotations. In: Ubiquitous Computing (UbiComp03). (2003)
19. Klemmer, S.R., Newman, M.W., Farrell, R., Bilezikjian, M., Landay, J.A.: The designers' outpost: a tangible interface for collaborative web site. In: Proceedings of the 14th annual ACM symposium on User interface software and technology, ACM Press (2001) 1–10
20. http://www.hitl.washington.edu/artoolkit: ARToolkit website (2003)
21. Billinghurst, M., Kato, H.: Collaborative augmented reality. Communications of the ACM **45** (2002) 64–70
22. Gong, Y.: Detection of regions matching specified chromatic features. Computer Vision and Image Understanding **61** (1995) 263–269
23. Jamzad, M., Sadjad, B.S., Mirrokni, V.S., Kazemi, M., Chitsaz, H., Heydarnoori, A., Hajiaghai, M.T., Chiniforooshan, E.: A fast vision system for middle size robots in robocup. In: RoboCup Symposium. (2001)
24. Schulz, D., Fox, D., Hightower, J.: People tracking with anonymous and id-sensors using rao-blackwellised particle filters. In: IJCAI. (2003)
25. Rekimoto, J., Ayatsuka, Y.: Cybercode: Designing augmented reality environments with visual tags. In: Proceedings of DARE. (2000)
26. http://www.mimio.com: Virtual Ink Mimio website (2003)

Augmenting Collections of Everyday Objects: A Case Study of Clothes Hangers as an Information Display

Tara Matthews[1], Hans-W. Gellersen[2], Kristof Van Laerhoven[2], and Anind K. Dey[3]

[1] EECS Department, University of California, Berkeley
[2] Computing Department, Lancaster University
[3] Intel Research Berkeley, Intel Corporation
tmatthew@cs.berkeley.edu

Abstract. Though the common conception of human-computer interfaces is one of screens and keyboards, the emergence of ubiquitous computing envisions interfaces that will spread from the desktop into our environments. This gives rise to the development of novel interaction devices and the augmentation of common everyday objects to serve as interfaces between the physical and the virtual. Previous work has provided exemplars of such everyday objects augmented with interactive behaviour. We propose that richer opportunities arise when collections of everyday objects are considered as substrate for interfaces. In an initial case study we have taken clothes hangers as an example and augmented them to *collectively* function as an information display.

1 Introduction

The graphical user interface, commonplace on desktops and mobile computers, has been criticized as being too detached from the architectural spaces and physical artefacts around which people's activities evolve. Under headings such as ubiquitous computing, pervasive computing, and ambient intelligence, alternative approaches are investigated that move the interface from the desktop into the environment. One emerging possibility is to augment physical artefacts that already surround us so that they become the interface to an otherwise invisible computing infrastructure. Previous work has produced a range of examples in which everyday artefacts have been augmented with interactive behaviour while retaining their original purpose. These include a picture frame augmented as a context display [1], a table enabled as pointing device [2], and a chest of drawers affording digital lookup of physical content [3].

In this paper, we go beyond the consideration of individual artefacts as interface objects and propose a new type of interface that is based on sets of everyday, physical artefacts that collectively provide interactive behaviour. There are two main reasons to explore this approach. First, many artefacts naturally exist in collections (*e.g.,* CDs, cutlery) and in arrangements (*e.g.,* chairs around a table, pictures on a mantel), a fact that can be taken advantage of for composing interfaces. Second, collections afford additional interactions such as adding, removing, and arranging component artefacts.

Interfaces composed of individual physical entities are of course not a new consideration – interfaces spanning multiple devices have been studied widely in traditional

A. Ferscha and F. Mattern (Eds.): PERVASIVE 2004, LNCS 3001, pp. 340-344, 2004.

user interface research, and there is also a range of examples of tangible interfaces that use the arrangement of physical components as an interaction mechanism (*e.g.* Triangles [4] and the Urp interface [5]). However, the distinct novelty in considering collections of everyday artefacts as interfaces is that the component artefacts are individually meaningful entities to start with. This gives rise to a host of questions, such as when is an artefact part of an interface and when not, and how can an interface be made robust against any changes in composition that may result from everyday use of component artefacts.

To explore this new interface concept we have focussed on a concrete case study as a starting point. Our approach was to go through the complete process of designing an example interface, of implementing the design as working prototype, and of engaging users with the built prototype to collect formative feedback in addition to our design experience. The example that we have investigated is a *Hanger Display*. We learned some lessons designing and formatively evaluating the Hanger Display: that existing structures can be useful in defining the interface scope of collections, and that interfaces composed of everyday artefacts must be resilient to change.

2   Design and Implementation of the Hanger Display

The Hanger Display is a generic system for portraying information at the periphery of attention. As shown in Figure 1, the Hanger Display consists of a row of hangers, each augmented with a display element (*e.g.* a light emitting diode, LED), placed on a rod (*e.g.* in a wardrobe). Collectively the hangers function as a simple display defined by the array of display elements arranged along the rod. The hangers have individual meaning in everyday activity, in the sense that individual hangers are used independently of the state of other hangers (*e.g.* to hang up a coat, or to move it from one wardrobe to another). But as display elements, they are only meaningful as part of the collection of hangers placed together on the rod.

From an interaction design perspective we think of the Hanger Display as being useful for portraying information that may be relevant to the activities involving hangers (such as hanging and selecting clothes, morning rituals, and getting ready for the day). For example, the hanger display could portray a sense of the day's expected high temperature: given a $0 - 100°$ F range, if 7 of 10 hangers were lit this would represent $70°$ F. However, the Hanger Display does not prescribe how information is mapped onto the variable collection of display elements; it is entirely open-ended for different display purposes. For instance, the hangers may convey any relative quantities (as in the temperature example), progress over time (*e.g.* in a reminder display

Fig. 1. *Design Sketch of the Hanger Display.* Each of the hangers is augmented with a display element, and all hangers arranged on the rod together function as a coherent display.

that counts down days before an important event), or trend information (*e.g.* analogous to how a barometer displays trends in weather conditions).

We have implemented a working prototype of the Hanger Display based on the following set of components:

- Ordinary hangers augmented with very low-cost embedded hardware (an LED, and a switch to control the LED).
- Hanger rods augmented with a physical network medium to provide connectivity to the hangers.
- Software to manage the display system, *i.e.* determining which hangers are present on a rod and controlling communication in the system.
- Software mapping an input to the LED output of the Hanger Display, *i.e.* determining which hangers to turn on and sending commands.

Two aspects are particularly noteworthy in this implementation. First, rods have a distinct role in the design as defining the scope of a display (*i.e.* determining which artefacts are part of the display and which are not). This is directly reflected in the implementation by foreseeing rods as a connection medium for hangers, in which physical arrangement is overlaid with digital connectivity. Second, the implementation allows the display to be dynamic. Hangers can be removed or inserted without causing the display to break. Instead, the displayed information is re-mapped when the physical composition of the display changes.

Figure 2 shows a photo of the implemented prototype and an illustration of its hardware components. The rod has two strips of aluminium along its length that act as two wires, enabling its use as a network medium. One strip serves as ground and the other provides data and power transmission, using the MicroLAN protocol [6] and a similar networking concept as presented by the Pin&Play project [7]. In order to connect a hanger to this medium, we put two loops of wire around its hook so that separate contacts are made to the two aluminium strips on the rod. The balance of a hanger, with or without clothing, generally causes it to naturally fall into the correct place on the rod to make these contacts. The hardware on each hanger includes a switch that can be controlled over MicroLAN and an LED whose output is directly connected to the state of the switch. The Hanger Display software is written in Java. Based on the MicroLAN protocol, it provides core functionality to control the display

 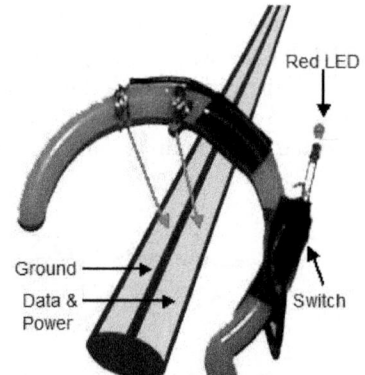

Fig. 2. A wardrobe augmented with two built-in Hanger Displays (the top and bottom rods). As detailed on the right, hangers have two wires added that contact the rod through which data and power are provided to the LED and the switch (hidden under black tape).

system, such as detecting which hangers are present on the rod and addressing hangers in order to control the LEDs.

Note that the hardware components are all very small and do not compromise the form-factor of the hangers in any significant way. In addition, despite the fact that only very simple and inexpensive hardware is used for each hanger, the collection of hangers is quite expressive in the way it conveys information.

3   Formative User Study and Discussion

In order to obtain formative user feedback we implemented two basic application scenarios in which weather-related information was displayed on the Hanger Display. The applications mapped temperature forecast and the probability of rain to the two rods of our prototype, based on local weather data parsed from the weather.com web site. The intent of this application setup was not to evaluate how effective weather information might be conveyed by such a display, but to engage users in a more general discussion on the use of multiple everyday artefacts as an interface.

To this end, we conducted a formative, in-lab user study with the two applications. We set up a wardrobe as in Figure 2: the top rod displaying the predicted high temperature (the range is $0 - 30°$ C and the figure shows 4 of 8 hangers are lit, so it will be $15°$ C), and the bottom rod showing the chance of rain (2 of 3 hangers lit, indicating a 66% chance of rain). Seven researchers from other groups at our department came to the lab for one individual evaluation session each with one of the authors. Participants were given a brief description of the applications and asked to explore and interact with the interface with no specific task. An interview followed, in which the participants were asked for qualitative feedback on the interface. Although feedback was influenced by the participants' previous computing experience, discussions primarily centred on domestic rather than work experience.

The feedback from the formative evaluation showed that there is a realistic use for displays and more generally interfaces composed of multiple everyday, physical artefacts. Every user had ideas for other applications that they would find useful in their own homes. For example, a collection of CDs could be augmented to give you a sense of which CDs you have not listened to recently. When you pick out music, you may enjoy selecting a CD you have not listened to for a long time, information that would be readily available without you reading the CD labels. Users suggested many other everyday objects that could be augmented similarly: books, videos, food containers, cutlery, dishes, shoes, household decorations, files folders, and furniture.

Users also suggested applications that involve richer interactions than the current peripheral display. These suggestions were based on the affordances of the physical collection, including adding, removing, and arranging the hangers. For example, the hangers could send messages to other people when rearranged. This could be useful in a clothing shop: hangers could indicate things like sale items or items to be moved by employees. By moving a certain clothing item to the "sale rack," an employee could cause all hangers holding the same item to light up. Employees could quickly find and gather all of these items onto the sale rack, at which time the LEDs would turn off.

The exciting prospect for examples like these is that they point to new, creative, and useful applications of ubiquitous computing that support activities *in situ*. Several users pointed out the benefit of an everyday object like hangers displaying additional information: "I like having information in the place where it is useful, so it is nice to get weather information when you are picking out clothes." With further exploration, other collections of everyday objects might suggest new ways to provide the benefits of technology without sacrificing the ease and naturalness of everyday activities.

The formative feedback also helped confirm some design considerations for interfaces composed of multiple everyday, physical artefacts. The first consideration is that existing structures can be useful in defining interface scope. For the Hanger Display, the natural affordances of the rod made it easy for users to understand the interface and to determine which hangers were part of it.

The second consideration is that interfaces composed of everyday artefacts must be resilient to change. Aside from some problems with the prototype implementation (*e.g.*, hangers not always making a connection with the rod), the users asserted that the embedded technology did not interfere with natural interaction. The Hangers show we can design to deal with physical re-composition, but future work is needed to investigate how this can be supported in a general way.

4 Conclusion

In this paper, we have presented a case study of an interface composed of a collection of everyday artefacts, which attempts to provide information in the context of everyday activities and environments. The design of and formative feedback on the Hanger Display shows that this approach is interesting for two main reasons: many artefacts naturally exist in collections; and collections afford additional interactions such as adding, removing, and arranging of component artefacts. When designing such interfaces, we learned that existing structures can be useful in defining interface scope, and that interfaces composed of everyday artefacts must be resilient to change.

References

1. Mynatt, E.D., *et al.* Digital family portraits: Providing peace of mind for extended family members. *Proc. of ACM CHI'01*, pp. 333-340.
2. Schmidt, A., *et al.* Ubiquitous interaction - Using surfaces in everyday environments as pointing devices. *Proc. 7th Workshop on User Interfaces for All*, 2002, pp. 263-279.
3. Siio, I., *et al.* Finding Objects in 'Strata Drawer.' *Extended Abstracts CHI '03*, pp. 982-983.
4. Gorbett, M.G., *et al.* Triangles: Tangible interface for manipulation and exploration of digital information topography. *Proc. of ACM CHI'98*, pp. 49-56.
5. Underkoffler, J. and Ishii, H. Urp: A luminous-tangible workbench for urban planning and design. *Proc. of ACM CHI'99*, pp. 386-393.
6. Dallas Semiconductors. Overview of 1-Wire Techn. & its Use. App Note 1796, Dec 3, '03.
7. Van Laerhoven, K., *et al.* Pin&Play: Networking Objects Through Pins. *Proc. of Ubicomp '02*, pp. 219-229.

MirrorSpace: Using Proximity as an Interface to Video-Mediated Communication

Nicolas Roussel, Helen Evans, and Heiko Hansen

Laboratoire de Recherche en Informatique & INRIA Futurs**
Bât 490, Université Paris-Sud XI
91405 Orsay Cedex, France
roussel@lri.fr, helen@hehe.org, heiko@hehe.org

Abstract. Physical proximity to other people is a form of non-verbal communication that we all employ everyday, although we are barely aware of it. Yet, existing systems for video-mediated communication fail to fully take into account these proxemics aspects of communication. In this note, we present MirrorSpace, a video communication system that uses proximity as an interface to provide smooth transitions between peripheral awareness and very close and intimate forms of communication.

1 Introduction

Physical proximity to other people is a form of non-verbal communication employed everyday by us all, although we are barely aware of it. We constantly use space and distance to define and negotiate the interface between private and public matter, particularly during the moments leading up to contact. By altering our physical distance from other people in a space, we communicate subtle messages such as our willingness to engage into dialogue with them, the desire for more intimacy or a lack of interest.

The term *proxemics* refers to the study of spatial distances between individuals in different cultures and situations. It was coined by E.T. Hall in 1963 when he investigated man's appreciation and use of personal space. Hall's model lists four distances which Northern Americans use in the structuring of personal dynamic space [1]: *intimate* (less than 18 inches), *personal* (between 18 inches and 4 feet), *social* (between 4 and 12 feet) and *public* (more than 12 feet). For each communication situation, there is a distance within these four categories that we find appropriate, based on our cultural background and on the particular context of the situation. If the perceived distance is inappropriate, we become uncomfortable and we usually adjust it by physically moving closer or further away, or even simply turning our head or looking in another direction.

Existing systems for video-mediated communication fail to take into account the proxemics aspects of communication. Although some of the people who designed the systems understood the importance of these aspects, they failed to fully provide the support they require. In this note, we present MirrorSpace, a video communication system that uses proximity as an interface to provide smooth transitions between peripheral awareness and very close and intimate forms of communication.

** projet In Situ, Pôle Commun de Recherche en Informatique du plateau de Saclay, CNRS, Ecole Polytechnique, INRIA, Université Paris-Sud

A. Ferscha and F. Mattern (Eds.): PERVASIVE 2004, LNCS 3001, pp. 345–350, 2004.

2   Related Work

Most video communication systems are based on a glass pane metaphor. VideoWindow [2] probably best illustrates this concept, displaying remote people as life-sized images on a large vertical surface, making them appear as if they were seen through a virtual window. The glass pane metaphor provides a sense of shared space and supports gesture-based communication. However, even with life-sized images, the psychological distance to someone at the other end of the system is still greater than that in a comparable face-to-face situation. In particular, the distance between the camera and the image of a remote person's eyes can make eye contact and gaze awareness a real challenge. A number of solutions to these problems have been proposed for specific contexts. ClearBoard [3], for example, supports both eye contact and gaze awareness in close collaboration situations based on shared drawing.

As a cultural artifact, the mirror has a prominent position in the creation and expression of esthetics. Throughout Western culture narratives such as the Narcissus myth, *Snow White* or *Through the Looking Glass*, it has come to many different meanings including vanity, deception, identity or a passage to another world. A number of interactive art installations, such as Liquid Views [4], have picked up on these meanings and taken advantage of the universal and irresistible fascination for self-image. A mirror metaphor offers an interesting potential to attract people to a video-based system [5]. It also helps reduce the psychological distance between local and remote participants by displaying them side-by-side, as if they were all in one room [6].

No matter the metaphor, the interpersonal distance perceived by participants determines in great part the suitability of a video communication system for a particular context. ClearBoard, for example, creates the impression of standing about one meter away from the other person, which corresponds to the personal distance of Hall's classification [3]. Although perfectly suited for use with friends and colleagues, this distance might seem too small for a formal meeting with a person of a higher rank. Another consequence is that while ClearBoard makes it easy to establish eye contact, it also makes it difficult to avoid. Users of VideoWindow experienced the same problem and "went to great lengths to avoid eye contact" when they wanted to avoid conversation [2].

ClearBoard authors suggest that the communication system could provide users with some control over the perceived interpersonal distance [3]. This distance is influenced by many factors such as the spatial distance from the display, the size and quality of the video images, backdrops or voice fidelity. The potential exists for proximity as a form of non-verbal communication to affect behavior in video-mediated interactions. Yet, very little work has been carried out on the control over perceived proximity [7].

3   MirrorSpace

While existing video communication systems create a shared space corresponding to a particular interpersonal distance, the goal of MirrorSpace is instead to create a continuum of space, to allow a variety of interpersonal relationships to be expressed. Our work focuses on the understanding of how people's interactions can trigger smooth transitions between situations as extreme as peripheral awareness of remote activity and intimate situations.

MirrorSpace relies on the mirror metaphor. Live video streams from all the places it connects are superimposed on a single display on each site so that people see their own reflection combined with the ones of the remote persons. A real mirror is already perceived as a surface for mediating communication with its own rules and protocols. As an example, making eye contact with a stranger through a mirror is usually considered less intrusive than direct eye contact. Since the mirror is already associated to this idea of reaching out to other people and other spaces, we believe it is the ideal enabling metaphor for establishing a new communication experience.

As we aim to support intimate forms of communication, it felt important to us that people could actually look into each other's eyes, so the camera was placed right in the middle of the screen. This setup allows participants to come very close to the camera while still being able to see the remote people and interact with them. MirrorSpace also includes a proximity sensor that measures the distance to the closest object or person in front of it. A blur filter is applied on the images displayed to visually express a distance computed from the local and remote sensor values. Blurring distant objects and people allows one to perceive their movement or passing with a minimum involvement. It also offers a simple way of initiating or avoiding a change to a more engaged form of communication by simply moving closer or further away.

MirrorSpace was originally conceived as a prototype for the interLiving project[1] of the European *Disappearing Computer* initiative. A first video mock-up illustrating its design concept was made in August 2002. Several units were then created and presented to the public as an interactive video installation in four art exhibitions, in February, May, July and December 2003.

3.1 Hardware Configuration

Two MirrorSpace units were built for the first exhibition and slightly modified before the other ones. Each unit consists of a flat screen, a camera, a proximity sensor and a computer that runs dedicated software. These prototypes have been designed to minimize their technological appearance so they can discreetly blend in their environment. The computer and the wires are kept hidden from users. The screen and its attached sensors are placed into a wooden box, protected by a transparent glass partially covered with a real mirror film (Fig. 1).

The image sensor and the lens of a Philips ToUcam Pro have been placed in the center of the screen. The sensor is connected back to the logic board of the camera using hair thin isolated wires running over the screen surface. Informal tests quickly confirmed that the lens is hardly noticeable once placed onto the screen, since people are generally focused on the images displayed rather than the screen itself. The proximity sensor, a Devantech SRF04, has been placed at the bottom of the screen. It is connected to a Parallax BASIC Stamp chip, itself connected to the computer via a serial interface. The computers were initially Apple PowerMac Cubes. They were later replaced by 2.8GHz Pentium IV machines with 2GB of memory and an NVIDIA GeForce FX 5200. A 100 Mbits/sec Ethernet network was set up to connect them during the exhibitions.

---

[1] http://interliving.kth.se/

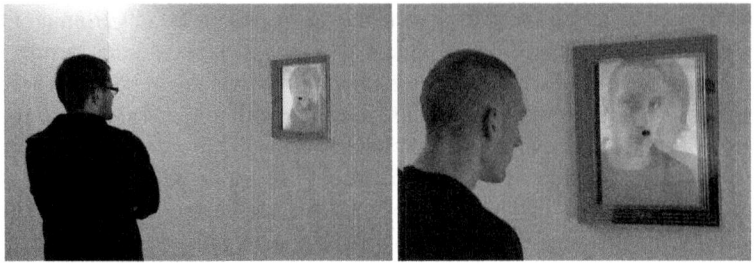

Fig. 1. MirrorSpace installation for the second exhibition

3.2   Software

MirrorSpace software is written in C++. It uses the videoSpace library [8] to capture SIF images from the camera in real-time and OpenGL to display a graphical composition created from these images and the proximity sensor values. Although only two were used for the exhibitions, the software doesn't make any assumption on the number of connected units. Proximity sensor values and images are sent on the network with a best-effort strategy (images are transmitted as JPEG data compressed to fit in a single datagram). The compositing process applies a blur filter on the image of each unit and superimposes them using alpha blending. The resulting composition is flipped horizontally before display to produce the expected mirror effect.

The blur effect is implemented with a two-pass incremental box filter. The size of the filter (i.e. the number of neighbors taken into account for one pixel) determines the blur level. The sensor values of all connected units are used to compute the size s of the filter to apply to each image. Three computation modes have been investigated so far. The first one (1) only takes into account the distance d, measured by the unit that captured the image. The two others (2 and 3) also take into account the distance d_{loc}, measured by the unit that displays the image:

$$s = f(d) \tag{1}$$

$$s = f(d_{loc} + d) \tag{2}$$

$$s = f(|d_{loc} - d|) \tag{3}$$

The software allows to choose a different mode for each unit. However, a strict WYSIWIS condition (*What You See Is What I See*) was imposed for the exhibitions.

4   Interacting with MirrorSpace

The first mode of operation of MirrorSpace (1) is quite intuitive: objects and people close to the mirror are better perceived than those far away. It is the one we used for all the exhibitions. It allows people to slowly get into focus as they move closer to the unit (Fig.2) and out of focus as they move away from it. The second mode introduces the notion of relative distance between participants. By moving forward or backward, people alter not only their own image but also the image of the remote persons. By

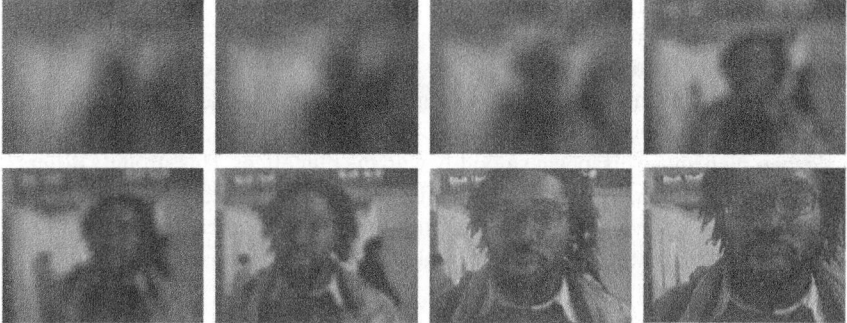

Fig. 2. Moving from peripheral awareness to focused communication by approaching the mirror

moving away from the mirror, one can still slowly disappear. However, in this case, the other people can follow that person to a certain extent. The third mode should allow multiple "islands" of communication aligned in front of the sensor. However, a lot of space and more than two units are needed, which is why it hasn't really been tested yet.

Almost all visitors of the exhibitions agreed on one point: interacting with MirrorSpace is fun. Proximity sensing helps creating an intimate relationship between users and the system. Many of them played with their own image and the blur effect. People didn't hesitate to make a fool of themselves and many took pictures or recorded video clips of themselves and other people interacting through the system. When they saw another person appearing next to them on the screen, many people turned over, looking for that person behind them. This shows that the superposition of the images creates a sense of sharing the same space. It also shows that MirrorSpace is perceived as a mirror and not as a remote video communication system. In fact, the majority of the people didn't think about the camera at all. Only after playing with the system for some time, they suddenly asked surprised "where is the camera?".

Fig. 3. Close and intimate communication through MirrorSpace

The superposition of the images allows not only to share space but also to become one. People who were visiting the exhibitions with friends or relatives immediately understood that and tried to overlay their faces (Fig. 3). Some went as far as kissing each other. At the same time, other persons were surprised and even disturbed to find strangers able to come so close to them. In that case, they simply backed away, which made their own image disappear smoothly with the blur effect. This strongly differs from systems such as ClearBoard or VideoWindow where eye contact is difficult to avoid. It shows that MirrorSpace can be used as an intimate communication device and, at the same time, supports at least part of the body language we are used to.

5   Conclusion

We hope that MirrorSpace will help researchers and practitioners realize the importance of the understanding of proxemics for the design of video-mediated communication systems. The design concept of this system as well as some details of its implementation have been described. We have also described some user reactions to presentations of the system that were made during several art exhibitions. These initial reactions show that MirrorSpace supports smooth transitions between peripheral awareness and very close and intimate forms of communication. We strongly believe that the use of proximity as an interface to computer-mediated communication is a promising research direction. We plan to continue this work on image-based communication and to apply the ideas described in this paper to other forms of communication as well.

References

1. Hall, E.: The Hidden Dimension: Man's use of Space in Public and Private. Doubleday, New York (1966)
2. Fish, R., Kraut, R., Chalfonte, B.: The VideoWindow system in informal communication. In: Proceedings of ACM CSCW'90 Conference on Computer-Supported Cooperative Work, ACM Press (1990) 1–11
3. Ishii, H., Kobayashi, M., Grudin, J.: Integration of interpersonal space and shared workspace: ClearBoard design and experiments. ACM Transactions on Information Systems **11** (1993) 349–375
4. Fleischmann, M., Strauss, W., Bohn, C.: Liquid views: rigid waves. In: ACM SIGGRAPH 98 Electronic art and animation catalog, ACM Press (1998) 21
5. Roussel, N.: Experiences in the design of the well, a group communication device for teleconviviality. In: Proceedings of ACM Multimedia 2002, ACM Press (2002) 146–152
6. Morikawa, O., Maesako, T.: HyperMirror: toward pleasant-to-use video mediated communication system. In: Proceeding of ACM CSCW'98 Conference on Computer-Supported Cooperative Work, Seattle, Mass., ACM Press (1998) 149–158
7. Grayson, D., Anderson, A.: Perceptions of proximity in video conferencing. In: CHI'02 extended abstracts on Human factors in computer systems, ACM Press (2002) 596–597
8. Roussel, N.: Exploring new uses of video with videoSpace. In Little, R., Nigay, L., eds.: Proceedings of EHCI'01, the 8th IFIP International Conference on Engineering for Human-Computer Interaction. Volume 2254 of Lecture Notes in Computer Science., Springer (2001) 73–90

SearchLight – A Lightweight Search Function for Pervasive Environments

Andreas Butz, Michael Schneider, and Mira Spassova

Department of Computer Science, Saarland University,
Stuhlsatzenhausweg, Bau 36.1, 66123 Saarbrücken, Germany
{butz,mschneid,mira}@cs.uni-sb.de

Abstract. We present a lightweight search function for physical objects in instrumented environments. Objects are tagged with optical markers which are scanned by a steerable camera and projector unit on the ceiling. The same projector can then highlight the objects when given the corresponding marker ID. The process is very robust regarding calibration, and no 3D model of the environment is needed. We discuss the scenario of finding books in a library or office environment and several extensions currently under development.

1 Introduction

Ubiquitous computing landscapes extend our physical surroundings by a computational layer providing new functionalities. One such functionality can be the capability of objects to make themselves known in order to be noticed or found by humans. This functionality was already proposed in the original Ubiquitous Computing vision [8]. A search function for physical environments would alleviate the need to keep track of all of the things in our environment.

One obvious application is keeping track of books in our office, in a library or a book store. Let's think of a library with ubiquitous display capabilities. Let's assume there will still be a conventional inquiry terminal to find out about books. The inquiry interface on the computer terminal in a library could then just provide an additional "show this book" button for the selected book, which prompts the environment to highlight its position on the shelf.

Let's think of an instrumented office in which there are many books and other things, either neatly sitting on the shelf where they belong, or left somewhere in the room on a pile. A physical search functionality will find and highlight missing objects for us, no matter where we left them. In this paper we describe such a search function for physical objects, and how it was implemented in our instrumented environment.

2 The SUPIE Environment

The Saarland University Pervasive Instrumented Environment (SUPIE) is an instrumented room with various types of displays, sensors, cameras, and other

A. Ferscha and F. Mattern (Eds.): PERVASIVE 2004, LNCS 3001, pp. 351–356, 2004.

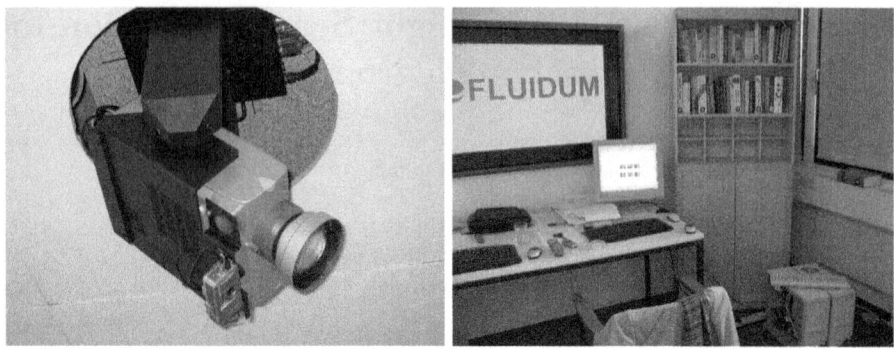

Fig. 1. The steerable projector mounted on the ceiling (left) and one corner of SUPIE with a book shelf (right)

devices. By means of a steerable projector, the room has continuous display capabilities which are limited in resolution and temporal availability (only one area at a time). The various other displays provide islands of higher resolution and interactivity within this display continuum.

Fig. 2. Images taken during the scanning process (left) and an example of an AR Toolkit marker (right)

The projector can project arbitrary light patterns onto all surfaces in the room with line of sight to its position in the center of the ceiling. An attached camera can take high resolution pictures from (almost) the same position and scan them for objects and markers. It also provides a low resolution video stream in real time for the recognition of movements.

3   Implementation of *SearchLight*

3.1   General Thoughts

All that is needed to implement *SearchLight* is the steerable projector with its camera and the capability to recognize optical markers. The only determining factors for finding objects and highlighting them, are the pan and tilt angles of the unit moving the projector. If a marker is found at given pan and tilt angles, these can be stored and used directly for projecting a highlight area around the corresponding object. Thus, not even a 3D model of the room is needed. Absolute registration errors of the camera and projector will cancel each other out since they are mounted in one unit and moved together.

3.2   Devices

The steerable projector is mounted on a moving yoke produced by a stage equipment manufacturer, controlled via a DMX interface, and mounted in the center of the ceiling of the room. The unit carries a 3.000 ANSI lumen projector with a lens of long focal length, in order to provide sufficient contrast and brightness in daylight. The camera attached to the projector (see figure 1 right) is a regular Digicam with an image resolution of four Megapixels. The high resolution pictures are triggered and read out over USB with the camera's freely available SDK. Each object that should be found by *SearchLight* is tagged by an AR Toolkit marker (see [1]). Each marker has a large black fringe and an individual small black symbol on a white background. (See figure 2) These markers are recognized by the Java version of AR Toolkit, the *JAR Toolkit*.

3.3   Implementation

SearchLight performs two main tasks: It *scans* the room for markers and memorizes the corresponding angles, which are used later to *show* searched objects.

Scan.   The room is scanned by taking slightly overlapping pictures in all horizontal and vertical directions (see Fig. 2 left). Each picture is analyzed using JAR Toolkit. Marker IDs are stored in a list together with their pan and tilt angle, derived from their position and orientation in the picture and the orientation of the moving yoke when taking the picture.

Show.   After the room has been scanned, the user can search for marked objects. The request to show a specific book in a library can, for example, be given from the library enquiry system through a "show this book" button. The object's Marker ID is then looked up and the projector unit moves to the position where this marker was detected during the last scan and projects a bright spot around the searched object. Due to the speed of the steerable projector, highlighting an arbitrary object takes less than a second, thus providing almost instant feedback to the search request.

Fig. 3. A book was found on the shelf (left) and another one on the window sill (right)

4 Related Work

The use of a steerable projector to transform environments into continuous displays has first been proposed by Claudio Pinhanez [3] at IBM. In their Siggraph 2001 emerging technologies demo, where colored M&Ms were composed by visitors to form large pixel images, the Everywhere Display Projector was used to highlight trays with different colors on a shelf. The positions of these trays were, however, not acquired automatically.

Additional applications are presented in [4], including a ubiquitous product finder and an interactive shelf. The product finder guides people towards products in a store environment, using blank projection surfaces as individual direction signs and interaction spaces. The interactive shelf uses similar surfaces for the display of product-related information. While these prototypes are strongly related to the work presented here, both of them require dedicated projection space in the environment.

Raskar et al. present in [5] a method for simultaneous acquisition of room geometry and use of the acquired surfaces as an output medium for mobile projectors. They also use markers for object recognition and annotation. The FindIT Flashlight [2] can find objects instrumented with responsive elctronic tags. Users scan the environment with an electronic flashlight emitting a digital code, and the corresponding tags respond when hit by the flashlight's beam. Both systems rely on the user pointing to the right direction, while in our demo, the environment actively controls the projector.

In [7] another augmented library environment is described, where books or their requested positions can be highlighted on the shelf through a head-mounted display. While this approach is more general in terms of scalability and topology of the environment, it requires an instrumentation of the user and an explicit 3D model of the environment, which we were able to avoid.

5   Current Results and Future Work

One immediate application scenario is a public library where each book has a marker on its front and back side and on the spine. In this way a book can be located even if it is not at its usual place in the shelf. When the library is closed, *Search Light* can scan all shelves and tables and take inventory. During opening hours, only certain areas must be re-scanned as described above. With our experimental setup in SUPIE (room size 5x6m, shelf on the wall, 4 Megapixel steerable camera with 3x optical zoom in the center of the ceiling) we were able to reliably recognize markers down to a size of 10mm in the whole room. Initial scanning of the room took roughly 1 hour, mostly due to the slow transmission of pictures over the USB 1.1 link to the camera. Currently we are improving *SearchLight* in various respects.

5.1   Continuous Model Update

In the current demo, scanning is done only once when *SearchLight* is started. In the future we will use idle times of the projector to systematically re-scan the room for changes. This process will also prioritize regions where changes are more likely by using additional sensors, such as RFID tags or motion detection with other cameras. On the down side, including external sensors will require the use of at least a simple 3D model of the room in order to relate locations of sensor events to pan and tilt angles of the projector. This model might, however, be acquired automatically by methods similar to those described in [6]. Upcoming versions of *SearchLight* may use the camera even when the projector is used for other tasks by just analyzing images as they appear rather than actively steering the camera to certain positions for scanning.

5.2   Extension to Other Markers

In theory, existing bar codes on many products could be used for the recognition process. In the case of books, OCR could even eliminate the need for markers altogether, since book spines and covers are designed to clearly identify books. Currently, we are working on integrating RFID tags by watching the region around the RFID antenna. When the antenna registers a change in its field, a new image is taken and the exact position of the new resp. removed object is identified from a difference image. As a side effect the exact silhouette of the object might be obtained for highlighting by the projector. To overcome the visibility problems inherent to the use of optical markers, radio tags might even fully replace them. This will, on the other hand, make a 3D model of the environment necessary, in order to relate object positions to pan and tilt angles of the projector.

5.3   Fuzzy Search Results

If the exact position or dimension of a searched object is unknown, a fuzzy search result can be visualized as a spot whose diameter reflects the amount of uncertainty. If, in addition, we know the probabilistic distribution over possible locations, we can adjust the brightness of the spot to reflect this distribution. In the extreme case this means highlighting a whole area of the room, which would signify that the object must be "somewhere in that area".

6   Conclusions

We have discussed an approach to implementing a search functionality for physical environments by a steerable camera-projector unit, and we have shown that it is possible to implement this functionality without an explicit 3D model of the environment. One big advantage of our approach is, that no calibration between the projector-camera-unit and the environment is needed. One obvious limitation is that only objects within sight from the unit can be searched and found. This limits the topology of the environments in which our approach can be used.

Acknowledgements. The work described here was funded by "Deutsche Forschungsgemeinschaft" under a young investigator award for the project FLU-IDUM. The SUPIE environment is shared with the REAL project. We thank our anonymous reviewers for helpful suggestions and pointers to additional related work.

References

1. M. Billinghurst, S. Weghorst, and T. Furness. Shared space: An augmented reality approach for computer supported collaborative work. *Virtual Reality*, 3(1):25–36, 1998.
2. H. Ma and J. A. Paradiso. The findit flashlight: Responsive tagging based on optically triggered microprocessor wakeup. In *Proceedings of UbiComp 2002, Gothenburg, Sweden*, 2002.
3. C. Pinhanez. Using a steerable projector and a camera to transform surfaces into interactive displays. In *Proc. ACM CHI 2001*, Seattle, Washington, USA, March 31 - April 5 2001. ACM Press.
4. C. Pinhanez, R. Kjeldsen, A. Levas, G. Pingali, M. Podlaseck, and N. Sukaviriya. Applications of steerable projector-camera systems. In *Proceedings of the IEEE International Workshop on Projector-Camera Systems at ICCV 2003*, Nice Acropolis, Nice, France, October 12 2003. IEEE Computer Society Press.
5. R. Raskar, J. van Baar, P. Beardsley, T. Willwacher, S. Rao, and C. Forlines. ilamps: Geometrically aware and self-configuring projectors. In *ACM SIGGRAPH 2003 Conference Proceedings*. ACM Press, 2003.
6. R. Raskar, G. Welch, M. Cutts, A. Lake, L. Stesin, and H. Fuchs. The office of the future: A unified approach to image-based modeling and spatially immersive displays. In *Proc. ACM SIGGRAPH '98*, pages 179–188, 1998.
7. G. Reitmayr and D. Schmalstieg. Location based applications for mobile augmented reality. In *Proceedings of the 4th Australasian User Interface Conference*, pages 65–73, Adelaide, Australia, Feb. 2003.
8. M. Weiser. The computer for the 21st century. *Scientific American*, 3(265):94–104, 1991.

Author Index

Lecture Notes in Computer Science

For information about Vols. 1–2896

please contact your bookseller or Springer-Verlag

Vol. 2963: R. Sharp, Higher Level Hardware Synthesis. XVI, 195 pages. 2004.

Vol. 2962: S. Bistarelli, Semirings for Soft Constraint Solving and Programming. XII, 279 pages. 2004.

Vol. 2961: P. Eklund (Ed.), Concept Lattices. IX, 411 pages. 2004. (Subseries LNAI).

Vol. 2960: P.D. Mosses (Ed.), CASL Reference Manual. XVII, 528 pages. 2004.

Vol. 2958: L. Rauchwerger (Ed.), Languages and Compilers for Parallel Computing. XI, 556 pages. 2004.

Vol. 2957: P. Langendoerfer, M. Liu, I. Matta, V. Tsaoussidis (Eds.), Wired/Wireless Internet Communications. XI, 307 pages. 2004.

Vol. 2954: F. Crestani, M. Dunlop, S. Mizzaro (Eds.), Mobile and Ubiquitous Information Access. X, 299 pages. 2004.

Vol. 2953: K. Konrad, Model Generation for Natural Language Interpretation and Analysis. XIII, 166 pages. 2004. (Subseries LNAI).

Vol. 2952: N. Guelfi, E. Astesiano, G. Reggio (Eds.), Scientific Engineering of Distributed Java Applications. X, 157 pages. 2004.

Vol. 2951: M. Naor (Ed.), Theory of Cryptography. XI, 523 pages. 2004.

Vol. 2949: R. De Nicola, G. Ferrari, G. Meredith (Eds.), Coordination Models and Languages. X, 323 pages. 2004.

Vol. 2948: G.L. Mullen, A. Poli, H. Stichtenoth (Eds.), Finite Fields and Applications. VIII, 263 pages. 2004.

Vol. 2947: F. Bao, R. Deng, J. Zhou (Eds.), Public Key Cryptography – PKC 2004. XI, 455 pages. 2004.

Vol. 2946: R. Focardi, R. Gorrieri (Eds.), Foundations of Security Analysis and Design II. VII, 267 pages. 2004.

Vol. 2943: J. Chen, J. Reif (Eds.), DNA Computing. X, 225 pages. 2004.

Vol. 2941: M. Wirsing, A. Knapp, S. Balsamo (Eds.), Radical Innovations of Software and Systems Engineering in the Future. X, 359 pages. 2004.

Vol. 2940: C. Lucena, A. Garcia, A. Romanovsky, J. Castro, P.S. Alencar (Eds.), Software Engineering for Multi-Agent Systems II. XII, 279 pages. 2004.

Vol. 2939: T. Kalker, I.J. Cox, Y.M. Ro (Eds.), Digital Watermarking. XII, 602 pages. 2004.

Vol. 2937: B. Steffen, G. Levi (Eds.), Verification, Model Checking, and Abstract Interpretation. XI, 325 pages. 2004.

Vol. 2936: P. Liardet, P. Collet, C. Fonlupt, E. Lutton, M. Schoenauer (Eds.), Artificial Evolution. XIV, 410 pages. 2004.

Vol. 2934: G. Lindemann, D. Moldt, M. Paolucci (Eds.), Regulated Agent-Based Social Systems. X, 301 pages. 2004. (Subseries LNAI).

Vol. 2930: F. Winkler (Ed.), Automated Deduction in Geometry. VII, 231 pages. 2004. (Subseries LNAI).

Vol. 2929: H. de Swart, E. Orlowska, G. Schmidt, M. Roubens (Eds.), Theory and Applications of Relational Structures as Knowledge Instruments. VII, 273 pages. 2003.

Vol. 2926: L. van Elst, V. Dignum, A. Abecker (Eds.), Agent-Mediated Knowledge Management. XI, 428 pages. 2004. (Subseries LNAI).

Vol. 2923: V. Lifschitz, I. Niemelä (Eds.), Logic Programming and Nonmonotonic Reasoning. IX, 365 pages. 2004. (Subseries LNAI).

Vol. 2919: E. Giunchiglia, A. Tacchella (Eds.), Theory and Applications of Satisfiability Testing. XI, 530 pages. 2004.

Vol. 2917: E. Quintarelli, Model-Checking Based Data Retrieval. XVI, 134 pages. 2004.

Vol. 2916: C. Palamidessi (Ed.), Logic Programming. XII, 520 pages. 2003.

Vol. 2915: A. Camurri, G. Volpe (Eds.), Gesture-Based Communication in Human-Computer Interaction. XIII, 558 pages. 2004. (Subseries LNAI).

Vol. 2914: P.K. Pandya, J. Radhakrishnan (Eds.), FST TCS 2003: Foundations of Software Technology and Theoretical Computer Science. XIII, 446 pages. 2003.

Vol. 2913: T.M. Pinkston, V.K. Prasanna (Eds.), High Performance Computing - HiPC 2003. XX, 512 pages. 2003. (Subseries LNAI).

Vol. 2911: T.M.T. Sembok, H.B. Zaman, H. Chen, S.R. Urs, S.H. Myaeng (Eds.), Digital Libraries: Technology and Management of Indigenous Knowledge for Global Access. XX, 703 pages. 2003.

Vol. 2910: M.E. Orlowska, S. Weerawarana, M.M.P. Papazoglou, J. Yang (Eds.), Service-Oriented Computing - ICSOC 2003. XIV, 576 pages. 2003.

Vol. 2909: R. Solis-Oba, K. Jansen (Eds.), Approximation and Online Algorithms. VIII, 269 pages. 2004.

Vol. 2908: K. Chae, M. Yung (Eds.), Information Security Applications. XII, 506 pages. 2004.

Vol. 2907: I. Lirkov, S. Margenov, J. Wasniewski, P. Yalamov (Eds.), Large-Scale Scientific Computing. XI, 490 pages. 2004.

Vol. 2906: T. Ibaraki, N. Katoh, H. Ono (Eds.), Algorithms and Computation. XVII, 748 pages. 2003.

Vol. 2905: A. Sanfeliu, J. Ruiz-Shulcloper (Eds.), Progress in Pattern Recognition, Speech and Image Analysis. XVII, 693 pages. 2003.

Vol. 2904: T. Johansson, S. Maitra (Eds.), Progress in Cryptology - INDOCRYPT 2003. XI, 431 pages. 2003.

Vol. 2903: T.D. Gedeon, L.C.C. Fung (Eds.), AI 2003: Advances in Artificial Intelligence. XVI, 1075 pages. 2003. (Subseries LNAI).

Vol. 2902: F.M. Pires, S.P. Abreu (Eds.), Progress in Artificial Intelligence. XV, 504 pages. 2003. (Subseries LNAI).

Vol. 2901: F. Bry, N. Henze, J. Ma luszyński (Eds.), Principles and Practice of Semantic Web Reasoning. X, 209 pages. 2003.

Vol. 2900: M. Bidoit, P.D. Mosses (Eds.), Casl User Manual. XIII, 240 pages. 2004.

Vol. 2899: G. Ventre, R. Canonico (Eds.), Interactive Multimedia on Next Generation Networks. XIV, 420 pages. 2003.

Vol. 2898: K.G. Paterson (Ed.), Cryptography and Coding. IX, 385 pages. 2003.

Vol. 2897: O. Balet, G. Subsol, P. Torguet (Eds.), Virtual Storytelling. XI, 240 pages. 2003.

GPSR Compliance

The European Union's (EU) General Product Safety Regulation (GPSR) is a set of rules that requires consumer products to be safe and our obligations to ensure this.

If you have any concerns about our products, you can contact us on ProductSafety@springernature.com

In case Publisher is established outside the EU, the EU authorized representative is:

Springer Nature Customer Service Center GmbH
Europaplatz 3
69115 Heidelberg, Germany

Batch number: 09490862

Printed by Printforce, the Netherlands